INTRODUCTION TO GEOMAGNETIC FIELDS

Introduction to Geomagnetic Fields is a textbook for advanced undergraduate and graduate students of geophysics. It explains the natural magnetic fields in and surrounding the Earth that arise from a variety of electric currents. Such electric currents exist within atomic structures, in the Earth's liquid outer core, in the ionized upper atmosphere, and in the Earth's space environment during solar–terrestrial disturbances (magnetic storms).

The author clearly presents these different components of the Earth's magnetic field with a minimum of mathematical complexity. Variations in the geomagnetic field over a range of time-scales are discussed, including reversals of the Earth's main dipolar field, disturbances of the magnetosphere caused by particles and fields radiating from the Sun, and daily changes caused by the tidal and wind motion of the ionosphere. Readers are also introduced to the techniques and instrumentation for measuring geomagnetic fields, and to the range of applications for which these measurements are used.

This second edition has been fully revised to include many of the most recent advances in this subject area. It has been designed as a textbook for use with semester courses in geomagnetism and includes student exercises at the end of each chapter. Special appendices review relevant mathematical techniques and direct the reader to various journals, books, organizations, and websites where the latest computer programs for geomagnetism may be downloaded. Solutions to the exercises can be found at http://www.cambridge.org/9780521822060.

WALLACE CAMPBELL graduated with a Ph.D. in Physics from the University of California, Los Angeles, in 1959. Following a year with the Geophysical Institute in the University of Alaska, he accepted a position for geomagnetic field studies with the National Bureau of Standards Laboratory in Boulder, Colorado, that subsequently became the Environmental Research Laboratory of NOAA. He remained there until 1973 when a federal reorganization transferred his group to the United States Geological Survey in Golden, Colorado, for studies in geomagnetic applications. He retired from the USGS in 1996 and since then has worked as a Guest Scientist with the Solar–Terrestrial Physics Division, National Geophysical Data Center of NOAA. Dr Campbell is the author of 128 geomagnetism publications. His research subjects include ionospheric currents, deep-Earth electrical conductivity, geomagnetic storms, geomagnetic pulsations, quiet-time field variations, and geomagnetic field applications. He is also the coeditor (and contributing author) of the textbook *Physics of Geomagnetic Phenomena* (1967) and the author of *Earth Magnetism: a Guided Tour Through Magnetic Fields* (2001).

INTRODUCTION TO GEOMAGNETIC FIELDS

Second Edition

WALLACE H. CAMPBELL

CAMBRIDGE
UNIVERSITY PRESS

CAMBRIDGE UNIVERSITY PRESS
Cambridge, New York, Melbourne, Madrid, Cape Town, Singapore, São Paulo

Cambridge University Press
The Edinburgh Building, Cambridge CB2 8RU, UK

Published in the United States of America by Cambridge University Press, New York

www.cambridge.org
Information on this title: www.cambridge.org/9780521822060

First published 1997
Second edition 2003

A catalogue record for this publication is available from the British Library

ISBN 978-0-521-82206-0 hardback
ISBN 978-0-521-52953-2 paperback

Transferred to digital printing 2007

Contents

Preface

This second edition of *Introduction to Geomagnetic Fields* has been redesigned as a classroom textbook for a semester course in geomagnetism. Student exercises have been added at the end of each chapter. Outdated figures and tables are replaced with more modern equivalents. Recent discoveries, field information, and references have been added along with special websites and computer programs. The basic structure of the original edition remains, providing a condensed and more readable coverage of geomagnetic topics than is afforded by existing textbooks.

My intention has been to focus upon the basic concepts and physical processes necessary for understanding the Earth's natural magnetic fields. When mathematical presentation is required, I have tried to remove the mystery of the scientists' special jargon and to emphasize the meanings of important equations, rather than obscure the relationships with complex formulas. Because some formulas are needed to appreciate geomagnetism, I have included, in an appendix, a succinct review of the required mathematical definitions and facts. For those readers who are approaching the subject of Earth magnetic fields for the very first time it may be helpful to start with the small layman's presentation, devoid of all mathematical equations, that I provided as *Earth Magnetism: A Guided Tour Through Magnetic Fields*, Academic Press, San Diego, 151 pp, 2001.

The student reader is expected to have a familiarity with the elementary scientific concepts identified by words of specific meaning, such as "force, velocity, energy, temperature, heat, charge, light waves, and fields of electric, magnetic, and gravitational nature". Excellent help is available if you are among those readers who have somehow missed receiving an explanation of these terms in your schooling. Albert Einstein and Leopold Infeld realized this need back in 1938 and wrote for you a small book, *The Evolution of Physics* (republished by Simon and Schuster, New York, 1961), which is as applicable today as it was over sixty years ago. That book uses remarkable simplicity of logic and language to reveal the fundamental concepts and terms now in use by physical scientists. In particular, with no mathematical formulas, their first two chapters not only provide the necessary basics for the science novice but also can reawaken an appreciation of the physical world made dormant by a schooling overdose of isolated facts and mathematical gymnastics.

To be called a proper "Introduction to Geomagnetic Fields" my book must contain the particular subjects that I have grouped into five chapters as follows. In Chapter 1 we explore the Earth's main field, consider its dipole representation, examine the vast extension of the fields into space, decipher the modeling methods of compact field representation, locate the many magnetic poles, and discuss paleomagnetic field reversals due to the source currents in the deep liquid core of the Earth. In Chapter 2 we find the reasons for quiet-time regular daily and seasonal field variations arising as electric currents in our Earth's upper atmosphere. We separate out the secondary currents that are induced to flow in the Earth itself – but find application in determining Earth conductivity profiles. In Chapter 3 we consider the major disturbances to the ordered particle and field configuration about the Earth. These geomagnetic storms have their origin in solar outbursts and act as monitors of changes in our space environment. In Chapter 4 we discover how magnetic field sensors (magnetometers) function, discuss some observation techniques, and look at measurement methods on the ground and in space. Finally, in Chapter 5 we consider the many useful applications of our knowledge of the main geomagnetic field and its temporal changes.

The book also contains three appendices. Appendix A is a review of mathematical concepts that the reader will encounter. Appendix B provides a guide to the major international organizations, a geomagnetism bibliography, and the useful source and website addresses for geomagnetic information. Appendix C gives the description of free special computer programs that are designed to help the beginning student in geomagnetism.

I have included more than the typical amount of text illustrations. They have been carefully selected to clarify concepts for the reader, rather than to burden him with a tedious overabundance of data reproductions. A great number of significant website addresses have been included. Although I have tried to guide the reader to known non-volatile websites, neither I nor the publisher can be held responsible if a site listed in this book is modified or disappears. The book index has been limited to those special words that I have considered to be of particular importance for an introduction to geomagnetic fields; don't expect to find place names, etc. This is a beginning textbook about geomagnetism; for full details on each of the chapter topics please borrow, from your library, the excellent reference books listed in Appendix B.

For the interpretation of phenomena, I have tried to stay with the current scientific consensus. However, in the future, many aspects of geomagnetism that are still being explored will undoubtedly change some of my viewpoints. Perhaps you can be the one to make such a contribution.

WHC

Acknowledgments

This small book had its origin, organization, and testing in my tutorial lectures at US Geological Survey; the Space Environment Laboratory of the National Oceanic and Atmospheric Administration; the World Data Center of the National Geophysical Data Center; the Australian Geological Survey Organisation (now Geoscience Australia); the Australian IPS Radio and Space Services; the Colorado School of Mines; the University of Colorado; the Academy of Sciences, Beijing; and the University of Cairo, Egypt.

I give special thanks for the guidance provided by researchers whose works are listed in the Bibliography and Reference sections, all of whom have made important contributions to the study of geomagnetic fields. In addition I have relied heavily on my personal scientific publications; the unreferenced figures are my own.

I appreciate the help of many scientists, teachers, special friends, and excellent technical reviewers who provided suggestions and material for improving various sections of the book. In particular, for assistance in preparing this second revision, I give special thanks to J. Quinn and D. Herzog (both at the US Geological Survey, Golden, USA), S. McLean, H. Coffey, E. Kihn, and B. Poppe (all at the National Oceanic and Atmospheric Administration, Boulder, USA), C. Barton, P. McFadden, and P. Milligan (all at Geophysics Australia, Canberra, Australia), D. Cole (at IPS Radio and Space Services, Sydney, Australia), D. Boteler (at the Geological Survey of Canada), G. Rostoker (at the University of Alberta, Canada), M. Barreto (at Observatorio Nacional, Rio de Janeiro, Brazil) and A. Adam (at the Geodetic and Geophysical Research Institute, Sopron, Hungary). I also thank my wife Beth, who assisted in editing the manuscript and endured my many hours spent at the computer.

WHC

Chapter 1
The Earth's main field

1.1 Introduction

The science of geomagnetism developed slowly. The earliest writings about compass navigation are credited to the Chinese and dated to 250 years B.C. (Figure 1.1). When Gilbert published the first textbook on geomagnetism in 1600, he concluded that the Earth itself behaved as a great magnet (Gilbert, 1958 reprint) (Figure 1.2). In the early nineteenth century, Gauss (1848) introduced improved magnetic field observation techniques and the spherical harmonic method for geomagnetic field analysis. Not until 1940 did the comprehensive textbook of Chapman and Bartels bring us into the modern age of geomagnetism. The bibliography in the Appendix, Section B.7, lists some of the major textbooks about the Earth's geomagnetic field that are currently in use.

For many of us the first exposure to the concept of an electromagnetic field came with our early exploration of the properties of a magnet. Its strong attraction to other magnets and to objects made of iron indicated immediately that something special was happening in the space between the two solid objects. We accepted words such as *field, force field*, and *lines of force* as ways to describe the strength and direction of the push or pull that one magnetic object exerted on another magnetic material that came under its influence. So, to start our subject, I would like to recall a few of our experiences that give reality to the words *magnetic field* and *dipole field*.

Toying with a couple of bar magnets, we find that they will attract or oppose each other depending upon which ends are closer. This experimentation leads us to the realization that the two ends of a magnet have

Figure 1.1. The Chinese report that the compass (Si Nan) is described in the works of Hanfucious, which they date between 280 and 233 B.C. The spoon-shaped magnetite indicator, balancing on its heavy rounded bottom, permits the narrow handle to point southward, to align with the directions carved symmetrically on a nonmagnetic baseplate. This photograph shows a recent reproduction, manufactured and documented by the Central Iron and Steel Research Institute, Beijing.

Figure 1.2. Diagram from Gilbert's 1600 textbook on geomagnetism in which he shows that the Earth behaves as a great magnet. The field directions of a dip-needle compass are indicated as tilted bars.

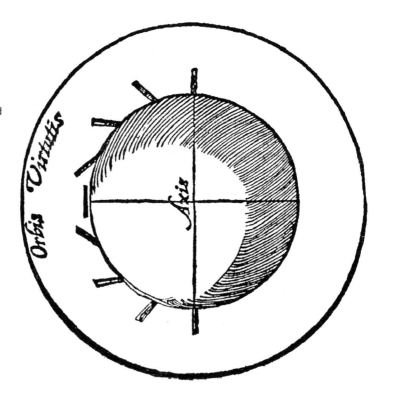

oppositely directed effects or *polarity*. It is a short but easy step for us to understand the operation of a compass when we are told that the Earth behaves as a great magnet.

In school classrooms, at more sophisticated levels of science exploration, many of us learned that positive and negative electric charges also have attraction/repulsion properties. The simple arrangement of two charges of opposite sign constitute an *electric dipole*. The product of the charge size and separation distance is called the *dipole moment*; the pattern of the resulting electric field is called *dipolelike*.

The subject of this chapter is the dipolelike *magnetic field* that we call the "Earth's main field." I will demonstrate that this magnetic field shape is similar in form to the field from a pair of electric charges with opposite signs. Knowing that a loop of current could produce a dipolelike field, arguments are given to discount the existence of a large, solid iron magnet as the Earth-field source. Rather, the main field origin seems to reside with the currents flowing in the outer liquid core of the Earth, which derive their principal alignment from the Earth's axial spin.

To bring together the many measurements of magnetic fields that are made on the Earth, a method has been developed for depicting the systematic behavior of the fields on a spherical surface. This spherical harmonic analysis (SHA) creates a mathematical representation of the entire main field anywhere on Earth using only a small table of numbers. The SHA is also used to prove that the main field of the Earth originates mostly from processes interior to the surface and that only a minor proportion of the field arises from currents in the high and distant external environment of the Earth. We shall see that the SHA divides the contributions of field into dipole, quadrupole, octupole, etc., distinct parts. The largest of these, the dipole component, allows us to fix a geomagnetic coordinate system (overlaying the geographic coordinates) that helps researchers easily organize and explain various geophysical phenomena.

The slow changes of flow processes in the Earth's deep liquid interior that drive the geomagnetic field require new sets of SHA tables and revised geomagnetic maps to be produced regularly over the years. Such changes are typically quite gradual so that some of the past and future conditions are predictable over a short span of time. A special science of paleomagnetism examines the behavior of the ancient field before the Earth assumed its present form. Paleomagnetic field changes, for the most part, are not predictable and give evidence of the magnetism source region and the Earth's evolution.

Also, in this chapter we will look at the definitions of terms used to represent the Earth's main field. We will see how the descriptive maps and field model tables are obtained and appreciate the meaning that these numbers provide for us.

1.2 Magnetic Components

A typical inexpensive compass such as a small needle dipole magnet freely balanced, or suspended at its middle by a long thread, will align itself with the local horizontal magnetic field in a general north–south direction. The north-pointing end of this magnet is called the *north pole*; its opposite end, the *south pole*. Because opposite ends of magnets, or compass needles, are found to attract each other, the Earth's dipole field, attracting the north pole of a magnet toward the northern arctic region, should really be called a south pole. Fortunately, to avoid such confusion, the convention is ignored for the Earth so that geographic and geomagnetic pole names agree. Other adjectives sometimes given are *Boreal* for the northern pole and *Austral* for the southern pole. We say that our compass points northward, although, in fact, it just aligns itself in the north–south direction. The early Chinese, who first used a compass for navigation (at least by the fourteenth century) considered southward to be the important pointing direction (Figure 1.1). Naturally magnetized magnetite formed the first compasses. Early Western civilization called that black, heavy iron compound *lodestone* (sometimes spelled *loadstone*) meaning "leading stone." It is believed that the word "magnet" is derived from Magnesia (north-east of Ephesus in ancient Macedonia) where lodestone was abundant.

By international agreement, a set of names and symbols is used to describe the Earth's field components in a "right-hand system." Figure 1.3 illustrates this nomenclature for a location in the Northern Hemisphere where the total field vector points into the Earth. The term *right-hand system* means that if we aligned the thumb and first two fingers of our right hand with the three edges that converge at a box corner, then the x direction would be indicated by our thumb, the y direction by our index (pointing) finger, and the z direction by the remaining finger. We say these are the three *orthogonal directions* along the X, Y, and Z axes in space because they are at right angles ($90°$) to each other. When a measurement has both a size (magnitude) and a direction, it can be drawn as an arrow with a particular heading that extends a fixed distance (to indicate magnitude) from the origin of an orthogonal coordinate system. Such an arrow is called a *vector* (see Section A.6). Any vector may be represented in space by the composite vectors of its three orthogonal components (projections of the arrow along each axis).

A magnetic field is considered to be in a positive direction if an isolated north magnetic pole would freely move in that field direction.

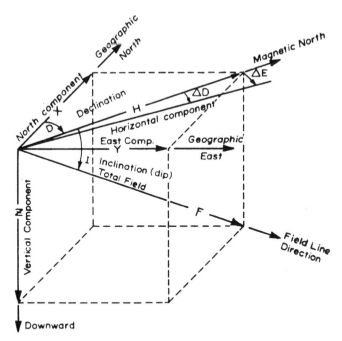

Figure 1.3. Components of the geomagnetic field measurements for a sample Northern Hemisphere total field vector **F** inclined into the Earth. An explanation of the letters and symbols is given in the text.

Observers prefer to describe a vector representing the Earth's field in one of two ways: (1) three orthogonal component field directions with positive values for geographic northward, eastward, and vertical into the Earth (negative values for the opposite directions) or (2) the horizontal magnitude, the eastward (minus sign "–" for westward) angular direction of the horizontal component from geographic northward, and the downward (vertical) component. The first set is typically called the X, Y, and Z (*XYZ-component*) representation; the last set is called the H (horizontal), D (declination), and Z (into the Earth) (*HDZ-component*) representation (or sometimes *DHZ*). In equations, a boldface on a field letter (e.g., **H**) will be used to emphasize the vector property; without the boldface we will just be interested in the size (magnitude).

In the early days of sailing-ship navigation the important measurement for ship direction was simply D, the angle between true north and the direction to which the compass needle points. Ancient magnetic observations therefore used the *HDZ* system of vector representation. By simple geometry we obtain

$$X = H\cos(D), \quad Y = H\sin(D). \tag{1.1}$$

(See Section A.5 for trigonometric functions.) The total field strength, F (or T), is given as

$$F = \sqrt{X^2 + Y^2 + Z^2} = \sqrt{H^2 + Z^2}. \tag{1.2}$$

The angle that the total field makes with the horizontal plane is called the *inclination, I,* or *dip angle*:

$$\frac{Z}{H} = \tan(I). \tag{1.3}$$

The quiet-time annual mean inclination of a station, called its "Dip Latitude", becomes particularly important for the ionosphere at about 60 to 1000 km altitude (Section 2.3) where the local conductivity is dependent upon the field direction.

Although the *XYZ* system provides the presently preferred coordinates for reporting the field and the annual INTERMAGNET data disks follow this system, activity-index requirements (Section 3.13) and some national observatories publish the field in the *HDZ* system. It is a simple matter to change these values using the angular relationships shown in Figure 1.3. The conversion from *X* and *Y* to *H* and *D* becomes

$$H = \sqrt{(X^2 + Y^2)} \text{ and } D = \tan^{-1}(Y/X). \tag{1.4}$$

On occasion, the declination angle *D* in degrees ($D°$) is expressed in magnetic eastward directed field strength *D* (nT) and obtained from the relationship

$$D(\text{nT}) = H \tan(D°). \tag{1.5}$$

Sometimes the change of *D* (nT) about its mean is called a magnetic eastward field strength, ΔE. (For small, incremental changes in a value it is the custom to use the symbol Δ.)

In the Earth's spherical coordinates, the three important directions are the angle (θ) measured from the geographic North Pole along a great circle of longitude, the angle (ϕ) eastward along a latitude line measured from a reference longitude, and the radial direction, *r*, measured from the center of the Earth. On the Earth's surface (where *x, y,* and *z* correspond to the $-\theta, \phi$, and $-r$ directions) the field, **B**, in spherical coordinates becomes

$$B_\theta = -X, \quad B_\phi = Y, \quad \text{and} \quad B_r = -Z. \tag{1.6}$$

Originally, the *HDZ* system was used at most world observatories because the measuring instruments were suspended magnets and there was a direct application to navigation and land survey. Usually, only an angular reading between a compass northward direction and geographic north was needed. In the *HDZ* system, the data from different observatories have different component orientations with respect to the Earth's axis and equatorial plane. The $\theta\phi r$ system is used for mathematical treatments in spherical analysis (of which we will see more in this chapter). The *XYZ* coordinate system is necessary for field recordings by many high-latitude observatories because of the great disparity in the geographic angle toward magnetic north at polar region sites. The XYZ system is becoming the preferred coordinate system for most modern digital observatories. Computers have made it simple to interchange the digital field representation into the three coordinate systems. Figure 1.3 shows the angle of *inclination (dip)*, I, and the *total field* vector, **F**.

The unit size of fields is a measurable quantity. We can appreciate this fact when we consider the amount of force needed to separate magnets of different strengths or the amount of force that must be used to push a compass needle away from its desired north–south direction. Let us not elaborate on tedious details of establishing the unit sizes of fields. What will be called "field strength" results from a measurement of a quantity called "magnetic flux density," B, that can be obtained from a comparison to force measurements under precisely prescribed conditions. The units for this field strength have appeared differently over the years; Table 1.1 lists equivalent values of B.

Table 1.1. *Equivalent magnetic field units*

$B = 10^4$ Gauss
$B = 1$ Weber/meter^2
$B = 10^9$ gamma
$B = 1$ Tesla

At present, in most common usage, the convenient size of magnetic field units is the gamma, or γ, a lower-case Greek letter to honor Carl Friedrich Gauss, the nineteenth-century scientist from Göttingen, Germany, who contributed greatly to our knowledge of geomagnetism. The International System (SI) of units, specified by an agreement of world scientists, recommends use of the Tesla (the name of an early pioneer in radiowave research). With the prefix nano meaning 10^{-9}, of course, one gamma is equivalent to one nanotesla (nT), so there should be no confusion when we see either of these expressions. To familiarize the reader with this interchange (which is common in the present literature), I will use either name at different times in this book.

Geomagnetic phenomena have a broad range of scales. The main field is nearly $60,000$ (6×10^4; see scientific notation in Section A.3) gamma near the poles and about $30,000$ (3×10^4) gamma near the equator. A small, 2 cm, calibration magnet I have in my office is 1×10^8 gamma at its pole (about 10,000 times the Earth's surface field in strength). Quiet-time daily field variations can be about 20 gamma at midlatitudes and 100 gamma at equatorial regions. Solar–terrestrial disturbance–time variations occasionally reach 1,000 gamma at the auroral regions and 250 gamma at midlatitudes. Geomagnetic pulsations arising in the Earth's space environment are measured in the 0.01 gamma to 10 gamma range at surface midlatitude locations. In Chapter 4 we will see how this great 10^6 dynamic range of the source fields is accommodated by the measuring instruments.

The magnetic fields that interest us arise from currents. Currents come from charges that are moving. Much of the research in geomagnetism concerns the discovery (or the use of) the source currents responsible for the fields found in the Earth's environment. Then, we ask, what about the fields from magnetic materials; where is the current to be found? A simple "Bohr model" (with planetarylike electrons about a sunlike nucleus) suffices in our requirements for visualizing the atomic structure. In this model the spinning charges of orbital electrons in the atomic structure provide the major magnetic properties. Most atoms in nature contain even numbers of orbiting electrons, half circulating in one direction, half in another, with both their orbital and spin magnetic effects canceling. When canceling does not occur, typically when there are unpaired electrons, there is a tendency for the spins of adjacent atoms or molecules to align, establishing a domain of unique field direction. Large groupings of similar domains give a magnet its special properties. We will discuss this subject further in Section 4.2 on geomagnetic instruments. However, for now, we find consistency in the idea that charges-in-motion create our observed magnetic fields.

1.3 Simple Dipole Field

To many of us, the first exposure to the term "dipole" occurred in learning about the electric field of two point charges of opposite sign placed a short distance from each other. Figure 1.4 represents such an arrangement of charges, $+q$ and $-q$ (whose sizes are measured in units called "coulombs"), separated by a distance, d, along the z axis of an orthogonal coordinate system. We call the value (qd) by the distinctive name *dipole moment* and assign it the letter "p." The units of p are coulombmeters. Figure 1.5 shows the dipole as well as symmetric quadrupole and octupole arrangements of charge at the corners of the respective figures. The reason for introducing the electric charges and multipoles here is to help us understand the nomenclature of the magnetic fields, for which isolated poles do not exist, although the magnetic field shapes are identical to the shapes of multipole electric fields.

The point $P(x, y, z)$ is the location, for position x, y, and z from the dipole axis origin, at which the electric field strength from the dipole charges is to be determined (Figure 1.4). This location is a distance

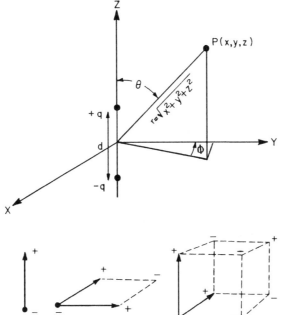

Figure 1.4. Electric dipole of charges, $\pm q$, and the corresponding coordinate system. An explanation of the letters and symbols is given in the text.

Figure 1.5. Charge distribution for the electric dipole, quadrupole, and octupole configurations.

dipole   quadrupole   octupole

$r = \sqrt{x^2 + y^2 + z^2}$ from the midpoint between the two charges and at an angle θ from the positive Z axis. We will call this angle the colatitude (colatitude $= 90° -$ latitude) of a location. The angle to the projection of r onto the X–Y plane, measured clockwise, is called ϕ. Later we will identify this angle with east longitude. Sometimes we will see the letter e, with the r, θ, or ϕ subscript, used to indicate the respective unit directions in spherical coordinates. Obviously, there is symmetry about the Z axis so the electric field at P doesn't change with changes in ϕ. Thus it will be sufficient to describe the components of the dipole field simply along r and θ directions.

Now I will need to use some mathematics. It is necessary to show the exact description that defines something we can easily visualize: the shape of an electric field resulting from two electric charges of opposite sign and separated by a small distance. I will then demonstrate that such a mathematical representation is identical to the form of a field from a current flowing in a circular loop. That proof is important for all our descriptions of the Earth's main field and its properties because we will need to discuss the source of the main dipole field, global coordinate systems, and main-field models. If the mathematics at this point is too difficult, just read it lightly to obtain the direction of the development and come back to the details when you are more prepared.

Let us start with a property called the *electric potential*, Ω, of a point charge in air from which we will subsequently obtain the electric field.

$$\Omega = \frac{q}{4\pi \in_0 r},\qquad(1.7)$$

where q is the coulomb charge, r is the distance in meters to the observation, $\in_0$ (the "inductive capacity of free space") is a constant typical of the medium in which the field is measured, and Ω is measured in volts.

For two charges with opposite signs, separated by a distance, d, the potential at point r at x, y, z coordinate distances becomes

$$\Omega = \frac{1}{4\pi \in_0}\left[\frac{q}{\sqrt{(z-d/2)^2 + x^2 + y^2}} + \frac{-q}{\sqrt{(z+d/2)^2 + x^2 + y^2}}\right]\qquad(1.8)$$

Now, for the typical dipole, d is very small with respect to r so, with some algebraic manipulation, we can write

$$\Omega = \frac{q}{4\pi \in_0 r}\left[\left(1 + \frac{zd}{2r^2}\right) - \left(1 - \frac{zd}{2r^2}\right)\right] + \Delta A,\qquad(1.9)$$

where ΔA represents terms that become negligible when $d = r$. Because $(z/r) = \cos(\theta)$, Equation (1.9) can be written in the form

$$\Omega = \frac{qd\cos(\theta)}{4\pi \in_0 r^2}.\qquad(1.10)$$

Now let us see the form of the electric field using Equation (1.7). We are going to be interested in a quantity called the *gradient* or *grad* (represented by an upside-down Greek capital delta; see Section A.8) of the potential. In spherical coordinates the gradient can be represented by the derivatives (slopes) in the separate coordinate directions:

$$\nabla\Omega = \mathbf{e}_r\left(\frac{\delta\Omega}{\delta r}\right) + \mathbf{e}_\theta\left(\frac{\delta\Omega}{r\delta\theta}\right) + \mathbf{e}_\phi\left(\frac{1}{r(\sin\theta)}\frac{\delta\Omega}{\delta\phi}\right),\qquad(1.11)$$

where the es are the unit vectors in the three spherical coordinate directions, r, θ, and ϕ. The electric field, obtained from the negative of that gradient, is given as

$$\mathbf{E}_r = -\frac{\delta\Omega}{\delta r} = \frac{p}{2\pi \in_0}\left(\frac{\cos\theta}{r^3}\right)\mathbf{e}_r\qquad(1.12)$$

and

$$\mathbf{E}_\theta = -\frac{1}{r}\frac{\delta\Omega}{\delta\theta} = \frac{p}{4\pi \in_0}\left(\frac{\sin\theta}{r^3}\right)\mathbf{e}_\theta,\qquad(1.13)$$

where p is the electric dipole moment qd.

Symmetry about the dipole axis means Ω doesn't change with angle ϕ, so $\mathbf{E}$ in the ϕ direction is zero. Equations (1.12) and (1.13) define the form of an electric dipole field strength in space. If we would like to draw lines representing the shape of this dipole field (to show the directions

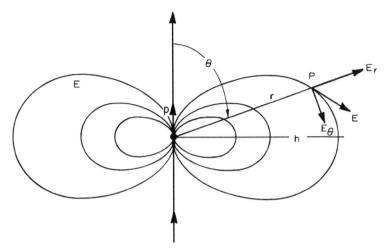

Figure 1.6. Electric dipole (of moment p) field configuration with directions for the components of electric field vectors **E** in the r and θ directions to an arbitrary observation point P; h is the equatorial field line distance.

that a charge would move in its environment), there is a convenient equation,

$$r = h \sin^2(\theta), \tag{1.14}$$

that can be used, in which h is the distance from the dipole center to the equatorial crossing (at $\theta = 90°$) of the field line (Figure 1.6).

These field descriptions (Equations (1.11) to (1.13)) come from scientists, mainly of the late eighteenth and early nineteenth centuries, who interpreted laboratory measurements of charges, currents, and fields to establish mathematical descriptions of the natural electromagnetic "laws" they observed. At first there was a multitude of laws and equations, covering many situations of currents and charges and relating electricity to magnetism. Then by 1873, James Clerk Maxwell brought order to the subject by demonstrating that all the "laws" could be derived from a few simple equations (that is, "simple" in mathematical form). For example, in a region where there is no electric charge, Maxwell's equations show that there is no "divergence" of electric field, a statement that mathematical shorthand shows as

$$\nabla \cdot \mathbf{E} = 0, \tag{1.15}$$

where the "del-dot" symbol is explained in Section A.8. But **E** is given as $-\text{grad}\,\Omega$ (which is titled "the negative gradient of the scalar potential"). Thus, for the mathematically inclined, it follows that

$$\nabla \cdot \mathbf{E} = -\nabla\nabla\Omega = 0 \tag{1.16}$$

or

$$\nabla^2\Omega = 0, \tag{1.17}$$

for which Equation (1.10) can be shown (by those skilled in mathematical manipulations) to be a solution for a dipole configuration of charges.

A simple experiment, often duplicated in science classrooms, is to connect a battery and an electrical resistor to the ends of an iron wire (with an insulated coating) that has been wrapped in a number of turns, spiraling about a wooden matchstick for shape. It is then demonstrated that when current flows, the wire helix behaves as if it were a dipole magnet aligned with the matchstick, picking up paper clips or deflecting a compass needle. If the current direction is reversed by interchanging the battery connections, then the magnetic field direction reverses.

Now let us illustrate with mathematics how a current flowing in a simple wire loop produces a magnetic field in the same form as the electric dipole. My purpose is to help us visualize a magnetic dipole, when there isn't a magnetic substance corresponding to the electric charges, so that we can later understand the origin of the main field in the liquid flows of the Earth's deep-core region.

Consider Figure 1.7, in which a current, $\mathbf{i}$, is flowing in the $X–Y$ plane along a loop enclosing area, A, of radius b, for which Z is the normal (perpendicular) direction. Let P be any point at a distance, r, from the loop center and at a distance, R, from a current element moving a distance, ds. The electromagnetic law for computing the element of field, dB, from the current along the wire element, ds, is

$$dB = \frac{\mu_0 i}{4\pi} \frac{ds\, R \sin(\alpha)}{R^3},\qquad (1.18)$$

where α is the angle between ds and R, so that dB is in the direction that

Figure 1.7. Coordinate system for a loop of current i, having radius b, area A, and enclosed perimeter of element length ds. The magnetic field vectors of $\mathbf{B}$ in the r and θ directions with respect to an orthogonal (right-hand) x, y, z coordinate system are shown.

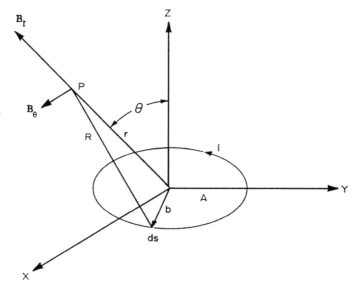

a right-handed screw would move when turning from ds (in the current direction) toward R (directed toward P). The magnetic properties of the medium are indicated by the constant, μ_0, called the "permeability of free space." We wish to find the r and θ magnetic field components of $\mathbf{B}$, at any point in space about the loop, with the simplifying conditions that $r \gg b$. Using the electromagnetic laws, we sum the dB contributions to the field for each element of distance around the loop, and after some mathematics obtain

$$B_r = \frac{\mu_0 i A \cos(\theta)}{2\pi r^3} \tag{1.19}$$

and

$$B_\theta = \frac{\mu_0 i A \sin(\theta)}{4\pi r^3}. \tag{1.20}$$

Comparing these two field representations with those we obtained for the electric dipole (Equations (1.12) and (1.13)), we see that the same field forms will be produced if we let the current times the area (iA) correspond to p, the electric dipole moment, qd. Thus, calling M the *magnetic dipole moment*,

$$M = iA \tag{1.21}$$

or

$$M = md, \tag{1.22}$$

where d becomes the equivalent separation of hypothetical magnetic poles of strength m.

We saw, in the parallel case of the electric dipole, that E was obtained from the scalar potential in a charge-free region. In a similar fashion, Maxwell's equations show that

$$\mathbf{B} = -\nabla V, \tag{1.23}$$

where V is called the magnetic scalar potential. Then, a person with math competence can write, for a current-free region (where curl $\mathbf{B} = 0$),

$$\nabla^2 V = 0 \tag{1.24}$$

and obtain a dipole solution

$$V = \frac{\mu_0 M \cos(\theta)}{4\pi r^2}. \tag{1.25}$$

To a first approximation, the Earth's field in space behaves as a magnetic dipole. At the Earth's surface we call $r = a$. Then

$$B_r = -Z = \frac{2[\mu_0 M \cos(\theta)]}{4\pi a^3} = Z_0 \cos(\theta) \tag{1.26}$$

and

$$B_\theta = -H = \frac{\mu_0 M \sin(\theta)}{4\pi a^3} = H_0 \sin(\theta), \qquad (1.27)$$

where constant $H_0 = Z_0/2$. The total field magnitude, F, is just

$$F = \sqrt{H^2 + Z^2}. \qquad (1.28)$$

On average, about ninety percent of the Earth's field is dipolar so we can use the approximation, $H_0 = 3.1 \times 10^4$ gamma, for rough field modeling. In Equations (1.26) and (1.27), recall (Section A.5) that $\sin(90°) = 1$, $\sin(0°) = 0$, $\cos(90°) = 0$, and $\cos(0°) = 1$. For the Southern Hemisphere, where $90° < \theta \leq 180°$, note that $\sin(180° - \theta) = \sin(\theta)$ and $\cos(180° - \theta) = -\cos(\theta)$. So the magnitude of the Earth's field at the equator ($\theta = 90°$) is just H_0 and at the poles ($\theta = 0°$ or $180°$) just $2H_0$.

For the dipole, the inclination, I (direction of the Earth's field away from the horizontal plane), at any θ is defined from

$$\tan(I) = \frac{Z}{H} = 2\cot(\theta). \qquad (1.29)$$

This is a valuable relationship for measurements of continental drift (see Section 5.10). It means that we can determine our geomagnetic latitude ($90° - \theta$) from field measurements of H and Z. Using ancient rocks to tell the field direction in an earlier geological time, the apparent latitude of the region can be fixed by Equation (1.29). Later, in Section 1.9, there will be more details regarding this paleomagnetism subject.

Conjugate points on the Earth's surface are locations P and P' that can be connected by a single dipole field line (Figure 1.8). The relatively strong Earth's field lines become guiding tracks for charged particles in the magnetosphere. The positions for conjugate points are used in studies of the Earth arrival of these phenomena from distant locations in space. The dipole field lines will extend out into the equatorial plane a distance, r_e. Up to about 65° geomagnetic latitude, θ' (in degrees), the length of this field line can be approximated by the relationship

$$\text{length} \approx 0.38\theta' r_e \qquad (1.30)$$

where the length and r_e are in similar units (e.g., kilometers or Earth radii). Figure 1.9 shows the relationship of latitude and field-line equatorial distance. As an illustration, at 50° geomagnetic latitude, read the appropriate x-axis scale; move vertically to the curve intersection, then read horizontally to the corresponding y-axis scale, obtaining 2.5 Earth radii for the distant extent of that field line. We shall see, in Chapter 3, that the outermost field lines of the Earth's dipole field are distorted

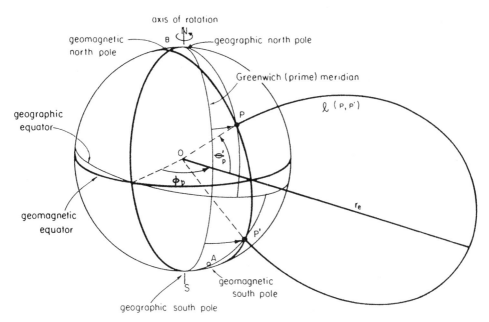

Figure 1.8. Geomagnetic locations, based on a spherical coordinate system aligned with respect to the dipole field, with latitude $\theta' = 90 - \theta$ (where θ is the colatitude) and longitude ϕ. A dipole field line of length l, connecting the conjugate points at P and P', extends to a distance r_e in the equatorial plane from the dipole center 0.

by a wind of particles and fields that arrive from the Sun; such change becomes quite noticeable above 60°.

The *magnetic shell parameter, L* shell, is an effective mean equatorial radius of a magnetic field shell, which, for a given field strength, B, defines the trapped-particle flux in the space about the Earth. Computation of the L shells for the Earth's field is complex. However, for a dipole field, the L-shell values may be considered almost equivalent to the number of Earth radii that the field line extends into space, r_e, and is a good approximation for all but the high latitudes. The *invariant latitude* (in degrees) of a location is obtained from L by the relationship

$$\cos(\text{invariant latitude}) = \frac{1}{\sqrt{L}} \qquad (1.31)$$

Figure 1.10 shows polar views of the L-shell contours for the two hemispheres, computed for the model, extremely quiet field of 1965. Many of the high-latitude geomagnetic phenomena are best organized when plotted with respect to their L shell or invariant latitude positions.

1.4 Full Representation of the Main Field

Now comes the most difficult part of this book, the representation of the main field by equations and tables. There is a considerable amount

Figure 1.9. Equatorial radial extent r_e (from the Earth's center) of a dipole field line starting from latitude θ' at the Earth's surface.

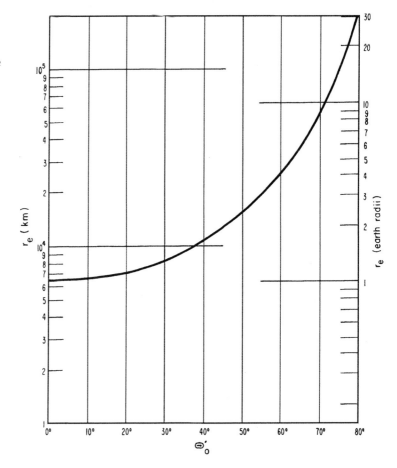

of mathematics here, with analysis techniques and shorthand math symbols that can frighten the casual reader. I will try to step gently in this section, but it is necessary for us to go through the details because there will be so many important physical results we can properly appreciate later if we understand their origin in the main field representation.

We will start with some of Maxwell's equations and show how the relationships appear in a spherical coordinate system. Then we will look for a solution of the equations of a type that will let us separate current sources that arise above and below a sphere's surface. Next, we will look at a method for fitting the measurements from a surface of observatory field values into the equations that produce our Earth's field models. It is important to know some of the strengths and failings of the methods so that we understand their successful application in geophysics. We will also find the main field representation important in the chapter on

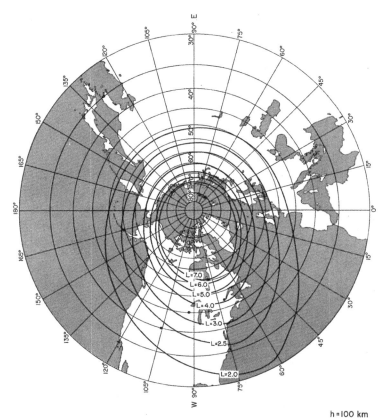

Figure 1.10. L-shell contours, computed for 100-km altitude, in the Northern (top) and Southern (bottom) Hemisphere regions. Geographic east and west radial longitude lines and circles of latitude (from 30° to the pole) are shown. These L-values were computed for the extremely quiet year, 1965, when there was a minimum distortion of the polar contours by solar wind.

h = 100 km

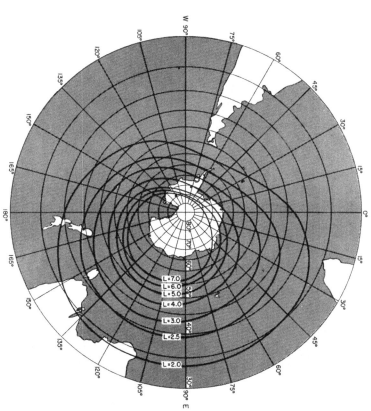

h = 100 km

quiet-field variations as well as in our discussion of solar–terrestrial disturbances. If you are not ready for the mathematics at this time, at least read through the steps lightly to focus upon what is being computed and the sequence followed.

Maxwell's great contribution to the understanding of electromagnetic phenomena was to show that all the measurements and laws of field behavior could be derived from a few compact mathematical expressions. We will start with one of these equations, adjusted for the assumptions that only negligible electric field changes occur and that the amount of current flowing across the boundary between the Earth and its atmosphere is relatively insignificant. Then, at the Earth's surface

$$\nabla \times \mathbf{B} = \mathbf{i}\left(\frac{\delta B_z}{\delta y} - \frac{\delta B_y}{\delta z}\right) + \mathbf{j}\left(\frac{\delta B_x}{\delta z} - \frac{\delta B_z}{\delta x}\right) + \mathbf{k}\left(\frac{\delta B_y}{\delta x} - \frac{\delta B_x}{\delta y}\right) = 0, \quad (1.32)$$

where $\mathbf{i}, \mathbf{j}, \mathbf{k}$ represent the three orthogonal directions and δ indicates that "partial" derivatives are used (see Section A.7). This equation is read "the curl of B equals zero" and requires that the field can be obtained from the "negative gradient of a scalar potential" so

$$\mathbf{B} = -\left[\mathbf{i}\frac{\delta V}{\delta x} + \mathbf{j}\frac{\delta V}{\delta y} + \mathbf{k}\frac{\delta V}{\delta z}\right] = -\nabla V. \quad (1.33)$$

The other Maxwell's equation that we will use is

$$\nabla \cdot \mathbf{B} = \left[\frac{\delta B_x}{\delta x} + \frac{\delta B_y}{\delta y} + \frac{\delta B_z}{\delta z}\right] = 0. \quad (1.34)$$

This equation is read "the divergence of the field is zero." Now, putting Equations (1.33) and (1.34) together, we obtain

$$\nabla \cdot \nabla V = \nabla^2 V = 0, \quad (1.35)$$

which is read as "the Laplacian of scalar V is zero." This *potential function* will be valid over a spherical surface through which current does not flow. In spherical coordinate notation, Equation (1.35) becomes

$$\frac{\delta}{\delta r}\left(r^2\frac{\delta V}{\delta r}\right) + \frac{1}{\sin\theta}\frac{\delta}{\delta\theta}\left(\sin\theta\frac{\delta V}{\delta\theta}\right) + \frac{1}{\sin^2\theta}\frac{\delta^2 V}{\delta\phi^2} = 0, \quad (1.36)$$

in which r, θ, and ϕ are the geographic, Earth-centered coordinates of the radial distance, colatitude, and longitude, respectively.

Now, the solution (i.e., solving the equation for an expression of V by itself) that is sought is one that is a product of three expressions. The first of these expressions is to be only a function of r; the second, only a function of θ; and the third, only a function of ϕ. That is what mathematicians call a "separable" solution of the form

$$V(r, \theta, \phi) = R(r) \cdot S(\theta, \phi), \quad \text{where } S(\theta, \phi) = T(\theta) \cdot L(\phi). \tag{1.37}$$

A solution of the potential function V for the Earth's main field satisfying these requirements has the converging series of terms (devised by Gauss in 1838)

$$V = a \sum_{n=1}^{\infty} \left[\left(\frac{r}{a} \right)^n S_n^e + \left(\frac{a}{r} \right)^{n+1} S_n^i \right], \tag{1.38}$$

where the $\sum$ means the sum of terms as n goes from 1 to an extremely large number, and for our studies, a is the Earth radius, R_e. The series solution means that for each value of n the electromagnetic laws are obeyed as if that term were the only contribution to the field. We will soon see that solving this equation for V allows us to immediately recover the strength of the magnetic field components at any location about the Earth.

There are two series for V. The first is made up of terms in r^n. As r increases, these terms become larger and larger; that means we must be approaching the current source of an external field in the increasing r direction. These terms are called V_e, "the external source terms of the potential function" (and our reason for labeling the S_n functions with a superscript e). By a corresponding argument for the second series, the $(1/r)^n$ terms become larger and larger as r becomes smaller and smaller, which means we must be approaching the current source of an internal field in the decreasing r direction. Scientists call these terms V_i, "the internal source terms of the potential function" (the reason for labeling the corresponding S_n functions with a superscript i).

The $S(\theta, \phi)$ terms of Equation (1.37) represent sets of a special class of functions called *Legendre polynomials* (see Section C.11) of the independent variable θ that are multiplied by sine and cosine function terms of independent variable ϕ. I shall leave to more detailed textbooks the explanation of what is called the required "orthogonality" properties and "normalization" and simply define the "Schmidt quasi-normalized, associated Legendre polynomial functions" that are used for global field analysis. Here I will abbreviate these as Legendre polynomials, $P_n^m(\theta)$, realizing they are a special subgroup of functions. The integers, n and m, are called *degree* and *order*, respectively; n has a value of 1 or greater, and m is always less than or equal to n.

When V is determined from measurements of the field about the Earth, analyses show that essentially all the contribution comes from the V_i part of the potential function expansion. For now, let us just call this

ASSOCIATED LEGENDRE POLYNOMIAL HARMONICS $P_n^m(\theta)$

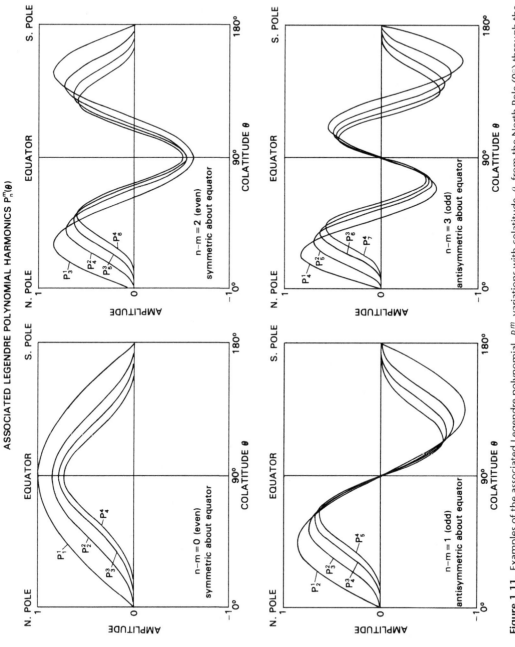

Figure 1.11. Examples of the associated Legendre polynomial, P_n^m, variations with colatitude, θ, from the North Pole (0°) through the equator (90°) to the South Pole (180°) for selected values of degree n and order m. The four sets are separated for similar values of $(n - m)$.

V and write the internal part with the Legendre terms

$$V = a \sum_{n=1}^{\infty} \left(\frac{a}{r}\right)^{n+1} \sum_{m=0}^{n} \left[g_n^m \cos(m\phi) + h_n^m \sin(m\phi)\right] P_n^m(\theta). \qquad (1.39)$$

The g_n^m and h_n^m are constants called the Gauss coefficients in recognition of Gauss' development of this analysis technique for geomagnetism. Let us pause here to look at some important characteristics of the Legendre polynomials that are used in this spherical fit.

The P_n^m are functions of colatitude θ only. They are quasi-sinusoidal oscillations having $n - m + 1$ waves (or just n waves if $m = 0$) as θ changes from $0°$ to $360°$ along a great circle of longitude. Figure 1.11 illustrates the variation of some of these waves (because of the obvious symmetry just the first half, to $180°$, need be displayed). Note how the even values of $n - m$ produce a symmetry about the equator (at $90°$), whereas the odd values of $n - m$ produce antisymmetric values about the equator. Also, as m increases in value, the "peaks" and "valleys" are sharpened from the more sinusoidal-appearing smooth shape.

For the mathematically inclined, let us pause a moment to show how the P_n^m are computed.

$$R_n^m = \sqrt{n^2 - m^2} \qquad (1.40)$$

$$P_0^0 = 1 \qquad (1.41)$$

$$P_1^0 = \cos(\theta); \quad P_1^1 = \sin(\theta) \qquad (1.42)$$

$$P_m^m = \sqrt{\frac{2m - 1}{2m}} \sin(\theta) P_{m-1}^{m-1} \quad \text{for } m > 1, n = m \qquad (1.43)$$

$$P_n^m = \left[(2n - 1)\cos(\theta)P_{n-1}^m - R_{n-1}^m P_{n-2}^m\right] / R_n^m \quad \text{for } n > m \qquad (1.44)$$

$$\frac{dP_n^m}{d\theta} = \left(n\cos(\theta)P_n^m - R_n^m P_{n-1}^m\right) / \sin(\theta) \text{ (except for } \theta = 0 \text{ or } 180), \qquad (1.45)$$

where Equation (1.45) is undefined at the poles ($\theta = 0°$ or $180°$). We call such formulas as these "recursion formulas" to indicate how the entire set, for different n and m, is obtained from the definition of the few earlier values in the set.

Now, looking back at our geomagnetic field potential function solution, Equation (1.39), we see that the functions multiplying the Legendre terms appear somewhat like a Fourier series: a harmonic series of cosine and sine terms that, when added, will produce the function they are to represent. Figure 1.12 illustrates how four Fourier harmonics ($m = 1, 2, 3,$ and 4) add to produce an oddly shaped $f(\phi)$ variation over $360°$. In mathematical form we would write for the Fourier representation

$$f(\phi) = A_0 + \sum_{m=1}^{M} \left[A_m \cos(m\phi) + B_m \sin(m\phi)\right] \qquad (1.46)$$

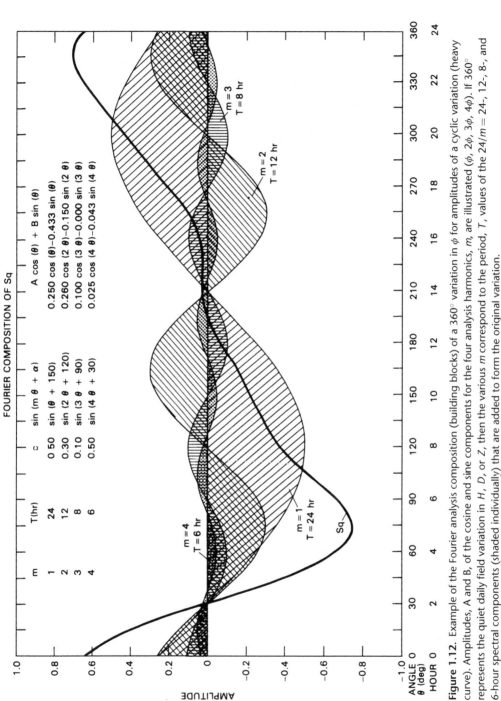

FOURIER COMPOSITION OF Sq

m	T(hr)	c	sin (m θ + α)	A cos (θ) + B sin (θ)
1	24	0.50	sin (θ + 150)	0.250 cos (θ) −0.433 sin (θ)
2	12	0.30	sin (2 θ + 120)	0.260 cos (2 θ) −0.150 sin (2 θ)
3	8	0.10	sin (3 θ + 90)	0.100 cos (3 θ) −0.000 sin (3 θ)
4	6	0.50	sin (4 θ + 30)	0.025 cos (4 θ) −0.043 sin (4 θ)

AMPLITUDE

| ANGLE θ (deg) | 0 | 30 | 60 | 90 | 120 | 150 | 180 | 210 | 240 | 270 | 300 | 330 | 360 |
| HOUR | 0 | 2 | 4 | 6 | 8 | 10 | 12 | 14 | 16 | 18 | 20 | 22 | 24 |

Figure 1.12. Example of the Fourier analysis composition (building blocks) of a 360° variation in φ for amplitudes of a cyclic variation (heavy curve). Amplitudes, A and B, of the cosine and sine components for the four analysis harmonics, m, are illustrated (φ, 2φ, 3φ, 4φ). If 360° represents the quiet daily field variation in H, D, or Z, then the various m correspond to the period, T, values of the 24/m = 24, 12-, 8-, and 6-hour spectral components (shaded individually) that are added to form the original variation.

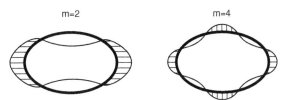

Figure 1.13. Positive (shaded) and negative (open) values of $m = 2$ and $m = 4$ harmonic oscillations represented about a latitude baseline (solid oval) viewed at a 45° angle.

or

$$f(\phi) = A_0 + \sum_{m=1}^{M} [C_m \sin(m\phi + \alpha)], \qquad (1.47)$$

where the A, B, and C are Fourier expansion constants that depend on the index m, with $C_m = \sqrt{A_m^2 + B_m^2}$; they are not SHA coefficients. The phase shift, α, is the angle whose tangent is A_n/B_n. The summation, Σ, is from $m = 1$ to M (the largest value of m).

For the Fourier analysis, it is assumed that the wave form being fitted repeats in every interval for both directions outside the region of interest. If we are instead studying some values along a latitude line of a sphere, then the fitted waves repeat by the natural closing around the circle of fixed latitude. Figure 1.13 shows Fourier harmonic waves for $m = 2$ and $m = 4$; the dark ovals represent a latitude line (viewed from a 45° angle in space), and the waves are shaded when values are positive and unshaded when the values are negative.

In spherical harmonic analysis, for the function being fit to the data (Equation (1.39)), the g and h are amplitude coefficients, the $\cos(m\phi)$ and $\sin(m\phi)$ are Fourier harmonic type sinusoidal oscillations about a latitude circle, and the $P_n^m(\theta)$ are the Legendre wave oscillations along a great circle of longitude.

The spherical harmonics that are used to produce the potential function (fitting the observed Earth surface values) are added in a manner similar to the way we added the sine and cosine waves in the last example. Figures 1.14 and 1.15 illustrate the pattern for various harmonics.

The nature of these polynomials requires that n be greater than or equal to m in size. There are m sine and cosine waves fitted around each latitude circle (called "sectoral harmonics"). Around each great circle of longitude there are "zonal harmonics" of Legendre polynomial waves: n of these if $m = 0$ (because a single half-cycle is located over each polar cap) and $n - m + 1$ of these waves if $m > 0$. Note the symmetry or asymmetry about the equator determined by the respective even or odd values of $(n - m)$. The finest-detail representation (shortest wavelength for fitting) along a latitude line is found by dividing 360° (degrees in the circle) by the largest value of m; along a longitude line it is found by dividing 360° by the largest value of n. With degree and order 12

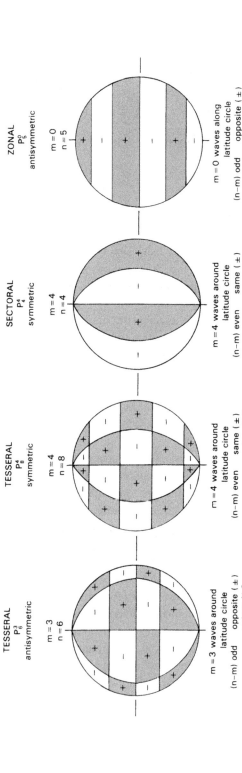

Figure 1.14. Tesseral, sectoral, and zonal forms of Legendre polynomial P_n^m spherical harmonics with values indicated by positive (shaded) and negative (clear) areas. The harmonics are symmetric or antisymmetric with respect to the equatorial plane. The four spheres show selections of n and m to determine the number of waves around latitude and longitude circles.

Figure 1.15. Legendre polynomial surface having $n = 12$ and $m = 6$. Polynomials such as this one are added using appropriate coefficients, to shape the surface potential function in a manner similar to the Figure 1.12 Fourier fitting. See the **SPH** program in Section C.11 for other displays of this type. Figure from programs by P. Swartztrauber and C. Ridley of NCAR.

the shortest wavelengths are about $30°$. Such wavelengths are important in providing some idea of the resolution we can expect in field models obtained from the SHA fittings. The computer program **SPH** described in Section C.11 is a graphic demonstration of the various forms taken by the spherical harmonics.

The analysis pole is selected by the person performing the analysis; for the main field, the Earth's spin axis is usually used. We shall see later that in the quiet-field variation analysis the geomagnetic pole axis is selected. If the analysis pole is the dominant feature by which the data is naturally organized, then the mathematical series representing the function will converge quickly (i.e., the amplitude of the higher numbered terms will diminish rapidly). If the analysis pole is far from the natural field pole, then many high-degree and high-order harmonics are needed to represent the observed field. Of course, with a large number of terms, the final field representations would be indistinguishable no matter where the analysis pole is selected. Computation of the SHA terms is usually truncated at that point where further field contribution is negligible or where the values of higher-degree and high-order harmonics are considered to be of crustal (rather than main field) origin.

Now, just how does one go about computing the Gauss coefficients of the SHA when what we have are the measurements of X, Y, and Z fields about the Earth? There are many routines presently used to accommodate observatory, temporary survey, and satellite recordings; to adjust for an elevation of the sample location; and to establish a common analysis time period (epoch). To understand the basic SHA method, let us follow through an early technique, in use when all data were from measurements taken at an irregular distribution of fixed locations on a spherical Earth surface.

First of all, some extrapolation and smoothing arrangements are followed so that the three-component field observations are represented at evenly spaced latitude and longitude locations all over the Earth. Next, at each colatitude, θ, a Fourier analysis along the latitude line provides m sine and cosine coefficients to represent this field for each X, Y, and Z component separately. We call the cosine coefficients X_c^m and the sine coefficients X_s^m and in a similar way name the Y and Z field components. With these values we compute some intermediate values:

$$a_n^0 = \frac{2n+1}{2n(n+1)} \int_0^{180} X_c^0 \frac{dP_n^0}{d\theta} \sin(\theta) d\theta \quad \text{for } n > 0, m = 0 \tag{1.48}$$

$$a_n^m = \frac{2n+1}{4n(n+1)} \int_0^{180} \left[X_c^m \frac{dP_n^m}{d\theta} \sin(\theta) + Y_s^m m P_n^m \right] d\theta \quad \text{for } m > 0 \tag{1.49}$$

$$b_n^m = \frac{2n+1}{4n(n+1)} \int_0^{180} \left[X_s^m \frac{dP_n^m}{d\theta} \sin(\theta) - Y_c^m m P_n^m \right] d\theta \quad \text{for } m > 0 \tag{1.50}$$

$$c_n^m = \frac{2n+1}{4} \int_0^{180} Z_c^m P_n^m \sin(\theta) d\theta \quad \text{for } m > 0 \tag{1.51}$$

$$\left(\text{but if } m = 0, c_n^0 = 2c_n^m \right)$$

$$d_n^m = \frac{2n+1}{4} \int_0^{180} Z_s^m P_n^m \sin(\theta) d\theta \quad \text{for } m > 0 \tag{1.52}$$

$$c_n^0 = \frac{2n+1}{4} \int_0^{180} Z_c^0 P_n^0 \sin(\theta) d\theta \quad \text{for } n > 0, m = 0. \tag{1.53}$$

Because the integral sign $\int$ here means a summation (see Section A.7) over a θ range of 0 to 180°, in the computation we can replace the integral by summing over that range in finite increments of $\Delta\theta$ steps. The size of these steps is selected to be appropriate to the wavelength resolution that is to be accomplished by the SHA fitting. For example, our wavelength representation for degree and order 12 was 30°. We ask how many values we would want to resolve this wave. That would probably be twelve. If so, we consider 30°/12 and get between 2° and 3° for the Δ steps in the summation.

From the values of Equations (1.48) through (1.53) we can compute what will become the Legendre polynomial coefficients:

$$a_n^{me} = \frac{(n+1)a_n^m + c_n^m}{2n+1} \tag{1.54}$$

$$a_n^{mi} = \frac{na_n^m - c_n^m}{2n + 1} \tag{1.55}$$

$$b_n^{me} = \frac{(n+1)b_n^m + d_n^m}{2n + 1} \tag{1.56}$$

$$b_n^{mi} = \frac{nb_n^m - d_n^m}{2n + 1} \tag{1.57}$$

where e and i represent the external and internal values, respectively. For the Earth's main field analysis the internal coefficients are the most significant; these are then renamed the Gauss coefficients g_n^m (corresponding to a_n^{mi}) and h_n^m (corresponding to the b_n^{mi}). It is a table of Gauss coefficients that becomes the SHA main field model, providing a smooth fit to the observations.

At this point, to shorten the presentation, whenever e or i is omitted it can be assumed that either can be used in the equation for the corresponding external or internal computations.

Now, how do we recover the fields from the table of these coefficients? First, we compute the potential function from

$$V(\theta, \phi) = \sum_{m=0}^{M} \left[V_c^m(\theta) \cos(m\phi) + V_s^m(\theta) \sin(m\phi) \right] \tag{1.58}$$

with the cosine and sine parts of the summation given by

$$V_c^m(\theta) = R_e \sum_{n=m}^{N} a_n^m P_n^m \tag{1.59}$$

and

$$V_s^m(\theta) = R_e \sum_{n=m}^{N} b_n^m P_n^m. \tag{1.60}$$

The N and M are the largest values of n and m, and R_e is the radial location of the surface analysis for our field determination (e.g., the mean Earth radius $= 6,371.2$ km). From the potential function relationship it is an easy step to obtain the field components in spherical coordinates:

$$X(\theta, \phi) = -B_\theta = \frac{1}{R_e} \frac{\delta V}{\delta \theta} \tag{1.61}$$

$$Y(\theta, \phi) = -B_\phi = \frac{-1}{R_e \sin(\theta)} \frac{\delta V}{\delta \phi} \tag{1.62}$$

$$Z(\theta, \phi) = -B_r = \frac{\delta V}{\delta r}, \tag{1.63}$$

where X, Y, and Z correspond to our components of the main field $-B_\theta$, B_ϕ, and $-B_r$, respectively. Using our values of the potential function (Equations (1.58), (1.59), and (1.60)) in the above, we obtain for the

X and Y representations (using either the external or internal parts of a and b for separating the corresponding fields)

$$X(\theta, \phi) = \sum_{m=0}^{M} \left[X_c^m(\theta) \cos(m\phi) + X_s^m(\theta) \sin(m\phi) \right] \qquad (1.64)$$

with

$$X_c^m(\theta) = \sum_{n=m}^{N} a_n^m \frac{\delta P_n^m}{\delta \theta} \qquad (1.65)$$

and

$$X_s^m(\theta) = \sum_{n=m}^{N} b_n^m \frac{\delta P_n^m}{\delta \theta}. \qquad (1.66)$$

Also

$$Y(\theta, \phi) = \sum_{m=0}^{M} \left[Y_c^m(\theta) \cos(m\phi) + Y_s^m(\theta) \sin(m\phi) \right] \qquad (1.67)$$

with

$$Y_c^m(\theta) = \frac{-m}{\sin(\theta)} \sum_{n=m}^{N} b_n^m P_n^m \qquad (1.68)$$

and

$$Y_s^m(\theta) = \frac{m}{\sin(\theta)} \sum_{n=m}^{N} a_n^m P_n^m. \qquad (1.69)$$

For the Z component the form is similar to the above:

$$Z(\theta, \phi) = \sum_{m=0}^{M} \left[Z_c^m(\theta) \cos(m\phi) + Z_s^m(\theta) \sin(m\phi) \right] \qquad (1.70)$$

only there are separate equations to be used for the external part:

$$Z_c^m(\theta) = \sum_{n=m}^{N} n a_n^{me} P_n^m \qquad (1.71)$$

and

$$Z_s^m(\theta) = \sum_{n=m}^{N} n b_n^{me} P_n^m. \qquad (1.72)$$

Then for the internal part

$$Z_c^m(\theta) = - \sum_{n=m}^{N} (n+1) a_n^{mi} P_n^m \qquad (1.73)$$

and

$$Z_s^m(\theta) = - \sum_{n=m}^{N} (n+1) b_n^{mi} P_n^m. \qquad (1.74)$$

There is an adjustment that can be made for the elevation of the station, h, above the mean Earth radius, R_e. We use a new radius, $r = R_e + h$ and multiply each element of the summations in Equations (1.65), (1.66), (1.68), (1.69), (1.71), (1.72), (1.73), and (1.74) by the fraction $(R_e/r)^{n+2}$.

One might ask why we have to compute all these troublesome functions with Legendre polynomials. Why not represent the field by something less complicated such as a latitude and longitude grid of sine and cosine waves. Also, if we want a series of r^n and $(1/r)^n$ terms, why not just fit the fields directly to something simple with just these terms; why all the potential function bother; can't we just represent the external and internal fields with any functions we care to?

The answer is no. The reason is that we need equations that are solutions of Maxwell's equations for the field because those expressions contain the physical properties of the behavior of electromagnetic fields. To obtain valid electromagnetic results that separate the external from the internal fields, the area of the field measurements must completely surround the volume of the internal source fields, separating them from the region of external sources. Any other mathematical fit of the fields will be nothing more than a mathematical fit. Currents derived from such a fit will not be realistic representations, the internal–external "separation" would be invalid, the depiction of changing field strength with distance from the Earth's surface will be false, etc. However, if just some area representation of the field is needed, it is usually easier to use some computer-generated contouring program rather than the SHA.

We might ask whether we can represent something other than fields with these Gauss spherical harmonics. For example, could we take the Earth's mean horizontal surface wind vectors to be our x and y components and the surface barometric pressure to be the z component and fit these values with a spherical harmonic fitting function? Of course that could be done, but there is no special significance to the result and no meaning to the extracted external and internal parts.

What if we actually used an electromagnetic field for the analysis? For example, suppose we used the monthly mean value of the field at a latitude distribution of observatories along a great circle of longitude. And suppose we assigned the whole year to 360° of change, as if the Earth revolved on its axis once a year with respect to the Sun's direction, and we let each month's value of observatory field be assigned to a longitude fraction of the 360°. Then we could do a spherical harmonic analysis of the latitude–longitude grid of magnetic field values, resulting in SHA model Gauss coefficients, separation of supposed external and internal parts, and representation of current sources. Do the results have a real meaning? No; the analysis is just a representation that may be convenient for picture purposes, but it cannot be used to determine real current sources obeying physical laws. Those measurements of surface fields that were put into the equations cannot be considered to enclose a sphere separating two realistic source–current systems. In this case, the relationship of the internal and external parts has no meaning, because we violated Maxwell's equations when we introduced a false condition on the nature of the source situation.

We need to meet three conditions for the appropriate representation of geomagnetic fields: (1) We should use functions that satisfy Equations (1.33) to (1.36), (2) we should use fields that could be realistic representations of continuous current–source distributions, and (3) our observation area should separate the internal and external sources with negligible current flowing between them.

When we were examining the fields from the electric dipole, perhaps we also thought about a symmetrical arrangement of electric charges for a *quadrupole* (charges on the corners of a square with adjacent charges given the opposite sign) and for an *octupole* (charges on the corners of a cube with adjacent charges given the opposite sign) as in Figure 1.5. Well, just as there are corresponding terms for the magnetic potential for the dipole, the succeeding terms in the Gauss solution can represent the magnetic quadrupole, octupole, etc., parts of the main field representation. We say that the SHA produces a separation of the *multipole* parts of the magnetic field.

Let us consider the altitude projection feature of the SHA field representation. The relative importance of the polynomial terms in the computation of the Earth's internal field decreases with increasing order of the multipole. Recall that our expansion for the internal sources contained terms of $(1/r)^n$ with increasing n. So, the larger the n, the smaller the contribution. Thus, for a finite number of polynomial terms, we would expect that the matching of the Earth's field should improve with altitude (r is larger so $1/r$ is smaller). This is the reason that the dipole field is effective in understanding many of the magnetospheric processes. Unfortunately, the generalization is complicated slightly by the existence of currents within the magnetosphere and by the deformation of the distant magnetosphere from its encounter with a solar wind of particles and fields (a subject of Chapter 3).

By fitting the spherical harmonics to the observed fields, all data points about the Earth contribute to the determination of all the Gauss coefficients. Modern methods of analysis include all available satellite, land, and marine surveys of the field. The permanent observatories then serve to establish the global annual change pattern necessary to bring all data to a common study period (epoch). However, permanent observatories are primarily located in continental areas and within industrialized countries; the vast oceans and polar regions are poorly represented. Special statistical methods are presently used to find the best field fitting to the common-time observations and establish the global grid of values. There are three reasons now used for terminating the analysis at degree and order values of 10 or 12: (1) errors in observatory data may limit greater accuracy; (2) the higher-order terms may represent the Earth's crustal geologic features; and (3) with the analysis in geographic coordinates, the tilted magnetic dipole location introduces nondipole Gauss harmonics that vary with the distance between the two pole positions.

A problem with the models of low degree and order is that each spherical harmonic has its own special global symmetry. The pressure of fitting an observed large excursion of field in one region of the Earth must necessarily cause critical harmonics (wavelengths with scales appropriate to the size of that region) to have manifestations outside of that region; all this would have been compensated by taking a larger number of polynomial terms.

For example, if the southeastern Australia main field contribution is determined from two SHA models, each of which used the same Australian observatory data values but differed greatly in the number of data points selected from other parts of the Earth, there would be two differing versions of the main field for the region of interest. Also, recall that no regional field change with a scale smaller than that determined from the degree and order of the polynomial can be represented by the analysis.

A group of geomagnetic field modelers belonging to the International Association of Geomagnetism and Aeronomy periodically examines various field representations for accuracy in reproducing actual field determinations. They select the best Gauss coefficients to represent a particular epoch (usually every 5-year period) and title this the "International Geomagnetic Reference Field" (IGRF). Also, when special retrospective studies are completed, the IGRF is corrected or modified (if necessary) and retitled a "Definitive Geomagnetic Reference Field" (DGRF). Table 1.2 shows a sample of the Gauss coefficients (in nanoteslas), up to the degree and order 6, computed for DGRF or IGRF field models from 1900 to 2000 with projections to 2005. On a computer program (see Section C.12) the **ALL-IGRF.TAB** file is a table of all the DGRF and IGRF coefficients, to degree and order 10, for 5-year epochs from 1900 to 2000. The final column of the tables gives the rate of annual change (*secular variation* (sv)) of the coefficients projected from the last 5-year model to the year 2005. Readers may obtain from the World Data Center, Boulder (address in Section B.2) a personal computer program that will compute the fields for any location and date using the appropriate geomagnetic reference field. A sample of this program, **GEOMAG**,

Table 1.2. *Definitive and International Geomagnetic Reference Field Values*

g/h	n	m	DGRF 1900	DGRF 1920	DGRF 1940	DGRF 1960	DGRF 1980	DGRF 1990	IGRF 2000	sv nT/yr
g	1	0	−31543	−31060	−30654	−30421	−29992	−29775	−29615	14.6
g	1	1	−2298	−2317	−2292	−2169	−1956	−1848	−1728	10.7
h	1	1	5922	5845	5821	5791	5604	5406	5186	−22.5
g	2	0	−677	−839	−1106	−1555	−1997	−2131	−2267	−12.4
g	2	1	2905	2959	2981	3002	3027	3059	3072	1.1
h	2	1	−1061	−1259	−1614	−1967	−2129	−2279	−2478	−20.6
g	2	2	924	1407	1566	1590	1663	1686	1672	−1.1
h	2	2	1121	823	528	206	−200	−373	−458	−9.6
g	3	0	1022	1111	1240	1302	1281	1314	1341	0.7
g	3	1	−1469	−1600	−1790	−1992	−2180	−2239	−2290	−5.4
h	3	1	−330	−445	−499	−414	−336	−284	−227	6.0
g	3	2	1256	1205	1232	1289	1251	1248	1253	0.9
h	3	2	3	103	163	224	271	293	296	−0.1
g	3	3	572	839	916	878	833	802	715	−7.7
h	3	3	523	293	43	−130	−252	−352	−492	−14.2
g	4	0	876	889	914	957	938	939	935	−1.3
g	4	1	628	695	762	800	782	780	787	1.6
h	4	1	195	220	169	135	212	247	272	2.1
g	4	2	660	616	550	504	398	325	251	−7.3
h	4	2	−69	−134	−252	−278	−257	−240	−232	1.3
g	4	3	−361	−424	−405	−394	−419	−423	−405	2.9
h	4	3	−210	−153	−72	3	53	84	119	5.0
g	4	4	134	199	265	269	199	141	110	−3.2
h	4	4	−75	−57	−141	−255	−297	−299	−304	0.3
g	5	0	−184	−221	−241	−222	−218	−214	−217	0.0
g	5	1	328	326	334	362	357	353	351	−0.7
h	5	1	−210	−122	−33	16	46	46	44	−0.1
g	5	2	264	236	208	242	261	245	222	−2.1
h	5	2	53	58	71	125	150	154	172	0.6
g	5	3	5	−23	−33	−26	−74	−109	−131	−2.8
h	5	3	−33	−38	−75	−117	−151	−153	−134	1.7
g	5	4	−86	−119	−141	−156	−162	−165	−169	−0.8
h	5	4	−124	−125	−113	−114	−78	−69	−40	1.9
g	5	5	−16	−62	−76	−63	−48	−36	−12	2.5
h	5	5	3	43	69	81	92	97	107	0.1
g	6	0	63	61	57	46	48	61	72	1.0
g	6	1	61	55	54	58	66	65	68	−0.4
h	6	1	−9	0	4	−10	−15	−16	−17	−0.2

Table 1.2. (*cont.*)

g/h	n	m	DGRF 1900	DGRF 1920	DGRF 1940	DGRF 1960	DGRF 1980	DGRF 1990	IGRF 2000	SV nT/yr
g	6	2	−11	−10	−7	1	42	59	74	0.9
h	6	2	83	96	105	99	93	82	64	−1.4
g	6	3	−217	−233	−249	−237	−192	−178	−161	2.0
h	6	3	2	11	33	60	71	69	65	0.0
g	6	4	−58	−46	−18	−1	4	3	−5	−0.6
h	6	4	−35	−22	−15	−20	−43	−52	−61	−0.8
g	6	5	59	44	18	−2	14	18	17	−0.3
h	6	5	36	18	0	−11	−2	1	1	0.0
g	6	6	−90	−101	−107	−113	−108	−96	−91	1.2
h	6	6	−69	−57	−33	−17	17	24	44	0.9

for dates from 1995 to 2005 is provided with the program set described in Section C.2.

For $n = 1$ and $m = 0$ or 1, using the Gauss coefficient nomenclature, the "first-order" potential function becomes

$$V_1 = \left(\frac{R_e^3}{r^2}\right)\left[g_1^0 \cos(\theta) + g_1^1 \sin(\theta)\cos(\theta) + h_1^1 \sin(\theta)\sin(\phi)\right]. \quad (1.75)$$

If we aligned the analysis coordinate system along the tilted geomagnetic dipole axis rather than the Earth's rotation axis (substituting a for g), we would have the dipole form

$$V_1 = R_e a_1^0 \left(\frac{R_e}{r}\right)^2 \cos(\theta). \quad (1.76)$$

The field is obtained from the gradient of the potential function (Equations (1.61), (1.62), and (1.63)). Assuming R_e to be about equal to r, and using the dipole term only (Equation (1.76)), the new, X', Y', and Z' field components (orthogonal to the dipole direction) become

$$X' = -a_1^0 \sin(\theta) \quad (1.77)$$
$$Y' = 0 \quad (1.78)$$
$$Z' = -2a_1^0 \cos(\theta). \quad (1.79)$$

As an approximation, we can now read the g_1^0 values (from our Table 1.2 of field models) for a_1^0 and estimate the size of our dipole field that was described in Equations (1.26) and (1.27).

1.5 Features of the Main Field

To visualize the *geomagnetic coordinate system*, consider a latitude and longitude grid engraved on a plastic shell covering a globe representing the Earth. The poles of this grid can be shifted with respect to any desired location. Centered at the Earth's spin axis pole locations, this plastic grid becomes the geographic latitude and longitude. Centered at the geomagnetic (dipole) pole, the grid becomes the geomagnetic coordinate system. All we need to define the geomagnetic coordinates for a sphere is a dipole pole location (because of symmetry, either the North or South Pole will suffice) and selection of a longitude starting point. A typical grid map for reading geomagnetic locations is illustrated in Figure 1.16. A computer program, **GMCORD**, for determining the geomagnetic coordinates for a selected geographic location is described in Section C.2. At the polar regions where the magnetospheric deformation and special currents play an important role in the measured surface fields, the dipole coordinate system must be used with caution (Chapter 3) if we are trying to explain geophysical phenomena.

Figure 1.16. Geomagnetic coordinates of 2000 epoch, shown as curved lines superposed upon a global map plotted in equally spaced geographic latitude and longitude dimensions. Figure provided by J. Quinn, USGS.

The *geomagnetic latitude*, θ', is a dipole model location just equal to $90° - \theta$. Now we will call θ the geomagnetic colatitude. The *geomagnetic dipole equator* is the trace along the Earth's surface for which the dipole inclination $I = 0°$ (Equation (1.3)). This is usually a different

Geomagnetic Coordinates

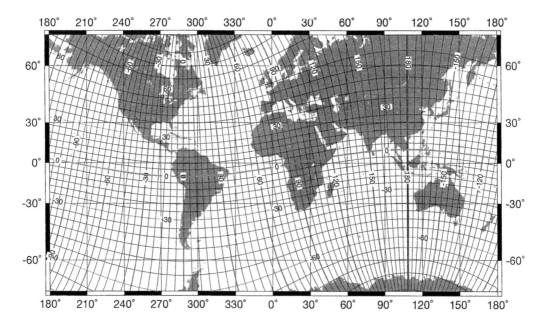

location from what we call the *dip equator* that corresponds to the $I = 0°$ obtained from the local measurements of H and Z (Equation (1.29)). Occasionally, in place of direct measurements, the dip equator is given by the locations where the full IGRF model field values indicate an inclination of $0°$. The difference between the geomagnetic and dip equators is quite large in the African continental region and small in the South American region (Figure 1.17).

The *geomagnetic longitudes*, ϕ, are defined as equal angular increments about the dipole axis, enumerated eastward from a zero value (prime meridian) that passes through the geomagnetic poles and also through the geographic South Pole (see Figure 1.8). *Geomagnetic poles* are the intersections of the axis line of the dipole field at the Earth's surface. At the present time these poles are inclined about $10.7°$ to the Earth's spin axis. In 1996 the geomagnetic north pole was near $79.3°$ north, $71.5°$ west. That location is not far from Thule, Greenland. The geomagnetic south pole at that time was at $79.3°$ south, $108.5°$ east, not far from Vostok Station, Antarctica.

The Locally Measured *Dip Pole* (called "magnetic pole" by cartographers) is the location where contours of the local fields indicate a most vertical situation, so that $I = 90°$ (Equation (1.3)). IGRF Geomagnetic Poles depend upon the representation of the Earth centered dipole components from a full set of the Earth's measurements. The Locally Measured Dip Pole location is different because of the greater dependence upon local effects. The two poles are not expected to correspond in location.

The Geomagnetic Pole location (colatitude θ_0 and longitude ϕ_0), called CD for *centered dipole*, can be obtained from the Gauss coefficient Table 1.2 using

$$\theta_0 = -\arctan\left[\frac{\sqrt{\left(h_1^1\right)^2 + \left(g_1^1\right)^2}}{g_1^0}\right] \qquad (1.80)$$

$$\phi_0 = \arctan\left(\frac{h_1^1}{g_1^1}\right). \qquad (1.81)$$

where the arctan means the angle whose tangent is the value determined by the bracketed ratio (see Section A.5). These poles are symmetrically located in the Northern and Southern Hemispheres. Figure 1.18 illustrates how the first three Gauss coefficients, used for computing the dipole, have changed over the years. The corresponding change in pole location is plotted on the maps of Figure 1.19. The program **POLYFIT**, described in Section C.7, has been used to extend the values beyond the model years.

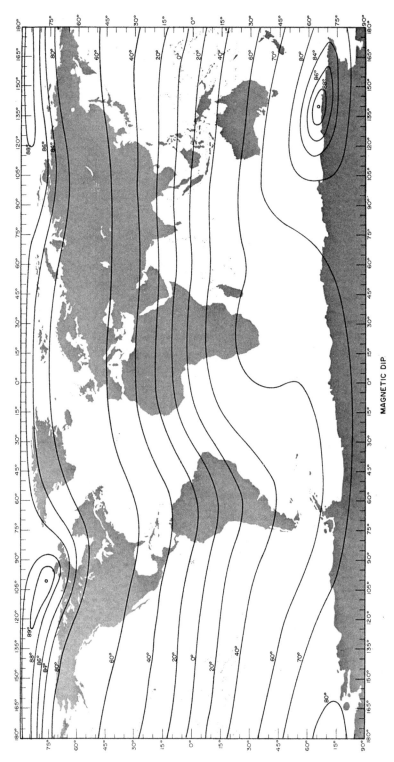

MAGNETIC DIP

Figure 1.17. Magnetic dip latitude curves superposed upon a global map plotted in equally spaced geographic latitude and longitude dimensions.

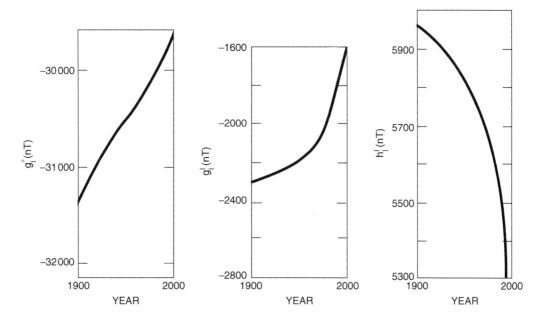

Figure 1.18. Behavior of first three main field Gauss coefficients from 1900 to 2000.

The *magnetic dipole moment M* is obtained from the Gauss coefficients with the equation

$$M = \left(4\pi/\mu_0\right) R_e^3 \sqrt{\left(g_1^0\right)^2 + \left(g_1^1\right)^2 + \left(h_1^1\right)^2} \text{ Ampere meter}^2, \qquad (1.82)$$

with R_e in meters and the permeability of free space $\mu_0 = 4\pi \times 10^{-7}$. The **GMCORD** program in Section C.1 gives a computation of the magnetic dipole moment for any selected date from 1940 to 2005.

If we had organized the field analysis about the geomagnetic dipole, rather than the spin axis, then the dipole moment would just be determined from g_1^0 because the other two Gauss coefficients of Equation (1.82) would be relatively insignificant in value. In other words, the first row of Table 1.2 shows us how the dipole moment is changing. Figure 1.20 is a graph of this variation since 1600. The Earth is now obviously in a period of rapidly declining field. Such decreases in the dipole moment have occurred in the past. On occasion, a dipole field reversal follows a minimum in the ratio of dipole to multipole field contributions.

A second series of Gauss coefficients may be obtained by tilting the analysis axis to align it with the geomagnetic axis. This process would enhance the dipole term at the expense of the higher multipoles. For a third series of coefficients we could allow not only the tilt but also the distance from the center of the analysis axis to the Earth center to vary. This "Eccentric Axis Dipole" analysis is carried out in such a way that

Figure 1.19. Changing locations of poles for CD, centered dipole (geomagnetic pole), ED, eccentric axis dipole, and locally measured dip pole. Southern Hemisphere is shown at top, Northern Hemisphere at bottom. Figure adapted from Fraser-Smith (1987).

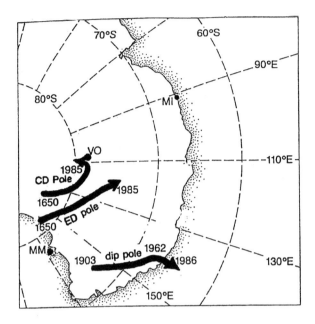

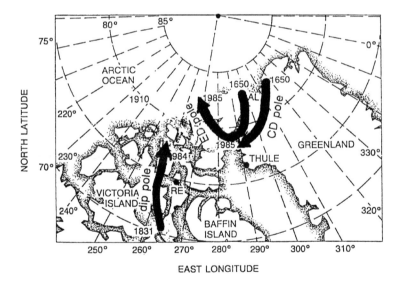

the quadrupole term is made to vanish, maximizing the dipole term. We can think of this process as ignoring the geodetic center of the Earth but allowing the measured field to fix the location of a dipole axis for a polynomial description such that the dipole term is maximized. In this configuration we have the *Eccentric Axis Dipole* (ED) model of the main field. Such a model can be useful for studies of the source current behavior, for tracking dip locations, and for understanding magnetospheric

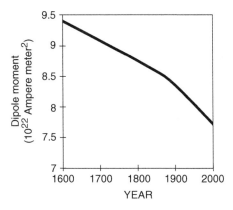

Figure 1.20. Dipole moment of the Earth's field from 1600 to 2000.

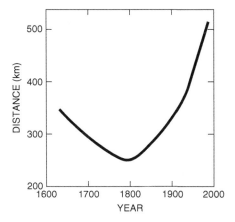

Figure 1.21. Distance from eccentric dipole midpoint to center of the Earth from 1600 to 2000. Figure redrawn from Fraser-Smith (1987).

particle dynamics. Charged particles, arriving from space, are guided by the eccentric dipole field.

Computations indicate that a displacement from the Earth's center of about 500 km is now required for the ED model. Figure 1.21 shows the variation of this distance since about 1600. The positions of the dipole poles, where the axis line penetrates the surface, are not symmetrically located in the Northern and Southern Hemispheres. Figure 1.19 shows the changing location of the ED dipole locations in the north and south polar regions. Because the eccentric axis center is offset from the Earth's center, the line of the axis exits the Earth at an angle to the surface. This fact gives rise to another set of poles (*Eccentric Axis Dip Poles*) for the pair of eccentric axis field lines that exit perpendicular to the Earth's surface.

Which poles are the most important? Well, it depends on what one wants to study. Remember that the observed dip pole and dip equator locations emphasize local measurements, whereas the other poles are fitted from values obtained all over the world. Later, as we look at specific phenomena, we will explore the suitable coordinates for each case.

At this point we will look at the meaning of "Geomagnetic Time". However, first recall the other times we use. *Universal Time* (UT) is the world accepted time at the 0° geographic meridian of longitude. Because this timing technique was originated at Greenwich, England, it is also called Greenwich Mean Time (GMT). Special atoms, whose natural frequencies are accurately determined, set an atomic clock time standard at the NIST laboratories in Boulder, Colorado, USA. (phone 303-499-7111). Each new year, necessary adjustments to the Earth's rotation rate can cause the UT to be changed by a second in time (leap second). Local Mean Times (LMT) are set and named, by each country, to roughly correspond to one or more dominant longitudes in that country. These countries may shift their clocks to one hour ahead in summer months to conserve local electrical power in the increased daylight period at middle and high latitudes. By an adjustment of UT to the actual local longitude, a Local (solar) Time (LT) is often used by geophysical scientists for understanding daily variations in their measurements.

A special time scale that takes into consideration the geomagnetic field is sometimes used for studies of phenomena that have a dependence upon the direction of the Sun with respect to the Earth's field. To explain this *Geomagnetic Time* let us use the north polar view of the Earth in Figure 1.22. To start with a simple example, assume that the Sun direction is toward the bottom of the figure and assume that the

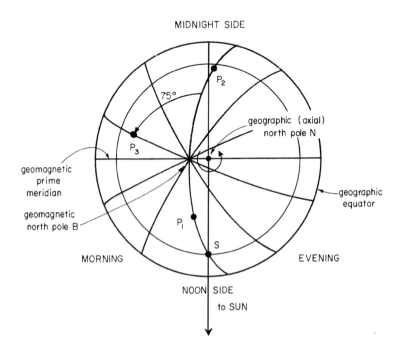

Figure 1.22. Configuration for explaining the computation of geomagnetic time at points P_1, P_2, and P_3 using the dipole field line at the subsolar point S. An explanation is found in the accompanying text.

day is the June solstice. In the figure, geomagnetic meridians of longitude are shown along with the two northern poles (geographic, N, and geomagnetic, B), locations P_1, P_2, P_3, and the subsolar point S (surface intersection of a line from the Earth center toward the Sun) at the Tropic of Cancer. The geomagnetic time of a station is obtained from the locations of the geomagnetic longitude of the subsolar point and the geomagnetic longitude of the station. Location P_1, on the same geomagnetic meridian as the subsolar point S, is at geomagnetic noon. Location P_2, 180° geomagnetic longitude from the subsolar point, is at geomagnetic midnight. Geomagnetic time is thus defined as the eastward hour angle between the geomagnetic meridian passing through the station and the geomagnetic meridian 180° from that passing through the subsolar point. Location P_3, which is 75° geomagnetic longitude from the subsolar meridian $+180°$, is said to be at 5 hr in geomagnetic time.

Three particular features should be noted: Geomagnetic time intervals vary slightly throughout the day. The relationship of geomagnetic time to local solar time changes a little during the year. The concept of geomagnetic time, like local solar time, becomes useless near the coordinate poles; in those regions, positions a very short distance apart can be on quite different longitudes and thereby have as much as 12 hr difference in assigned "time."

1.6 Charting the Field

Charting the changing Earth's surface field has been a necessary and regular function of the major nations of the world since the early days of global exploration. Typically, contour-type lines of equal (*isomagnetic*) increments of field are plotted. Magnetic field declination is best represented on the navigational charts that use Mercator projections (latitude and longitude lines cross at right angles; distances between parallels of latitude increase poleward from the Equator in the same proportion as the meridians of longitude are expanded) because such charts preserve azimuthal relationships. For consistency, the other field components are also displayed on these projections. Polar areas are particularly distorted in size by Mercator projections. Figures 1.23a to 1.23d illustrate the field contours of *H, D, Z,* and *F,* respectively, using the degree and order 12 Gauss coefficients of the 2000 IGRF. The South Atlantic/South American anomaly minimum in total field (Figure 1.23d) is a direct result of the eccentric axis midpoint offset from the Earth's geographic center. See Section 5.3 for the impact of this region on satellite operation. This anomalous region has been gradually moving westward over the years.

(a)

Horizontal Intensity - Main Field (H)

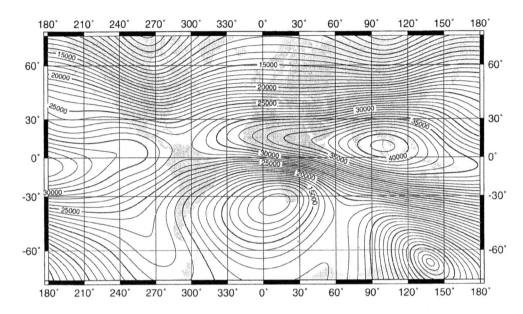

(b)

Declination - Main Field (D)

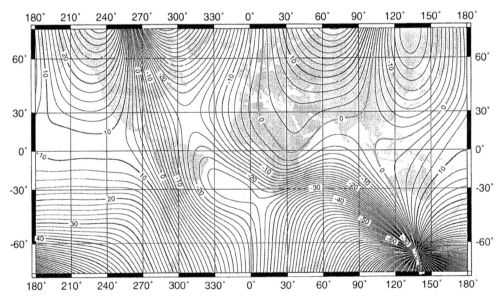

Figure 1.23. International Geomagnetic Reference Field (IGRF-2000) for (a) the horizontal, H, field component (nT). (b) the declination, D, field component (degrees eastward). (c) the vertical, Z, field component (nT). (d) the total intensity, F, of the field (nT). Figures provided by J. Quinn, USGS.

(c)

Vertical Component - Main Field (Z)

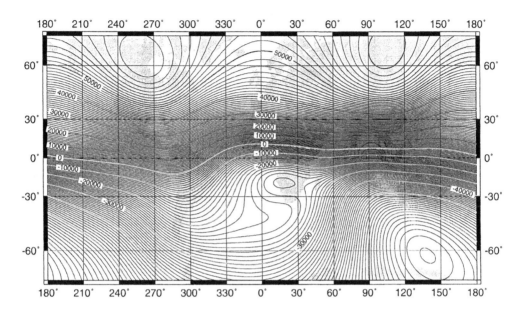

(d)

Total Intensity - Main Field (F)

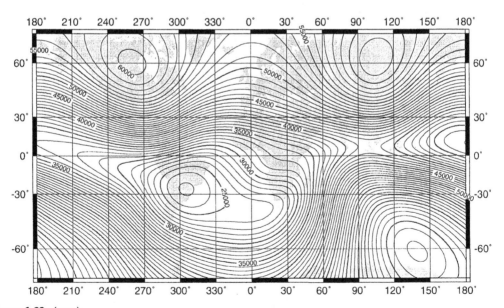

Figure 1.23. (*cont.*)

Through the years, several special adjectives have arisen to specify types of isomagnetic contour charts:

isoclinic– equal inclination (dip), I

isodynamic– equal intensity of field components, F, H, X, Y, Z

isogonic– equal declination, D ($0°$ is the "agonic line")

isoporic– equal secular variation of field component

When the published chart is a representation of some harmonic analysis of the field, the user should note the degree and order of the coefficients that have been used. This information is important for estimation of the spatial resolution of the fields shown on the chart. Also, the data epoch (date to which the measured fields have been adjusted) should be checked to consider whether a significant temporal change in field could have occurred since the map was prepared.

Because the principal Gauss dipole coefficients change gradually over the years, we can determine a smooth mathematical representation (see the polynomial fit program in Section C.7) to these values and reliably project the dipole characteristics to ten or more years past the last chart epoch. The **GMCORD** program (Section C.1) uses such polynomial fitting to determine the geomagnetic coordinates of a location for any day in the years from 1940 to 2005.

Regular changes of the field that are occurring over a period of years are called *secular variations* or secular changes. Figure 1.24 shows the change in H at Tucson observatory from 1958 through 1974. Charts of the magnetic field often include contours of the secular change (Figure 1.25). Secular changes can result from: (a) change in the magnitude of the principal (dipole source) current within the Earth; (b) motion of that current, causing a shift in the alignment of the dipole axis (the dipole north pole is now moving about 18 km northward and 5 km westward each year); and (c) change in the westward drifting, nondipole parts of the main field representation. The various harmonic components in the field model seem to be moving at different rates. Models of recent years indicate that the dipole component is moving rather slowly (nearly 0.1°/yr). Motion of the dipole about the spin axis is called the *precessional drift*.

With the dipole field components removed from a model, the remaining multipole parts, forming the *nondipole fields*, are illustrated in Figure 1.26. These contours presently seem to be drifting westward at a rate of about 0.2° or 0.3° each year, noticeably faster than the dipole part. Such drift would cause anomalies to circle the globe in 1000 to 2000 years, were it not for the fact that the form of the anomalies may be noticeably altered in less than 100 years. The drift may be faster at higher latitudes. On average, the nondipole components are changing by about 50 nanoteslas per year. There is some evidence that the anomalies are severely modified in form as they pass the Pacific Ocean region. The

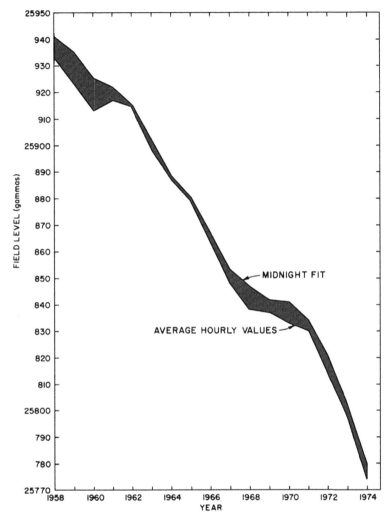

Figure 1.24. Secular change of geomagnetic field at Tucson, Arizona (United States) from 1958 to 1974. Upper boundary of trace represents *H* components determined from midnight values on quiet days of year. Lower boundary of trace represents *H* obtained from the annual averages of all hourly values. Large differences between the two values occur at sunspot maximum years.

nondipole components contribute an average of about 10% to the main field; in some places the proportion is less, but at other spots they account for as much as 30% of the field. The Northern Hemisphere contains fewer nondipole anomalies than the Southern Hemisphere. Accurate registration of the Earth's field has only been possible for the last 100 years, so our understanding of these regional features is quite limited.

Because the computation of the eccentric axis dipole location involves the multipole parts of the Gauss coefficients, the *westward drift* can be described by the westward motion of the eccentric axis center. The field change has also been represented as a westward drift of the individual multipoles. The several-thousand-kilometer dimension of the nondipole anomalies is said to be "regional" in size. A correlation

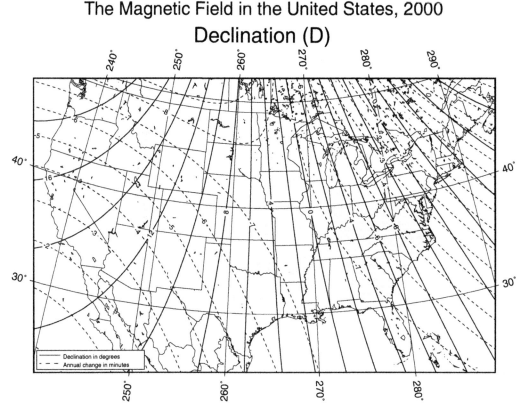

The Magnetic Field in the United States, 2000
Declination (D)

Figure 1.25. Magnetic declination D of the geomagnetic field in 2000 for the continental United States. The change of D in minutes per year is depicted by dashed contours. Figure provided by J. Quinn, USGS.

coefficient of 0.9 has been established between the change in westward drift and the change in Earth rotation speed that occurred several years earlier (1.0 represents perfect correlation; 0 represents no correlation; see Section C.10). The cause has been ascribed to differences in the rotation speed of the Earth's mantle, outer core, and inner core.

Some scientists believe that the time variations of the Gauss coefficients g_2^0, h_2^1, and h_3^1 showed a relatively sharp change in slope over a few months of 1969. This change, called a *geomagnetic jerk*, has been explained in terms of a disruptive process at the Earth's core–mantle boundary. Skeptics believe the "jerk" to be simply of solar–terrestrial disturbance origin and that outer core changes shorter than several years should be attenuated before the field reaches the surface.

The Earth-field models are usually not computed beyond degree and order 10 or 12 because, at about that point, the higher polynomials are thought to include a large portion of Earth crustal anomalies. Anomalies of small size, under about 100 km (*local anomalies*), are generally correlated with geological surface features. Such crustal fields come from the induced current flowing in conducting Earth materials as a response

Non-Dipole Total Intensity Field (F)

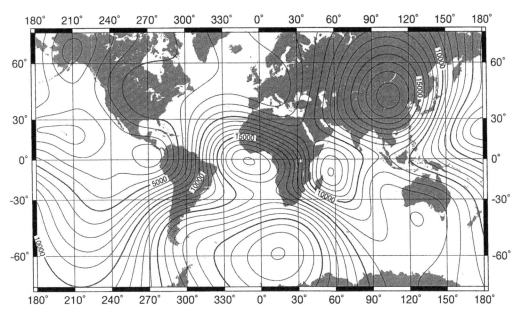

Figure 1.26. Nondipole field in year-2000 model for total intensity (nT). Figure provided by J. Quinn, USGS.

to current sources external to the Earth or from rock magnetization acquired in the region's early formation. The field effects of these anomalies usually disappear rapidly with altitude and typically make very little contribution in the space environment. However, there are places of intense magnetic fields that occur in the regions of major conducting and ferromagnetic deposits; these may give rise to a response in the higher coefficients of the model itself.

1.7 Field Values for Modeling

The annual mean values of H, D, and Z have traditionally been used for determining the Gauss coefficients. The annual means are affected by the solar–terrestrial activity level even in years with a greater than average number of quiet days (Figure 1.24). The idea of quiet and active geomagnetic times relates to the daily magnetic field analog recordings, which exhibit either smooth, gradually changing (quiet) traces (a subject of Chapter 2) or rapidly varying, irregular (disturbed) traces. In Chapter 3 we will discuss the solar–terrestrial disturbances that cause the irregular traces and describe indices to rate the daily field-level behavior. The year 1965 was a record year for quietness (in the time since global observations of the Earth's field started). Figure 1.27 illustrates, for 41 observatories, the differences between the field levels determined

Figure 1.27. Differences in the annual mean levels of the *H* component of field determined from values on selected quiet days minus the levels determined from values on all days. The data were gathered from 41 world observatories operating in 1965, a year of extremely low solar activity.

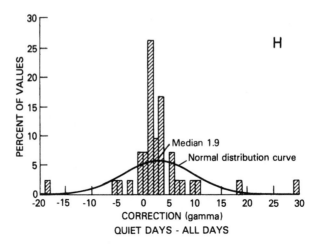

for all days of 1965 and field levels determined for selected very quiet days in 1965. The mean value of the difference was 2.9 γ, 0.04 min, and 0.3 γ for *H, D*, and *Z*, respectively. These values may be considered as a measure of the error that may be introduced into the spherical harmonic analysis by the analyzers' data selection procedure.

Buried in the records of geomagnetic observatories are annual and semiannual field changes. Figure 1.28 shows the distribution of midnight values of the *H* component of field on quiet days of 1973 for the Tucson, Arizona (United States) observatory. A clear annual and semiannual change in level is evident in the records. Figure 1.29 illustrates these seasonal changes (as external currents) determined from a North American distribution of observatories using the midnight field levels at quiet times. Representing such changes as a current system is convenient for visualizing both the surface field direction and the latitude and seasonal changes of the midnight field. However, it must be remembered that such currents have been derived with the false assumption of a continuous source current not far from the Earth with 360° representing one year of change; any relationship of external and internal parts is truly fictitious.

If the annual and semiannual changes are not removed by the main field modeler, then errors from the seasonal magnetospheric deformation and activity, as much as 20 γ in size, are entered into the computation of Gauss coefficients. We will see in Chapter 3 how these systematic changes can be caused by the annual distortion of the magnetosphere, by the arrival of particles and fields from the Sun, and by semiannual preferential months for disturbance activity interactions.

The separation of internal and external parts in main-field analysis helps remove many of the above distortions. However, induced currents

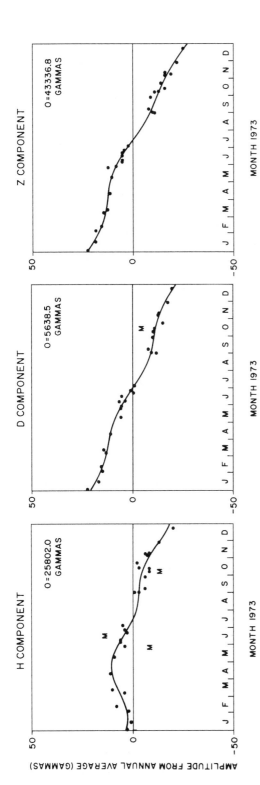

Figure 1.28. Tucson midnight geomagnetic field amplitudes for *H*, *D*, and *Z* components, using selected days when geomagnetic indices AE and Dst (Section 3.13) represented quiet conditions. *M* indicates an outlying value removed from the fitting. The letter of the month is indicated at the bottom of each diagram. The baseline value is given in the top right corner. Note the annual and semiannual variations.

Figure 1.29. Contours of fictitious external currents (in units of 1000 amperes) designed to visualize magnitude and direction of the midnight annual and semiannual surface field changes in the American sector. Arrows show the current-flow directions. Geomagnetic latitudes and months of 1965 are indicated. The pattern depicts the surface effect of seasonal field distortion by the nighttime magnetosphere.

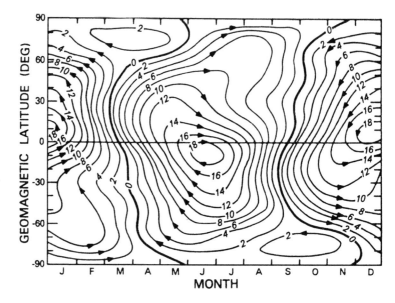

within the conducting Earth create problems with the exact identification of the Earth's core field. To demonstrate the dependence of the Gauss coefficients upon activity level, Langel (1987) computed the first three external coefficients for a 1980 main field model as a linear function of the solar–terrestrial activity levels (represented by the storm-time disturbance index, Dst; see Section 3.13):

$$g_1^0(\text{external}) = 18.4 - 0.63 \, \text{Dst}, \tag{1.83}$$

$$g_1^1(\text{external}) = -1.1 - 0.06 \, \text{Dst}, \tag{1.84}$$

and

$$h_1^1(\text{external}) = -3.3 + 0.17 \, \text{Dst}, \tag{1.85}$$

where the formulas apply to the quiet times of low Dst index values. He found that the induction from these external sources was significant enough to modify the internal field as well, so that

$$g_1^0(\text{internal}) = -29991.6 + 0.270 g_1^0(\text{external}). \tag{1.86}$$

The induction contributions to the main-field analysis during solar–terrestrial disturbances would be responsible for an 11-year solar-cycle fluctuation in the field model coefficients.

One major problem for the main-field analysis has been the irregular distribution of surface observations (Section 4.15). Satellite measurements of the Earth's field in space have helped considerably in this regard. Yet such measurements have introduced their own unique problems. Internal to the satellite orbit are ionospheric dynamo and equatorial electrojet currents (see Chapter 2) as well as high-latitude electrojet and field-aligned currents (discussed in Chapter 3), so the Gauss external–internal separation is slightly contaminated by these currents. In addition, satellite lifetimes are short with respect to surface observatories, thus limiting the ability of satellites to determine any long-term, secular change.

Special corrections to satellite data have to be made because one of the requirements for the application of our SHA is that no currents pass across the surface of analysis. Satellites for magnetic field measurements are placed at near-circular orbits at altitudes above 500 km, as low as possible for field sensitivity but high enough so that atmospheric drag will not pull the satellite from orbit. We will see, in Chapter 3, that at the higher latitude locations, strong currents often flow along the Earth's field lines through the analysis region. Presently, the ground and space data are best combined for the analysis epochs when satellite records are available.

1.8 Earth's Interior as a Source

The magnetic properties of iron and its abundance in the Earth led to early speculations that the Earth was a great magnet (Figure 1.2). One physical property immediately dispelled this hypothesis. All materials lose their magnetism at high temperatures. At this *Curie temperature*, characteristic of each ferromagnetic atom, the magnetic moment becomes randomly oriented. The magnetization of a material depends upon "exchange interactions" between the individual magnetic moments of each ferromagnetic atom, causing their co-alignment (see Section 4.2). The randomization of this co-alignment above a Curie temperature destroys the magnetization. That temperature is about $770°C$ for iron and $675°C$ for magnetite Fe_3O_4. The Earth's temperature increases with depth (Figure 1.30), and presently there seems to be as much heat generated in the interior as is lost at the surface. The Curie temperature isotherm for Earth materials is reached at an average depth of about 25 km. The Curie temperature does increase with pressure (which also increases with depth), but such change is quite gradual. At the boundary of the Earth's core, where the temperature can be close to $3,000°C$, the Curie temperature of iron would not be much more than $780°C$.

What about the Earth's crust that is shallower than the Curie temperature isotherm; could the source of the main field be there? When the materials known to exist at the crust are examined, it is found that their complete magnetization, at best, would provide an insignificant contribution to the dipole field.

The second important reason for dispelling the hypothesis of a great Earth-magnet source of our main field is that variations in dipole location,

Figure 1.30. Possible
temperatures of the Earth's
interior (uncertainty increases
with depth).

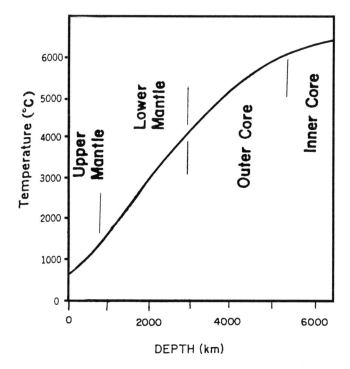

Figure 1.30. Possible temperatures of the Earth's interior (uncertainty increases with depth).

nondipole westward drift, and the evidence for ancient dipole field reversals all point to a liquid motion organized by the Earth's spin. Seismic studies have given us many details of the Earth's interior. Figure 1.31 illustrates the present density estimate obtained from the time of travel, refraction, and separation of the compressional and torsional seismic waves. This pattern of density with depth is used to name the interior regions of the Earth. We give the name *crust* to the solid outer layer which is about 5 km thick beneath the oceans but about 30 km thick for most continental regions. Tall mountain ranges cover *crust* thicknesses of up to 100 km. Below the *crust* begins the *upper mantle* whose solid top part is joined with the *crust* for the name *lithosphere*. The *upper mantle* becomes more plastic with depth to be called the *asthenosphere* in that region. The *upper mantle* extends to a transition zone near a depth of 600 km to 700 km, where the *lower mantle* begins. A *core–mantle (CM) boundary* is encountered at about 2,890 km, and a liquid *outer core* extends to about 5,150 km depth to surround a solid *inner core*. The Earth center is near 6,371 km (Figure 1.32).

Surface evidence of mantle activity is found in the existence of earthquakes; continental drift; volcanic activity; and the formation of oceans, continents, and volcanic islands. Mantle upwelling regions at spreading sites such as oceanic ridges and downwelling regions at subduction zones such as the east coast of New Zealand or the west coast of North and

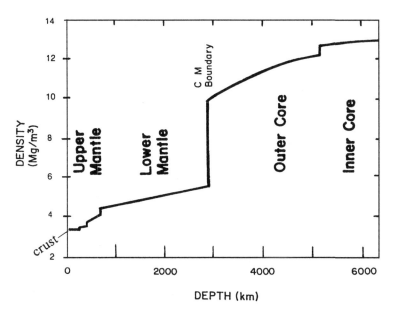

Figure 1.31. Density variation with depth within the Earth as computed from seismic waves. The region names are indicated.

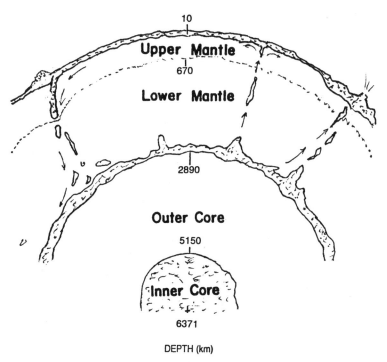

Figure 1.32. Schematic representation of regions within the Earth. Names and approximate depths are indicated. Figure adapted from Lay et al. (1990).

South America give evidence of mantle convection driven by gravitational energy and core-region heat sources (see Section 5.9). The mantle flow transports the heat, deforms the internal boundaries, and modulates the heat flow from the core. However, this sluggish convection is too slow to be responsible for the much faster geomagnetic main-field changes in dipole position or nondipole westward motion. We must look to the liquid outer core as the source-current region.

The primordial gravitational accretion processes that formed our spherical planet cause the density, pressure, and temperature to increase with depth. These properties gave rise to the modification and separa-tion of the original Earth-forming materials so that their character and behavior change in roughly concentric layers. A large composition con-trast occurs at the CM boundary; the two adjacent regions have distinct chemistries, dynamics, and physical states. The CM boundary has light products expelled from the core and heavy heterogeneities that have sep-arated from the mantle. Estimates of the conductivity, temperature, and composition of the mantle encourage some experimenters to extrapolate the Earth's surface nondipole field to this CM boundary. We must look to the liquid outer core for the massive ring of current whose flow causes the Earth's dipole field.

The best known hypothesis for generating a geomagnetic field is that the liquid outer core of the Earth maintains an electric current as a *"self-excited dynamo."* The first question researchers try to answer is, what power source drives this current? If it were the spin alone, the Earth would not be slowing at the present rate but rather spinning appreciably faster than the evidence allows. Radioactive decay of long-life isotopes within the Earth's interior is a principal source of heat leaving the Earth, but only a small fraction of the surface heat flux could come from the core region. Heating and chemical reactions are not presently thought to be responsible for geomagnetic current, but rather it is the flow caused by growth of the solid inner core.

To give us a mental picture of what is meant by a self-excited dynamo consider the following classical mechanical model. Recall that the "right-hand rule" means that when the current direction is indicated by the thumb pointing along a clockwise, horizontal, circular current flow, the resulting field, **B**, inside the circle, will be directed axially downward, as the fingers would point (Figure 1.33). Also, motion (at velocity **v**) of a condutor in a magnetic field will generate a current in the direction that a right-hand screw (Figure 1.33) would turn while moving the vector **v** into the vector **B**. Now let us look at the dynamo model.

Figure 1.34 illustrates the mechanical form of the simplistic self-excited dynamo. Starting with a small upward-directed field, **B**, and the rotating disk, at the brush connection (on the right) the right-hand rule describes the direction of a current, radially outward on the disk, that proceeds through the brush connection and then down the spiral wire encircling the dynamo axis. The current in the spiral wire then increases the strength of the field, causing the self-generation of more

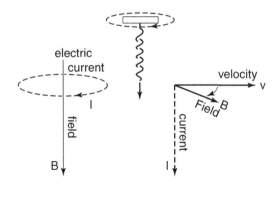

Figure 1.33. Right-hand screw (shown at top) correspondence for (at left) direction of **B** field from loop of current **I**, and (at right) direction of current **I**, from motion of conductor (moving at velocity **v**) in a magnetic field **B**.

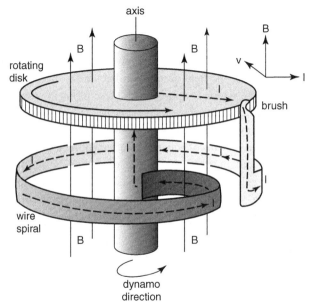

Figure 1.34. Self-excited dynamo current machine. The conducting disk is moved at constant velocity **v** in a small field **B**, creating current **I** through spiral of wire to generate an increase in field **B** and thereby increasing dynamo current **I**, and causing further field and current growth at expense of disk motion about the axis. The disk rotation direction, the winding direction of the spiral wire, and the initial direction of the small field are critical to the resulting increase in **B**.

current. The spin of the rotating disk drives the system to larger and larger **B**-field generation. When natural processes cause the dynamo to break up or decay and then reestablish, the dominant field direction is determined by the small initiating field direction in relationship to the dynamo flow. Reversals of the main field are explained by field decay and reestablishment in an opposite direction. Of course, the Earth-core dynamo has no insulation to define the currents and is certainly much more complex than this model.

We now believe that fluid motions, in the liquid outer core of the Earth, are driven by the gravitational energy released with the upward migration of lighter elements and growth of the inner core with the freezing-out of heavier components. This flow is organized by the Earth's spherical shape and forces delivered by the Earth's spin. Interactions at the CM boundary are believed to contribute to the nondipole field features. There is a correspondence of an order of magnitude (power of 10) decline of the Earth's field with an extensive continental plate rifting about 600 My (millions of years) ago. The coincidence is explained by the loss of core energy available to drive the field dynamo at that time.

Modelers of main-field sources are beset with difficulties. The chemistry of the outer core is not well known. Properties of the CM boundary are still speculative. Temperatures throughout the core are poorly represented. The viscosity of the outer core near the CM boundary seems to be only a few times that of Earth surface water, so the fluid motions can be exceedingly complex. Armed with a knowledge of the Earth's surface field and its variations over the last 300 years, the modeler of the core field must first determine the projections of field to the CM boundary and outer core. This is a difficult problem because the conductivity-depth profile of the Earth is not well known below about 500 km. The multipole terms of the field description become relatively more important as the source region is approached; these terms are the ones least well represented by our surface field analysis.

The solutions of the dynamo problem in the outer core involve highly complicated systems of equations; judgments (that often cannot be thoroughly tested) must be made to determine the relative importance of the many variables in order to limit the complication of the computations. In addition, the mathematical solutions involve time constants that may be many times larger than the record of reliable Earth-surface observations of the field. As a result, there are at present numerous competing current-source models and no obvious way of selecting between them.

Nevertheless, all liquid-core dynamo models of the Earth's field lead us to expect relationships between the field and the motion of the Earth. Courtillot et al. (1982) found a general agreement between the change in the Earth's spin (excess length of a day) and the secular change in declination from 1860 to about 1975; one millisecond per day change in time corresponded to about one minute angular change in the D component per year.

Variations in the temperature gradient across the outer core could critically alter core convection and cause the decay and reversal of the self-excited dynamo field. But such gradients are not presently measurable, nor do we know the history of mantle cooling from which the thermal performance of the core could be determined. Sudden changes in the mass distribution at the Earth's surface by the ice ages or a major asteroid impact could produce changes in the CM boundary that also might lead to major main-field modifications.

Seafloor spreading information (see Section 5.9) has verified that the Earth's field regularly undergoes a full change in dominant direction.

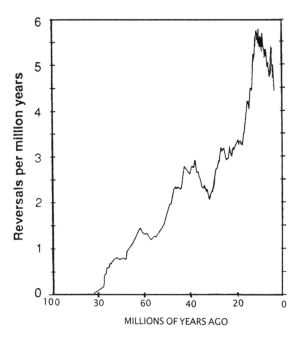

Figure 1.35. Geomagnetic field reversal rate (number per million years) for sample age. Figure adapted from Merrill and McFadden (1990).

MILLIONS OF YEARS AGO

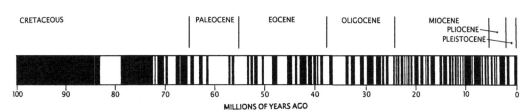

MILLIONS OF YEARS AGO

The last reversal occurred about 710 thousand years ago. During the past 100 million years the reversal rates have been generally increasing (Figure 1.35). We presently seem to be overdue for another major reversal. Although the figure shows reversal rates varying considerably, the Earth's dipole moment is presently rapidly declining in value (cf. Figure 1.20). Now we can expect a reversal about every 200 thousand years. Figure 1.36 illustrates the well-established reversal pattern back to the Cretaceous geological time when there seem to have been periods as long as 35 million years without reversals.

A major epoch of time (about fifty thousand to five million years) in which there is a dominance of one polarity or another is called a polarity *chron* and named after geomagnetic scientists. When an exceedingly long period of one polarity prevails it is titled a *super chron*. A reversal of shorter duration, when an opposite polarity occurs within a chron, is called a *subchron* and named after the region when it was first discovered. *Excursions* are short-lived divergences of the magnetic axis away from

Figure 1.36. Reversals of geomagnetic field recorded back about 100 My from present. Black areas indicate dipole field alignment similar to the present; open areas indicate reversed field. Some geological period names are shown. Figure adapted from Hoffman (1988).

the rotation axis of the Earth. This new direction must persist for an interval assumed to be ample for the magnetic pole to have circled the spin-axis pole of the Earth at least several times (about ten thousand years).

1.9 Paleomagnetism

The subject of field reversals brings us to the study of prehistoric field directions as evidenced by *remanent* (fossil) *magnetization* discovered in rocks whose dates of formation are established by independent techniques. Such magnetic studies come under the general heading of *paleomagnetism* (sometimes spelled *palaeomagnetism*). Three overlapping specialties are subdivisions of this topic. The first is the study of the physics of atomic, molecular, and grain structures responsible for rock magnetization. Such study also defines the corresponding specialized techniques of sample measurement to detect fossil magnetization. The second specialty is the study of geophysical processes producing remanent magnetization, geological rock sampling techniques, and the determination of dated field vector information that is representative of the Earth's fossil field environment. The third specialty is the application of observed remanence measurements in order to understand the evolution of both the Earth's main field and the Earth's structural change.

The first of these paleomagnetic topics covers a subject entitled *domain theory*. It concerns tiny regions (about 10^{-6} meters) called *domains* of field-aligned magnetic atoms, the energetics of valence (outer) orbital electrons about the atoms, the distributions of atomic and molecular magnetic dipoles, the response of rock grains and crystalline structure to magnetization, and the physics of the boundary conditions between domains. Although these are all interesting and important topics that are the basis for the application of rock sampling to paleomagnetic determinations, domain structure will not be described here because our chapter focus is upon the global field. We will visit domain theory briefly in Section 4.2 in order to understand the behavior of magnets.

When rocks that include tiny magnetic domains are formed in the Earth's crust, the magnetic field at the formation site may be imprinted in the rocks and thereby provide a mechanism for scientists to decipher the evolutionary history of our planet. The rock sample can pick up its *natural remanent magnetization* (NRM) either during its first formation or at a time when major physical or chemical geologic alteration took place; this is called *primary remanent magnetization*. If the magnetic remanence is acquired by later exposure of the rock to environmental fields, it is called *secondary remanent magnetization*. A major effort of laboratory processing of paleomagnetic rock samples involves the separation of primary and secondary types of NRM.

For each magnetized material there is a critical temperature (Curie temperature, T_c) above which magnetic properties are randomized by thermal energy so that no magnetization can be detected. As particles of rock cool to temperatures below T_c in the environment of the Earth's main field, a natural primary remanent magnetization, called *thermoremanent magnetization* (TRM) is acquired. Examples of TRM are obtained from cooled volcanic lavas, extruded Earth mantle magma at oceanic spreading centers, and kiln-baked pottery and bricks (see Section 5.9).

In addition to T_c, a second group of characteristic temperatures is important for paleomagnetic studies; they are called the *blocking temperatures, T_b*. As the temperature originally responsible for the TRM fell below the Curie point, the small magnetic grains of igneous rock reached steps in temperature, T_b. Each step corresponded to a characteristic grain size whose magnetization could no longer follow the temporal change in the local field (a process called *relaxation time*). At each T_b point a record of the ambient field was locked into that set of special grain sizes of the rock. The natural range of grain sizes provided the corresponding T_b values down to the final ambient (local) temperature level. The smallest grains have relaxation times of just a few seconds. Large igneous rock grains have relaxation times near 10^9 years.

Over long periods of time, the natural changes of pressure, temperature, and acidity in rocks can cause chemical changes in the particle structure of the magnetic grains such that the Curie temperature is raised from its former level and the rock acquires a *chemical remanent magnetization* (CRM). Metamorphic rocks, whose properties change over geologic time because of their unique local physical environment or chemical composition, lock in a memory of the Earth's field at the time of metamorphism.

Rocks formed as sediments deposited in rivers, lakes, estuaries, and oceans provide samples of *depositional remanent magnetization* (DRM). In natural water bodies there are fine grains of rock that have earlier acquired a TRM or CRM and become detrital fragments because of weathering. While settling in water, these free-falling detrital fragments take only a few seconds to align with the Earth's magnetic field. Then, below the sediment–water boundary, the grains lock in a fossil alignment with the Earth's field as they are naturally compacted into rock. Often a long, continuous record of the paleomagnetic field can be obtained from such DRM samples.

Unfortunately, the paleomagnetic records do not provide a continuous time representation of global evolution. There is a gradual turnover of the seafloor bottom (seafloor spreading), rising at the mid-oceanic ridges and subducting at oceanic trenches near continental margins, so that no seafloor sample is older than about 210 My. The best paleomagnetic sequence of fields comes from this period. Using the fossil magnetization of continental rock formations of earlier dates paleomagneticians have obtained an incomplete record of magnetization extending back in time about 2850 My. With the appearance of man on Earth came the fire baking of pottery and bricks. Carbon dating of the kilns together with a measurement of the TRM for the pottery objects produce the recent history of the Earth's field; this specialized subdiscipline of paleomagnetism is titled *archeomagnetism* (sometimes spelled archaeomagnetism). Such studies extend details of the Earth's main field back about 100,000 years.

Laboratory treatment of rock samples is concerned with the proper separation of primary and secondary remanent magnetization by making use of the difference in relaxation times of rock grains. The process involves a special type of demagnetization to remove the more recent additions to the remanence. Three popular analysis methods involve the application of a gradually decreased alternating field (*AF demagnetization*), thermal treatment, and chemical dissolution. The AF technique allows selective demagnetization of rock particles that are incapable of retaining remanence over periods longer than the rock's age. With *thermal demagnetization*, secondary fields in the rock sample are removed by step-wise use of the blocking temperature feature of the remanence described above. In *chemical demagnetization*, selectively dissolving rock grains by size in an acid treatment separates away remanence types.

The laboratory-measured TRM of a sample is the totality of thermal remanences imparted to a rock as it cooled from T_c to the ambient temperature. When the rock sample is later heated to a temperature T_i, the fraction of remanence acquired up to T_i is lost. With careful, step-wise laboratory heating and TRM-field measurements made at each step, the final paleomagnetic field acquired at the time of the rock's formation can be revealed.

Rock sample fields are determined in laboratories dedicated to the required highly specialized measurements. Such laboratories use the extremely sensitive astatic, spinner, or cryogenic magnetometers described in Sections 4.4, 4.7, and 4.12. Three typical rock dating techniques involve: (1) detection of the decay of radioactive potassium to argon for TRM samples from the past 5 My, (2) measurement of radioactive carbon decay (see Section 3.1) for samples dated after the emergence of human life, and (3) observation of the original stratigraphic emplacement of the sample with respect to other dated geological evidence. The K–Ar decay technique uses the fact that radioactive potassium (^{40}K), from crystals formed during a volcanic eruption, decay (change form) to argon (^{40}Ar)

as the lava cools. Using the potassium isotope half-life (time for half of the atoms to decay) of 1250 My, the fraction of decay product can be equated to the age of the volcanic rock sample.

Just as is the case for most geophysical studies, all separation of remanence is complicated. Unwanted secondary magnetization can be acquired, for example, due to exposure of the rock to lightning, mechanical and electrical laboratory equipment, or storage of the sample in a static-field environment. Other sources of remanence besides TRM, CRM, and DRM have been identified.

Only a sufficiently large number of rock samples can establish the representative time and remanent field from a site. The direction toward a magnetic pole, together with the magnetic latitude determined from the sample at the time of its primary magnetization, fixes a single pole location called the *virtual pole* position. Equation (1.29) gives the relationship of the magnetic inclination and colatitude for a dipole field measurement. For a paleomagnetic rock sample we call these values I_r and θ_r. Assuming that the north and east geographic coordinates of the sample site location are N and E, and that the rock sample measurements give a horizontal declination direction D_r, then the north and east virtual magnetic pole location, N_{vp} and E_{vp}, can be obtained from

$$N_{vp} = \sin^{-1}[\sin(N)\cos(\theta_r) + \cos(N)\sin(\theta_r)\cos(D_r)] \qquad (1.87)$$

and

$$E_{vp} = E + \alpha \quad \text{for } \cos(\theta_r) \geq \sin(N)\sin(N_{vp}) \qquad (1.88)$$

or

$$E_{vp} = E + 180 - \alpha \quad \text{for } \cos(\theta_r) < \sin(N)\sin(N_{vp}) \qquad (1.89)$$

with

$$\alpha = \cos^{-1}\left[\frac{\sin(\theta_r)\sin(D_r)}{\cos(N_{vp})}\right], \qquad (1.90)$$

where $\sin^{-1}() = \arcsin()$ and $\cos^{-1}() = \arccos()$, as explained in Section A.5.

Using a tightly defined sample region of the Earth, a sequence of virtual poles can be plotted. Dated sedimentary rocks give evidence that the Earth's geomagnetic pole has circled the geographic (spin axis) pole for about the past 10,000 years. Paleomagnetic studies show that the geomagnetic field has had a westward drift for the past 1,400 years, called the *secular change* or *secular variation* (SV). These records seem to show an average motion between 0.3° and 0.4° per year (see Figure 1.37). The spin-axis positions are calculated and the pole has moved with reference to the continent of measurement. The resulting geographic *apparent*

Figure 1.37. Computed path of northern virtual geomagnetic pole from 7000 BC to 1980 AD. Figure adapted from Dawson and Newitt (1982).

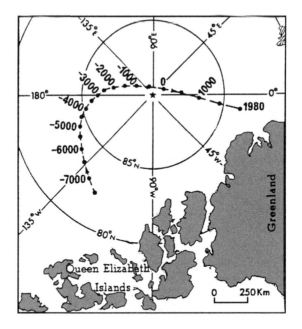

polar wander (APW) diagrams are drawn using sampling techniques that require a representative time period that is long with respect to a 1,000- to 2,000-year pole-circling time of geomagnetic secular variation motion. In Section 5.10 we will see how the apparent polar wander positions are used to reveal the gradual drift of the Earth's continents to their present location.

1.10 Planetary Fields

In recent years space probes have discovered that most of the Sun's planets have main magnetic fields. Only Venus, Mars, and the Moon are exceptions. Indications are that Mars and the Moon have patches of magnetization that most likely formed long ago before their interior core dynamo of liquid material disappeared.

The fields of the giant planets Jupiter, Saturn, Uranus, and Neptune are much stronger than that of the Earth. Their magnetic axes differ considerably. Jupiter has its magnetic axis oppositely directed to that of the Earth, although of similar tilt from its spin axis. The magnetic axis of Saturn seems about exactly aligned with its spin axis. Uranus has a magnetic axis aligned at about 60° to its spin axis, more nearly parallel to its orbit about the Sun. Neptune, more similar to Uranus, has its magnetic axis about 47° to its spin axis. All these giant planets have rings (the largest is that of Saturn) which modify the planets' extension of magnetic fields into space.

1.11 Main Field Summary

Let me now outline the principal things we have learned in this chapter. A field is a vector having both a magnitude and direction; exact definition requires three components. A magnetic field results from the flow of electric currents; the magnetic dipole field comes from a loop current and can be quantified by the value of a dipole moment. The Earth-centered dipole field defines geomagnetic coordinates that are tilted by about 11° with respect to the geographic coordinate system. Geomagnetic pole locations derived from whole-Earth measurements, are different from the dip pole locations derived from local measurements. When the dipole field is modeled with an offset from the Earth's center, there are eccentric-dipole pole locations different in position from the geomagnetic or dip poles.

Now, if someone asks us "Where is the magnetic pole?" we can intelligently answer: "There are five pairs of magnetic poles. Which pair interests you?" The IGRF representation of the measured fields locates one pair where the influence of all global observations contributes to the position. The centered dipole components of the IGRF determines our Geomagnetic Poles for geomagnetic coordinates. Eccentric Axis Dipole Poles are obtained from the field modeling that maximizes the dipole term by moving the center and tilt of the dipole. One pair of field lines from the Eccentric Axis Dipole exits perpendicular to the Earth's surface to determine the Eccentric Axis Dip Poles. Finally, an exploratory measurement of the global field dip location (contaminated with local crustal and external magnetic effects) indicates locally measured Magnetic Poles.

The behavior of electromagnetic fields is completely described by the Maxwell equations. Given a single field vector, a source current direction can be found, but its magnitude can only be prescribed if the distance to the current is known. The field effect from a single current can be identical to that from an equivalent summation of currents; the only requirement is that the composite currents be physically realizable. Fields can be represented by a function of radial distance, r, colatitude, θ, and longitude, ϕ in the spherical coordinate system. When we consider field measurements made over a closed sphere that has currents outside and within, but no current flowing between the two locations, Maxwell's equations provide a method for distinguishing the direction of the two sources and their dependence upon distance.

Gauss found a way to represent a special solution of Maxwell's equations, for global magnetic field observations, that separates the field contributions in r, θ, and ϕ and is composed of two series of terms in ascending powers of r and $(1/r)$. These two series represent spherical harmonics that separate the observed surface field into the two parts that originate from the external and the internal currents. Paired external and internal terms of the series can be considered to result from individual contributing source currents. The terms of the internal series can be identified with field patterns expected of dipole, quadrupole, octupole, etc., field distributions. The field modeling contains special functions called Legendre polynomials in n and m (called "degree" and "order") indices. These polynomials define spherical harmonics having m waves around a circle of latitude and $n - m + 1$ waves around a great circle of longitude (or just n waves if $m = 0$). The significant constants for the internal analysis series of terms are tabulated as "Gauss coefficients" g and h with distinguishing n and m indices (see Table 1.2 and the computer file **ALL-IGRF.TAB** in Section C.12).

Following a collaboration of international modelers, the tables of these coefficients, called the IGRF, are published to represent the fields for a specific 5-year epoch. Later, when refinements are made in these models, the name DGRF is assigned. There are inherent difficulties in producing main-field models due to the distribution of field observations, correction for induction, and limitations of the analysis procedure.

Physical evidence dismisses many suggested locations for the source region of the main field and establishes the site of the source in the Earth's liquid outer core. Currents at that location seem to be driven by gravitational growth of the inner core and organized by the spin of the Earth. However, scientists are still far from agreement on the exact processes in the liquid outer core that produce the dipole and multipole currents. Paleomagnetic studies of the fossil geomagnetic fields have not only proved the existence of reversals but have also revealed how the main field has changed over time and how continents have moved as the Earth has evolved.

J. Quinn reports "The USGS has generated wall-sized geomagnetic IGRF charts for the D, I, F, H, and Z magnetic field components and their secular variations. They exhibit the whole world as large contour charts. In the upper left and right corners of the chart are smaller projections of the main field and secular field given as shaded contours. In the lower left and right corners are polar projections contours of the main field and secular field, with smaller shaded versions of these projections. Everything is on the front-side of the chart for each magnetic component. The controls provided by the website allow zooming in and out. Details about these charts and whom to contact about them are given at the following website: www.esri.com/mapmuseum/mapbook_gallery/volume16/geology2.html (note the underline between mapbook and gallery)."

With the background given in this chapter, three computer programs, described in Appendix C, can be intelligently explored by the reader. The

GEOMAG program will allow you to use an IGRF field model to compute the field at your desired location. The **SPH** program demonstrates the spherical harmonics that are fitted to describe the Earth's field. The program, **GMCORD**, is designed to compute the geomagnetic coordinates of any place on the Earth. The Global Paleomagnetic Database website provides detailed information on time scales, data bases, literature, and organizations at: www.ngu.no/dragon/Palmag/paleomag.htm

1.12 Exercises

1. Assume that you have obtained these July, 2002, quiet daily mean field values from the geomagnetic observatory at Alibag, India: Xmean = 3664 nT, Ymean = 827 nT, Zmean = 2142 nT. Use the information in Section 1.2 to determine the equivalent mean values of H, D, F, and I.

2. Discover the geographic latitude and longitude of your home city. Then download the GMCORD program from the website described in Appendix C and determine the geomagnetic coordinates of that city for 1 June 2003. Compare these results with the Geomagnetic Map of Figure 1.16. From your geomagnetic latitude, would you consider the location to be a polar, high, middle, low, or equatorial geomagnetic location?

3. Use Figure 1.9 to estimate the equatorial radial extent of a dipole field line that leaves the Earth at the geographic latitude that you computed in Exercise 2 above.

4. Use Equation 1.30 to determine the approximate length of the magnetic dipole field line connected to your above location.

5. Use Figure 1.10 to estimate the L-shell of the location discussed in Exercise 2 above. Then follow Equation (1.31) to determine the corresponding invariant latitude.

6. Use the website described in Appendix C to download program SPH and display the Legendre polynomial surfaces with coefficients m and n corresponding to those in Figure 1.14.

7. Use the website in Appendix C to download the GEOMAG program and compute the expected quiet-time field strength at your selected city for 1 June 2003.

8. Substitute the g_1^0 for a_1^0 from Table 1.2 into Equations (1.77) to (1.79) in order to estimate the surface X, Y, and Z centered dipole field components at the geomagnetic colatitude of your selected site in Exercise 2 above for a year of your choosing (call it YEAR1).

9. Use Table 1.2 and Equations (1.80) & (1.81) to determine the colatitude and longitude of the geomagnetic pole for YEAR1. Note that the columns in the table list Gauss coefficient values for 1 January of the given year. To find an intermediate date you need to interpolate between the listed coefficients.

10. Using Equation (1.82) determine the value (in Amperes/meter^2) of the magnetic centered dipole moment on YEAR1.

11. Go to website http://geomag.usgs.gov and select the highlighted "National Geomagnetic Information Center". Next, select from the left-side list "Magnetic Charts". Then select the "DOD World Magnetic Field Charts (Mercator)". Look at the color reproductions of the world magnetic field representations. Click onto any map to obtain a full screen picture and note the icon to print a copy for yourself. Next, back up to "DOD World Magnetic Field Charts" and select "Polar Grid Magnetic Variation and Magnetic Coordinates". Enlarge the "North Main Field Grid Variation" by clicking on its figure. Note the pole location and discuss why it differs from the centered dipole location found in Exercise 9

12. Use Table 1.2 with Equation (1.82) to compute the longitude of the centered dipole geomagnetic pole position for a time at least 10 years different from YEAR1 above; call this YEAR2. Then estimate the change in pole longitude per year and how long it will take that pole to circle the Earth at that rate of movement.

13. Assume that a new magnetic observatory is being established in your country and your expert advice is requested. You are asked how to use the first year's data from that new station to determine the most reliable values of the base-line geomagnetic field levels and secular variation. These are the values to be reported to the World Data Centers for inclusion in the next IGRF. What is your answer?

Chapter 2
Quiet-time field variations
and dynamo currents

2.1 Introduction

It is the nature of geomagnetic fields to not divulge their sources simply. The observatory magnetometers (Earth-field measuring devices described in Chapter 5) respond to all the fields reaching the local environment, add them together, and limit recording only by the frequency response designed into the instruments. A large part of research in geomagnetism concerns the dissection of field variation recordings to isolate the individual contributing sources, discover the physical processes that cause these currents, and thereby understand another feature of our global environment. Occasionally, a newly revealed feature becomes immediately important and useful to society's needs; usually, its utility is discovered only after many years. One of the first field sources to be discovered (Stewart, 1883; Schuster, 1889, 1908) was a current driven by tidal forces and winds in a conducting region above the Earth that was subsequently named the *ionosphere*. Such currents are indicated by a recurring field pattern on quiet-time daily recordings. The accurate determination of quiet-day field variation now finds utility in improvement of satellite main-field modeling, in profiling the Earth's electric conductivity, and in establishment of baselines from which magnetospheric disturbances are quantified.

The purpose of this chapter is to explain the origin and behavior of the regularly recurring field variations that have periods of a day or less. Because the principal source for these currents lies in a naturally ionized layer above the Earth, we will examine the basic features of this ionosphere. In the process we will need to develop a way to distinguish between quiet and active days. Revisiting our spherical harmonic analysis techniques will provide methods for picturing the source currents. We will look at innovative ways to separate high- and low-latitude quiet-time current system sources, expose the lunar-tidal effects hidden in the geomagnetic records, and examine some special field changes that occur during eclipses and solar flares.

2.2 Quiet Geomagnetic Day

The daily record of geomagnetic variations at any world location typically shows a multitude of irregular changes in the field that represents the superposition of many spectral components whose amplitudes generally increase with increasing period. Unique current sources in the upper atmosphere and magnetosphere have been identified as the origins of many of these spectral field variations. On occasion there are days when the magnetic records smoothly change with primarily 24-, 12-, 8-, and 6-hour period spectral components dominating the field composition and few of the irregularly appearing, shorter or longer period changes are present. On these days, the oscillations of the three orthogonal field components produce records that are predictably similar to others recorded many days earlier or later and follow a pattern of gradual change through the seasons of the year. Such records describe the *quiet daily geomagnetic variations*. Figure 2.1 illustrates the difference between active and quiet magnetic magnetograms. Figure 2.2 shows the spectral composition of quiet records from periods of five minutes to twenty eight days. The computer program **FOURSQ1**, described in Section C.8, allows the user to determine the spectral composition of field values from a quiet-day record.

When the small but persistent effects ascribed to the lunar-tidal current system (L) have been removed, the changes are called "Sq" for *solar-quiet fields*, referring to their local-time changes when solar–terrestrial disturbances (Chapter 3) are absent. Some authors, particularly those interested in Earth crustal conductivity studies, prefer to call Sq the *daily variations* or *diurnal variations*. More correctly, "diurnal variation" should be restricted to meaning only the 24-hr spectral component of a phenomenon; Sq fields are not limited to this period alone.

Solar activity identified with sunspot number, controls the percentage of magnetically quiet (Sq-producing) days in a year. The quietest geomagnetic levels usually occur on, or one to two years after, the minimum sunspot number (see Section 3.3). Figure 2.3 illustrates the number of quiet days found each year for several conditions of quietness as defined by the 0 to 9 scale (with three substeps +, 0, −) of magnetic activity index Kp (see Section 3.13). Because of the 11-year cycle in

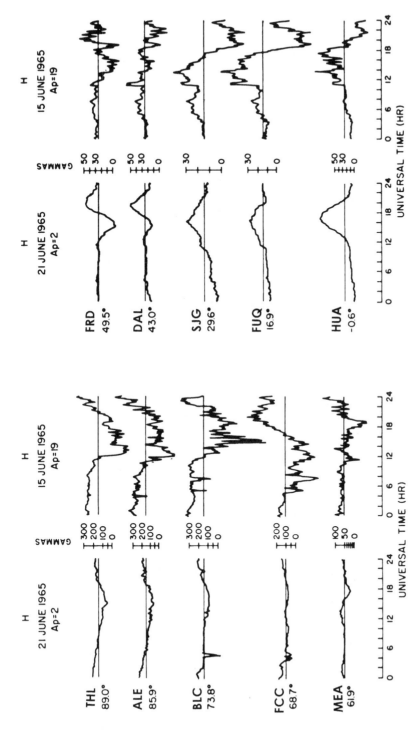

Figure 2.1. Comparison of quiet (21 June 1965) and active (15 June 1965) *H* field component magnetograms at 10 observatories (3–letter codes) having geomagnetic latitudes (under station codes) from the polar region to the equator in the American sector. The horizontal lines indicate the daily average field level. Note the changing scale size at the center of paired days. The daily geomagnetic activity index, Ap (Section 3.13), is indicated under the date.

Figure 2.2. Spectral composition (Fourier analysis amplitudes) of *D* component for the average of quiet days in March 1965 at the Tucson, Arizona, USA, observatory. (a) Variation periods from 28 days to 6 hours. (b) Variation periods from 4 hours to 5 minutes. Note the quiet spectral peaks at periods of 24, 12, 8, and 6 hours and the decreasing amplitude with decreasing period.

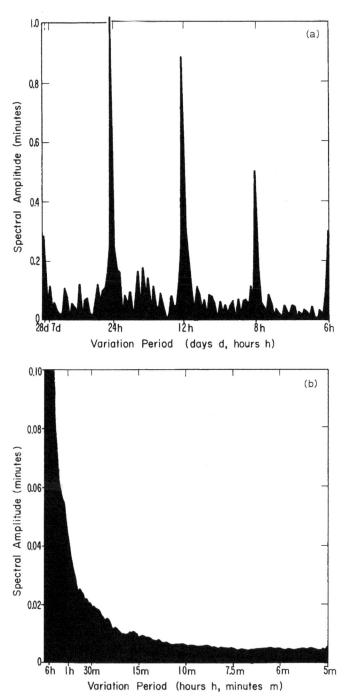

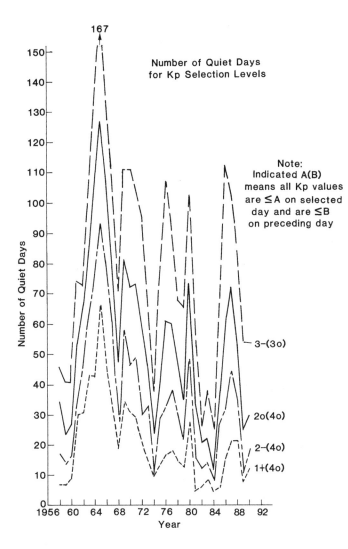

Figure 2.3. The number of quiet days for the years 1956 through 1990. Each trace gives a different selection of quietness defined by values of the activity index, Kp. Traces indicated to the right side as A(B) mean that all 8 daily Kp values were less than or equal to the value A and that all 8 Kp values of the preceding day were less than or equal to the value of B. The preceding day values are included to guard against long-period field disturbances not recorded by the Kp index.

solar activity, a similar cycle of geomagnetically quiet years occurs. The level of Kp $\leq 2+$ is often selected as a quietness level in order to provide enough days to determine the seasonal change of a phenomenon in even the most disturbed years. Conditions on the activity level of the "preceding day" are occasionally added to the quiet-day selection because the Kp index favors variations occurring with a period near 3 hours; on the day following high-Kp values there can be very long period field distortions that could be overlooked.

Figure 2.3 shows us that in recent history the quietest year occurred in 1965; the least quiet year was 1958. The World Data Centers (see Section B.2) provide a list of the *five quietest days* of each month. Some researchers select this set for definition of quiet-day phenomena.

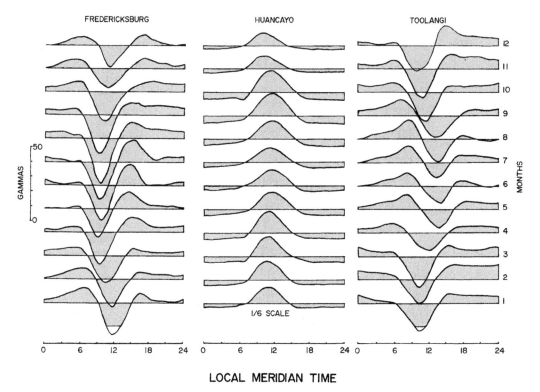

LOCAL MERIDIAN TIME

Figure 2.4. Average daily variation (local time 0 to 24 hours) of the Sq *H*-component at Fredericksburg, a northern midlatitude location; at Huancayo, a Southern Hemisphere dip equator location; and at Toolangi, a southern midlatitude location, for months of the year (1 to 12). Note the annual amplitude change at the midlatitudes and the semiannual change at the equator. Also observe the seasonal summertime annual shift to earlier times of maximum amplitudes. Baselines represent the daily mean field level.

Unfortunately, such "quietest days" may encompass considerable activity in a fully disturbed month; the first reports of a solar-cycle change in Sq were likely due to the inclusion of such disturbed data. However, with more careful selection of quiet days, some residual change in the behavior of Sq in active years is still observed. After we discuss the current generating mechanism of Sq, we shall see how such change can be anticipated.

The Sq field varies slowly in amplitude and phase (time of maximum) through the months of the year (Figure 2.4). At high latitudes (above about 60°), magnetospheric processes (discussed in Chapter 3) may completely dominate the recordings, and only on rare occasions can the true Sq be observed. Outside the polar regions the quiet-day variations about a mean daily level of field show a relatively flat appearance in the night hours. There is an obvious annual increase in amplitude during the summer months and a seasonal shift of the maximum (earlier in summer and later in winter) for Sq at middle- and low-latitude stations. The data from equatorial stations display a semiannual, equinoctial enhancement of amplitudes. At dip-equator locations the *H* component of the daily variation, Sq(*H*), attains an extremely large amplitude, whereas equatorial Sq(*Z*) and Sq(*D*) are relatively small (Figure 2.5). The daily pattern

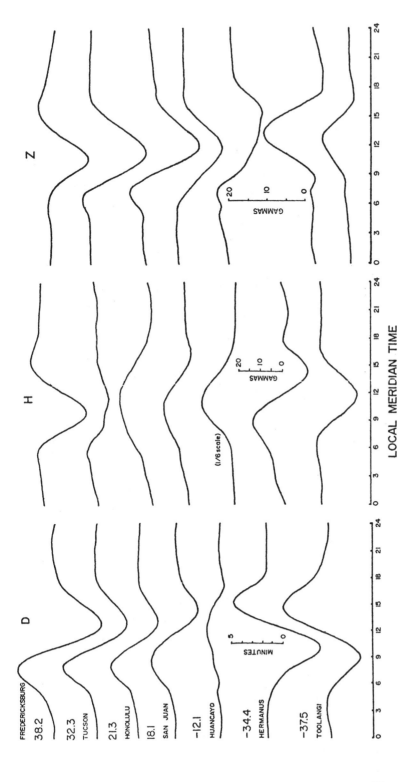

Figure 2.5. Daily variation (local time 0 to 24 hours) of H, D, and Z field components for averaged quiet days of March 1965 when all geomagnetic index values of Kp (Section 3.13) were less than or equal to 2+. Representative observatories (names to left) at mid-and low-geographic latitudes (values near the names) are selected to illustrate the pattern change of daily component variations between hemispheres. Note 1/6–amplitude scale for Huancayo.

73

for the H component of field seems to go through a transition in shape at middle latitudes, whereas the D components of the field are reversed in direction for the Northern and Southern Hemispheres. When the seasons are matched, there seem to be some Northern/Southern Hemisphere differences in the quiet variations; for equivalent geomagnetic latitudes the fields seem more intense in the north.

A computer program, **SQ1MODEL** (Section C.1), produces a representation of the Sq field components for any day at all world locations. These computer-generated values were made from extrapolations of global observatory records obtained in an extremely quiet year. In active years, the Sq fields are expected to be somewhat different.

2.3 Ionosphere

To understand the source of the quiet-time current behavior, we need to digress for a while and review the conditions in the source region for Sq, the ionosphere. Geophysical features having global dimensions that are related to the spherical shape of the Earth are typically given the suffix "sphere." An *ion* is an atom or bonded group of atoms that has become charged by the loss or gain of one or more electrons of negative charge. The electron density measurement of an ionized gas (called *plasma*) in the atmosphere can be used to indicate the degree of ionization. The name *ionosphere* is given to that region of the atmosphere from about 60 to 1,000 km above the Earth where the electron density is sufficiently high to affect the transmission of electromagnetic waves at radio frequencies.

The dynamics of ionosphere formation concerns the balance between production and removal of ionization in the upper atmosphere. The principal production source is the ultraviolet light and soft X rays ("soft" means less energetic) at the short wavelength end of the solar emission spectrum; the principal removal process is recombination of the positive and negative ions. Because our atmosphere becomes less dense and changes composition with altitude, the significant solar radiation encounters few atoms to ionize far above the Earth's surface where the atmosphere is very thin. Close to the surface, the atmosphere is so dense that there is little critical solar radiation remaining that has not been absorbed at higher altitudes. The sensitive molecules that are ionized by the Sun immediately recombine with oppositely charged particles that are nearby at these more dense lower altitudes. At ionospheric altitudes the solar radiation that maintains the breaking apart of molecules into ions and electrons has daily and seasonal changes that vary with location. Recombination, related to the frequency of collision between all local ionized species, dominates the ionosphere decay process.

Let us look at a simplified view of signal transmission using the ionosphere. *Radio waves* beamed from the Earth's surface to the ionosphere at proper frequencies cause the high-altitude atmospheric charged particles to try to vibrate with the local radio-wave field. The motions of these particles emit fields, altering the travel time of the wave in the region. Because the electron density in the lower ionosphere increases with height, the top portion of the radio wave moves faster than the lower part, with the effect that the wave bends downward sufficiently to cause a radio-wave reflection by the ionosphere. Bouncing of the radio wave between the conducting Earth and the ionosphere allows the radio signal to be transmitted to great distances.

In early ionospheric studies, the reflected radio-wave frequencies seemed to indicate specific layers of the ionosphere. These were given the letters D, E, and F (divided into a lower $F1$ and an upper $F2$) and still carry that naming. Presently, scientists prefer the term "regions" to be more appropriate than "layers" because the boundary transitions are not distinct. At high enough frequencies, the radio waves penetrate all the regions. The limiting value of a transmitted radio-wave frequency at which reflections just begin to disappear from any specific region is called the critical penetration frequency, f_0, of that region (e.g., f_0F2 for the critical frequency of the $F2$ region).

A special observatory instrument is used to monitor the changing ionospheric height-frequency profile from which electron density profiles can be determined. This instrument, which generates and receives a broad range of radio-wave frequencies, determines the reflection height from the delay time between an upward-directed specific frequency transmission and the corresponding down-coming return signal reception. Electron densities are exactly computed from the radio-wave frequency using special refraction formulas. The reflected frequency varies in a way similar to the square root of the electron density. The instrument "sounds" the ionosphere, hence the name *ionosonde*. Satellite instruments for a similar function, but which probe the ionosphere below their orbit, are called "topside sounders." A measure of the number of electrons of the ionosphere in a unit cross-section column above an observatory is called the *total electron content* (TEC) of the site at a particular time.

Extra ionization is caused by cosmic rays, solar flares, and the bombardment of particles associated with geomagnetic storms and auroras (Chapter 3). Solar eclipses give scientists opportunities to observe the ionosphere and compute its important variables from response patterns as the solar ionization is suddenly removed and restored at known eclipse times. The characteristic variables of the region are then compared to laboratory models to determine the exact physical processes.

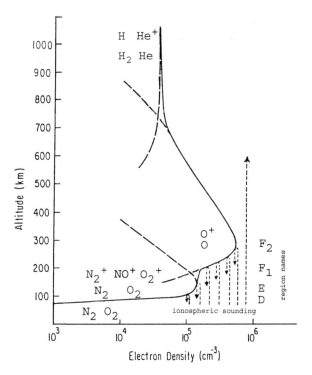

Figure 2.6. Typical summertime composition of the Earth's midlatitude ionosphere. Solid curves give the electron densities; the scale is at bottom. Long-dashed curves indicate the estimated regional density contribution. N_2, O_2, NO, O, H, and He represent atomic, molecular, and +ion composition. Ionospheric regions are named, from the bottom up, D, E, $F1$, and $F2$. Radio-wave soundings near the region maxima are illustrated with short dashes.

Figure 2.6 illustrates the height variation in electron density, principal components of the chemical composition, and the letter designations of the various ionospheric regions of the atmosphere. Each region has its own unique behavior. Most noticeable of these is the disappearance of the D and E regions during the nighttime at locations below latitudes where ionization processes from solar–terrestrial disturbances are important (see Chapter 3 regarding auroral zones and polar caps). The nightly disappearance is mostly due to rapid collisional recombination at lower ionospheric levels after sundown. Of course there are daily, seasonal, and solar-activity-level changes in this pattern. The $F2$ region slowly decays with nightfall; a minimum in electron density occurs there just before dawn (see Schunk, 1996).

Let us look at the simple equations that were first used to represent the change of electron density in the E region, of fundamental interest for the quiet-time currents. Here we will let the subscripts e and i refer to electrons and ions, respectively. It is first assumed that the production of ionization is dependent on the Sun's position; straight overhead, the Sun would produce P_0 ions per second; just below the horizon, it would produce none. Thus, we will call $P_0 \cos(\chi)$ the *production rate*, where χ is the angle from the zenith to the location of the Sun. Next, we call α_{ie} (or simply α) the *recombination coefficient*, a measure of the rate at which the ions, i, and electrons, e, recombine at the E region. Then the time, t, rate of change in the number, n, of ions, dn_i/dt and in the number of electrons, dn_e/dt, are given by

$$\frac{dn_i}{dt} = P_0 \cos(\chi) - \alpha_{ie} n_i n_e \qquad (2.1)$$

and

$$\frac{dn_e}{dt} = P_0 \cos(\chi) - \alpha_{ie} n_i n_e. \qquad (2.2)$$

Because the process changes slowly (χ varies slowly), we call it a "quasisteady-state" condition. In that situation there is no net electric charge so we can assume that $n_i = n_e$, which we simply call the number n. The rate of change of this number proceeds so slowly that we can assume $dn/dt = 0$. For this approximation the above equations may be written as

$$0 = P_0 \cos(\chi) - \alpha_{ie} n^2 \qquad (2.3)$$

or

$$n = \sqrt{\frac{P_0 \cos(\chi)}{\alpha_{ie}}} \qquad (2.4)$$

or

$$n = C\sqrt{\cos(\chi)}, \qquad (2.5)$$

where C is a constant equal to $\sqrt{P_0/\alpha}$.

Thus, to a first approximation, the E-region ionization changes as $\sqrt{\cos(\chi)}$. The name *Chapman function* is given to this variation in honor of Sydney Chapman, who made use of the relationship to explain the Sq changes. See the Section C.5 program **SUN-MOON** for a determination of the solar zenith angle, χ, and the Chapman function at any location. Figure 2.7 illustrates the variation of χ and $\sqrt{\cos(\chi)}$ for the midlatitude location at 40° north latitude. Figure 2.8 shows how the noon value of the Chapman function and the duration of sunup conditions change with northern geographic latitude.

Figure 2.9 is an example of the magnetic field variation on a quiet day superposed upon the daily variation of χ and the $\sqrt{\cos(\chi)}$. In this example, Sq(D) corresponds in onset and subsidence with the Chapman function change. If Sq originated in distant space, far outside the ionosphere, then the sudden local transitions at sunrise and sunset would not occur. Often there is a drift of the nighttime, quiet-field level from sunset to dawn; it is believed that this drift may represent the decay of small nighttime Sq currents flowing in the lower F region as the ionization decays toward dawn. Rocket-borne magnetometers have verified the presence of the principal Sq currents in the daytime E region. We

Figure 2.7. Daily variation of solar zenith angle χ (solid curves) and Chapman function $\sqrt{\cos(\chi)}$ (dashed curves) for summer solstice (SS), equinox (E), and winter solstice (WS) at Boulder (40° north). Local time at this site is Universal Time minus seven hours.

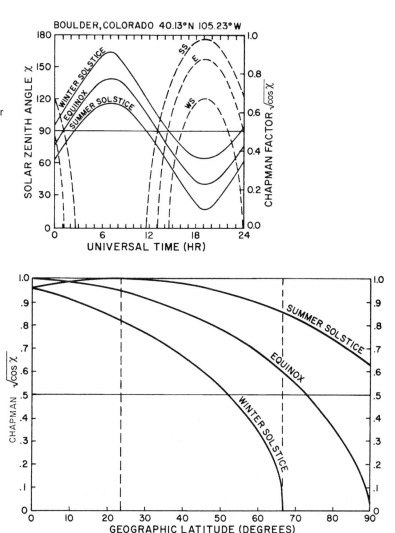

Figure 2.8. Chapman function (curves with scale to left) for sunup ($\chi < 90°$) times at northern geographic latitudes. Dashed vertical lines are the Tropic of Cancer (highest subsolar latitude) and the Arctic Circle (limit of sunup in winter). The presence of Sq currents can be identified with the presence of ionization defined by the Chapman function.

will see later that there is a small contribution to Sq from the distortion of the dipole field by the solar wind.

The property of a substance to allow the flow of current is called its *conductivity*. Recall that current results from the motion of electric charges. How easily these charges can move is a measure of the conductivity, σ, of the medium carrying the current density, J. For a conducting metal such as a wire we write

$$J = \sigma E, \tag{2.6}$$

where E here is the electric field driving the current (not to be confused with the E-region designation of the ionosphere). We call the reciprocal

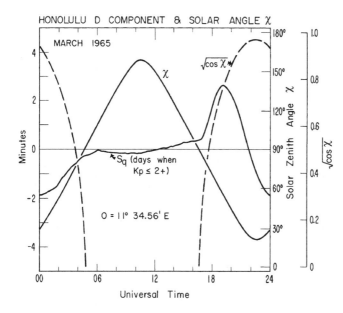

Figure 2.9. Comparison of the daily variation in Sq(*D*), the solar zenith angle χ, and the Chapman function $\sqrt{\cos(\chi)}$ at Honolulu observatory for the average of quiet days in March 1965. Sunrise is at 16:40 UT and sunset is at 4:41 UT. The baseline-level magnetic declination is 11°34.56′ east.

of the conductivity the electrical *resistivity*. Equation (2.6) is just another form of the familiar *Ohm's law*, in which voltage equals current times resistance.

The *E*-region *dynamo current* is given that special name because the ionospheric plasma (a conductor) that is moved through the Earth's field is similar, in a way, to the current that is generated in the wires of a hydroelectric plant dynamo as these wires are moved, by water turbines, through the field of a large permanent magnet. The major difficulty with this analogy is that the ionosphere is a plasma whose conductivity varies in time and is dependent upon the directions of the forces and fields in the region. The term *tensor conductivity* is employed to describe the directional conducting properties of the ionosphere.

We want to examine the characteristic equations of these tensor conductivities; but first, we need a few definitions. When a moving particle, with charge *q*, encounters a magnetic field with strength, B_0 (Tesla), perpendicular to its velocity vector (see Section A.8 for vectors), the particle is forced to move in a circular path of radius, *r*, so that the angular *gyrofrequency*, ω, is given by

$$\omega = \frac{|q|}{m} B_0, \qquad (2.7)$$

with ω (rdn/sec) $= 1/2\pi T$ (for *T* in seconds for one oscillation) and where *m* is the mass of the particle and $|q|$ represents the absolute value of *q*. For electrons, we call *q* the electronic charge *e*, and the value of $|e|/m_e$ (where m_e is the electron mass) is about 1.76×10^{11}; the gyrofrequency is then called the *Larmor frequency*. The *Larmor radius* of the gyrating electron is

$$r = \frac{m_e v_p}{e B_0}, \tag{2.8}$$

where v_p is the velocity perpendicular to the field B_0. Because the Earth's field direction and strength change with altitude, we expect the gyrofrequency of electrons to be altitude dependent.

In the ionosphere, where the electron motion can be impeded by collisions with other particles, the *collision frequency* becomes a necessary part of the representation of conductivity. The atmospheric density, composition, and temperature vary with altitude so the characteristic collision frequencies also vary. For our field studies the most significant collision frequency, v, for the E region is between electrons and neutrals (non-ionized air particles).

If there is an electric field $\mathbf{E}$ directed along the magnetic field, then the electric conductivity for a similarly oriented current is given approximately by

$$\sigma_0 = \frac{n e^2}{m v}, \tag{2.9}$$

where n is the concentration of electrons and σ_0 is called the *longitudinal conductivity* (which is the usual electric conductivity).

Two separate conductivities need to be considered when the electric field is perpendicular to a component of the Earth's magnetic field $\mathbf{B}$. For a current in the direction parallel to the electric field a *Pedersen conductivity* (or transverse conductivity), σ_1, applies. For the E region this conductivity becomes

$$\sigma_1 = \sigma_0 \left(\frac{v^2}{\omega^2 + v^2} \right) \approx \sigma_0 \left(\frac{v}{\omega} \right)^2 \tag{2.10}$$

For a current perpendicular to the electric field a *Hall conductivity* must be used. In such a case we have

$$\sigma_2 = \sigma_0 \left(\frac{v \omega}{\omega^2 + v^2} \right) \approx \sigma_0 \left(\frac{v}{\omega} \right) \tag{2.11}$$

These last two approximation equations (at the right side of (2.10) and (2.11)) are allowed in the E-region ionosphere because the collision frequency, v, in that region is relatively small with respect to the gyrofrequency, ω.

In summary, for $\mathbf{B}$ and $\mathbf{E}$ parallel, we use Equation (2.9) for σ_0. With $\mathbf{B}$ and $\mathbf{E}$ perpendicular we have either (a) the current (particle velocity) parallel to $\mathbf{E}$ so Equation (2.10) for σ_1 holds or (b) the current (particle velocity) perpendicular to $\mathbf{E}$ so Equation (2.11) for σ_2 holds. Of course, the typical field direction of $\mathbf{E}$ or $\mathbf{B}$ will be at some angle, so its parallel

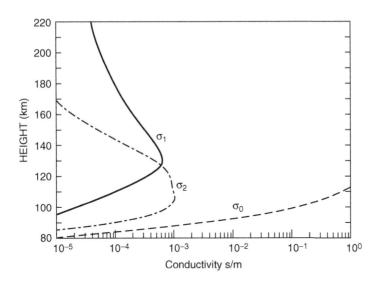

Figure 2.10. Parallel (σ_0), Pederson (σ_1), and Hall (σ_2) conductivities at midlatitude noon in summer with an Earth's magnetic field strength of 2.9×10^4 nT.

and perpendicular components will cause all of the three conductivities to come into play simultaneously. For more accurate computations the equations become more complex with the inclusion of the positive ion effects.

Figure 2.10 illustrates the relative magnitudes of the three conductivities (units are Siemens/meter) versus altitude for summer noontime at midlatitude. The Hall conductivity dominates the Pedersen conductivity in the dynamo region where the current is carried by the electrons. Therefore, the current flows most favorably in a direction perpendicular to the orthogonal electric and magnetic fields. Dynamo currents in the lower F region may also be important at night hours. At such times the solar production of electrons ceases, the recombination processes have essentially depleted the D- and E-region ionization, and the F region decays slowly because of the low density of ions present for recombination. The F-region minimum occurs just before dawn.

What I have described above is a quiet-time ionosphere. When the Sun is active, there is an increase in E-region ionization at high latitudes associated with the bombardment of the atmosphere by electrons and protons. A major modification of the global F region results from thermospheric processes that will be discussed in Chapter 3. Figure 2.11 shows some of the contributors to the tensor conductivity of our upper atmosphere.

2.4 Atmospheric Motions

We need to look at some features of the atmosphere that provide the daily motions to drive our quiet-time dynamo current. The temperature profile

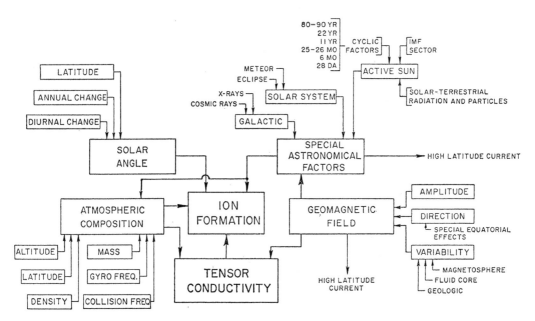

Figure 2.11. Diagram of some contributors to the behavior of tensor conductivity in the ionosphere.

defines the naming of characteristic regions. Solar heating tells us how the atmosphere expands and contracts. Global wind systems are found at ionospheric heights. We will now look at some of these atmospheric characteristics as a prelude to understanding the generation of quiet-time currents in the upper atmosphere.

The altitude–temperature plot for the Earth's atmosphere defines the regions *troposphere, stratosphere, mesosphere,* and *thermosphere.* Their boundaries are fixed by characteristic profile maxima and minima, called *pauses,* that occur at altitudes where the temperature momentarily ceases to change with altitude (Figure 2.12). Between regions separated by these pauses there is a minimum of atmospheric mixing. It is easy to understand the causes of this temperature–altitude profile. Solar radiation warms the Earth, heating the atmosphere nearest the surface; we recognize the difference in temperature between sea level and mountain tops. The drop in temperature with increasing altitude would continue with increasing height were it not for *ozone heating.*

Principally at altitudes from about 15 to 35 km, atmospheric oxygen molecules are broken into oxygen atoms by the penetrating ultraviolet (UV) radiation in the solar spectrum. Higher in altitude, there is too small a concentration of oxygen for UV effects; lower in altitude, the UV has already been dissipated. The oxygen atoms recombine with other oxygen molecules to form ozone, O_3, which is most concentrated near 25 to 35 km. The ozone is subsequently removed upon recombination with other oxygen atoms to form the oxygen molecules. Presence of some

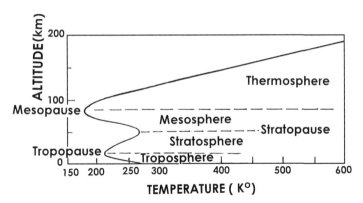

Figure 2.12. Air temperature in degrees Kelvin ($^\circ$C + 273) as a function of altitude. Regions of the atmosphere that derive their names from the temperature profile are indicated.

other specific molecules in the upper atmosphere (natural or man-made) can stimulate (by catalytic action) ozone formation or depletion.

The chemical processes associated with ozone formation provide a heat source that modifies our temperature–altitude profile (Figure 2.12) from the tropopause all the way to the mesopause. The release of man-made chlorofluorocarbons into the atmosphere has begun to destroy the natural ozone layer through the interaction of chlorine with the ozone. A special Antarctic springtime (October–November) condition at the stratosphere releases the elemental chlorine for photocatalytic destruc-tion of the ozone, creating a polar-cap region of depleted ozone called the "ozone hole." The resulting temperature anomaly there could modify the local thermospheric wind systems responsible for dynamo currents.

To understand the temperature increase in the thermosphere, we must first review the difference between temperature and heat. The tempera-ture of a gas is a measure that indicates the speed of motion of the gas particles. Depending on the speed and number of these particles hitting us, our skin warms, giving our body heat calories. Close to the Earth's surface, the density of air (determining the number of air molecules that hits our skin) changes so little, that we associate temperature (re-lated to gas particle velocity) with heat (calories) and we interchange the two meanings. At -273° Centigrade, particle motion ceases; we call this temperature "absolute zero." Using the same scale divisions as the Centigrade scale (100° between the mean sea level temperatures at which water freezes and boils), we set a Kelvin scale with 0 at absolute zero; water freezes at +273 K. The atmospheric temperatures of Figure 2.12 are in Kelvin units.

Gravity is responsible for holding the atmospheric molecules about the Earth; the atmospheric density decreases rapidly with altitude (Figure 2.13). The air we breathe at the Earth's surface is almost eighty percent nitrogen and twenty percent oxygen. With increasing altitude,

Figure 2.13. Atmospheric density decrease with altitude.

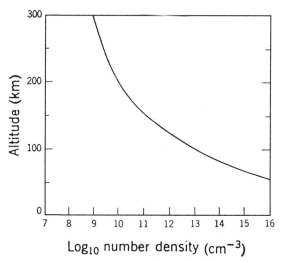

Figure 2.13. Atmospheric density decrease with altitude.

physical processes gradually change the atmospheric composition to the lighter molecules of helium and hydrogen (Figure 2.6). In the thinner atmosphere, the collisions decrease and the mean speed of the neutral molecules increases, resulting in a temperature rise that stabilizes at about 1,500°C near 600 km. The heat imparted to our space-walking astronauts is small, however, because so few of the fast molecules hit their spacesuits. Solar heating by solar radiation is a separate effect.

The thermal-tidal (*thermotidal*) motions that are responsible for the principal features of the quiet-day currents depend on the daily expansion of the Earth's heated atmosphere. Figure 2.14 shows the expected *solar radiation* through the year for all latitudes. There are obvious annual differences in the radiation. A semiannual overhead transit of the Sun occurs near the equator. The Southern Hemisphere shows a larger summer–winter contrast in radiation than the Northern Hemisphere because the Earth is slightly closer to the Sun in northern winter and farther away in northern summer. The heating forces responsible for the thermotidal motions in the atmosphere are slightly different from what is shown in Figure 2.14, mainly because of special effects at a large solar incidence angle and the influence of atmospheric water and ozone molecules. Near the Earth's surface, heat accumulation slows the time of maximum atmosphere heating to the post-noon hours and delays the date of maximum daily heating by one or two months past the summer solstice.

There is a linear relationship between selected wavelengths of the solar energy flux and the sunspot number (a measure of the Sun's activity; see Section 3.3). Therefore the solar energy flux shows a distinct solar cycle (approximately 11 years) change. Average daytime temperatures

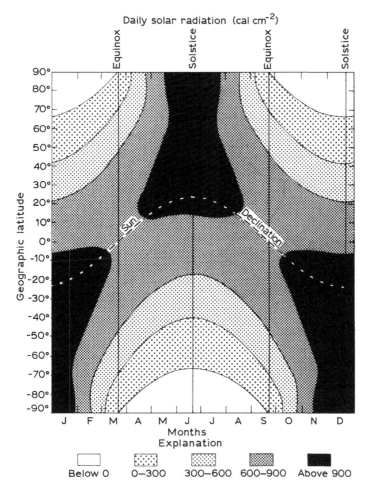

Figure 2.14. Solar radiation at top of the atmosphere as given in the Smithsonian Meteorologic Tables. The solar declination and seasonal demarcations are also shown. Darker shading indicates higher radiation.

at our atmospheric thermopause levels can increase by 1,000°C from sunspot-number minimum to maximum years.

Solar gravitational forces act upon the atmosphere, but these effects are minor compared to the daily atmospheric expansion and contraction due to heating. Sizable lunar-tidal forces are also present in the ionosphere and give rise to a lunar-time component of the dynamo current, L (Section 2.7) that can be extracted from the quiet-day records.

Recent thermospheric modeling techniques have provided a global picture of the horizontal wind systems at E-region altitudes. Figure 2.15 is a model for the solstitial period in quiet, average, and active geomagnetic conditions. Such winds vary with season and with the special heating that occurs as part of intense currents associated with auroras in active periods (Chapter 3). The thermospheric winds together with the thermotidal motions then drive the ions in the Earth's main field to

Figure 2.15. Schematic diagram of the average circulation in the Earth's thermosphere (arrows indicate wind direction) during solstice for various levels of auroral activity: (a) extremely quiet geomagnetic activity, (b) average activity, and (c) geomagnetic substorm. Figure from Roble (1977).

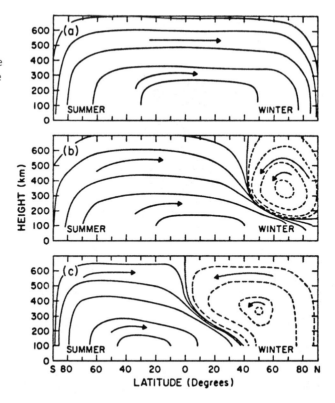

generate the observed quiet-time currents. Figure 2.16 is a diagram of the many geophysical processes affecting such current.

2.5 Evidence for Ionospheric Current

We have already seen that a large dipolelike field permeates the space about the Earth, that there are charged particles existing in regions of the atmosphere, and that there are winds and daily tidal motions that can drive dynamolike currents in these regions. Significant evidence of these currents comes from daily observatory recordings that show field variations exhibiting a sunup, Chapman-function dependence (Figure 2.9). The field patterns change from large to small in an annual cycle, in keeping with the ionization changes from summer to winter. At the equator, where the Sun passes overhead twice a year (at the equinoxes), a corresponding semiannual change in Sq amplitudes is found.

The most convincing direct proof of ionospheric currents comes from magnetic field sensors aboard rockets that have been sent through the ionosphere. An increasing field in the daytime E region always shows the location of a strong current as the rocket approaches the critical Sq altitude. Figure 2.17 shows the ionospheric current density observed for

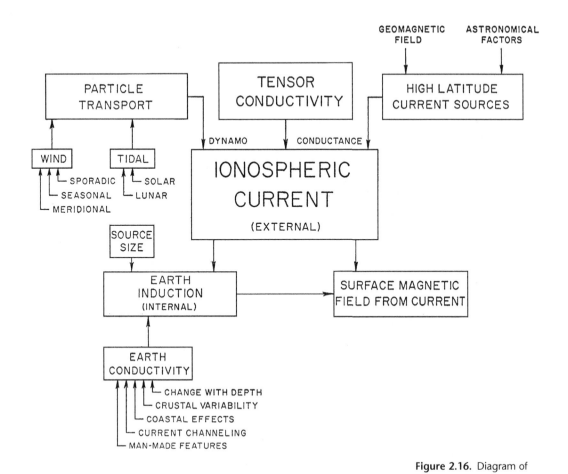

Figure 2.16. Diagram of some contributors to the ionospheric current and its field.

rocket flight near the equinox and close to the dip equator. Note the current maximum near 106-km altitude in the E region.

The H component of field becomes large and positive in a very narrow latitude region about the dip equator (Figure 2.18). It is difficult to envision a source more distant than the ionosphere that could cause so sudden a change of amplitude in such a short north–south distance. The physics of the ionosphere requires that a special current condition be established at the magnetic dip equator, where the Earth's field is directed magnetic south to north in a horizontal plane. There, the Hall current, inhibited by the ionospheric boundaries perpendicular to the flow, causes a polarization field to be established in opposition to the flow. Under those special equatorial conditions, the effective conductivity parallel to the boundary (perpendicular to the Earth's magnetic field) is increased above the usual Pedersen conductivity. The net result is a considerable enhancement of the east–west conductivity that concentrates the flow of any arriving E-region ionospheric current.

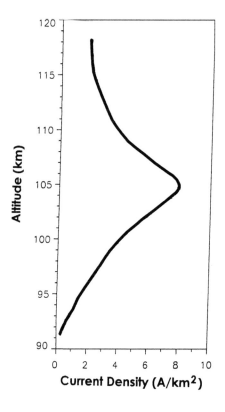

Figure 2.17. Example detection of equatorial *E*-region ionospheric current system. Rocket flight from Alcantara, Brazil, September 21, 1994. Figure adapted from Pfaff et al. (1997).

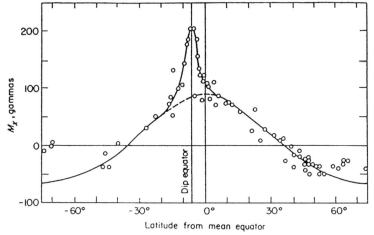

Figure 2.18. Daily range of *X* component of field at different latitudes measured at 76° W longitude. Note location of dip equator for this South American study. Figure adapted from Onwumechilli (1967).

At the Earth's surface, measurements show us that this "equatorial electrojet" effect is about 450 to 550 km wide (about 4° to 5° in latitude). Using the midnight field levels as a baseline (or main-field) position, the equatorial electrojet appears as a large daytime increase in the *H* component of field that maximizes just before local noon. Near Huancayo, Peru, the main field is smaller than at other dip-equator observatory locations (see Figure 1.23d); such a field alters the E-region gyrofrequency, which ultimately produces a larger than average electrojet effect at that location. The dip equator moves with the westward drift of the main field so that the position is presently moving southward in Peru (and northward in India).

On occasion, in the morning and/or evening hours depressions of the equatorial field appear that have been credited to the existence of an oppositely directed (*counter electrojet*, CEJ) current. Researchers have found a reversal in the local mean winds at E-region heights concurrent with the counter electrojet. Such occasions of CEJ seem to be related to global disturbances of thermospheric winds and tides (see Section 3.10 for Joule heating and thermospheric wind effects).

We can infer a general shape of the global pattern of the quiet-time ionospheric current from a closer view of the magnetic records and using the reliable *right-hand rule* (gripping the current with the right hand, thumb along the current direction, our fingers curled in the directions of the field). Figure 2.19 illustrates conclusions about the source current shape that can be ascertained from surface field component records. The positive northward (H) equatorial electrojet field therefore means an eastward ionospheric current exists in that region. The consistently recurring pattern of the D component at almost all latitudes in the Northern Hemisphere means that there is a southward current direction at the first part of the day and a northward current in the afternoon. The reversal of the D patterns for the Southern Hemisphere indicates an oppositely directed current system there.

Our knowledge of dynamo effects lets us expect that a reversal of the main field impressed upon the dynamo would reverse the current. The Earth's dipole-like field is directed away from the Earth in the Southern Hemisphere and into the Earth in the Northern Hemisphere. The reversal of the D-component patterns in the two hemispheres (Figure 2.19) is consistent with an ionospheric generation of the Sq currents. The Z-component fields, not shown in the figure, could be similarly diagrammed.

The H component of field is positive near midday at low latitudes and then reverses its pattern in the middle latitudes to become negative near midday at higher latitudes (Figure 2.19). The midday amplitude ratio D/H becomes smaller in the region of the equator. This means that as the equator is approached the current direction becomes more clearly aligned east–west. Taken all together, our directional information on Sq fields indicates that (viewed from above the ionosphere) the daytime Northern Hemisphere has a counterclockwise vortex of current and the daytime Southern Hemisphere has a clockwise vortex (Figure 2.20).

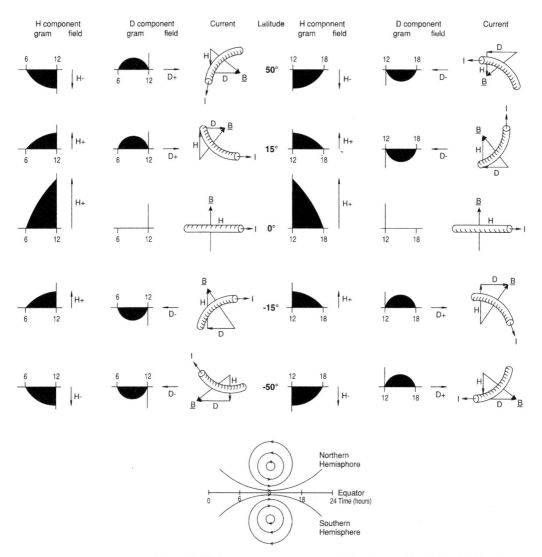

Figure 2.19. Sq current vortex patterns inferred from *H* and *D* field directions below 50° geomagnetic latitude. Morning patterns (0600–1200 local time) are shown on magnetograms to the left of the central latitude column; afternoon patterns (1200–1800 local time) are shown to right. Shaded variations for *H* and *D* fields for hours before and after 1200 are indicated with arrows. Current in above-surface wire segments equivalent to cause the observed *H* and *D* fields are sketched to right of each *H* and *D* picture along with the **B**-field vectors. The fields in the two hemispheres, at daytime, would be consistent with the Sq ionospheric current vortices depicted at the bottom of the figure.

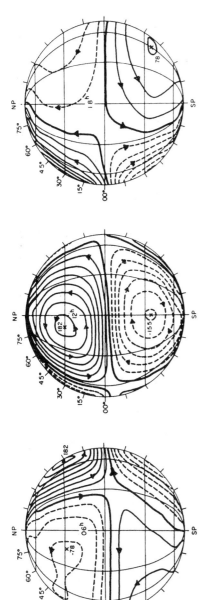

Figure 2.20. Yearly averaged Sq equivalent current system for 1958 viewed toward the 06-, 12-, and 18-hr meridians. Arrows indicate current-flow direction, clockwise contours are dashed, and counterclockwise currents are solid. Each contour represents a step of 30×10^3 Amperes flowing between the contour lines in the direction of the arrows. Figure from Matsushita (1967).

Still, there are questions about (1) what is happening in the polar region, (2) how much of the observed field is a result of currents induced in the conducting Earth, and (3) how we can detail exactly the Sq ionospheric current system. All these features can be resolved by revisiting spherical harmonic analysis.

2.6 Spherical Harmonic Analysis of the Quiet Field

For current sources such as Sq that we now know arise from thermal–tidal motions in the ionosphere, a secondary current is induced to flow in the conducting Earth. The depth of this induction depends upon how rapidly the source varies and upon the conductivity patterns within the Earth. Most of us have experienced the fading or blackout of our radio reception in a car traveling through a tunnel. Some motorists traveling the same route on different occasions may have noticed tunnels in which their FM radio signals are lost completely but the AM radio signals remain, although weaker than outside the tunnel. This is direct experience of the deeper penetration of lower-frequency (longer wavelength) AM signals compared to the higher-frequency (shorter wavelength) FM signals.

Figure 2.2 illustrated the spectral composition of the quiet-day field change. Note the strong contributions of the 24-, 12-, 8-, and 6-hr components; smaller 3- and 2-hr components may also appear. One-hour and shorter-period oscillations seem to be random, noiselike fluctuations. The program **FOURSQ1** described in Section A.8 allows the user to determine the spectral components of any daily field variation by entering the measured sampled field values for the desired day.

In Figure 1.12 we saw how Fourier analysis dissects a daily variation signal such as Sq into its harmonic sine (and cosine) wave parts. Effects of the longest wavelengths of Sq penetrate to a depth of 400 to 600 km, into the upper mantle of the Earth. This penetration means that currents are induced to flow through the conducting crust and mantle (see Section 5.8). At the Earth's surface, magnetometers measure the sum of fields from the source and the induced currents. From our mathematical exercise in Chapter 1, recall that spherical harmonic analysis (SHA) allows us to separate the contributions from sources above and beneath the Earth's surface.

To perform a SHA we need a global distribution of Sq measurements as quiet-day field variations about a baseline level selected to be either (a) the mean for the day (easy to determine from an average of all values), (b) the value at local midnight (when the ionospheric E-region should not support current) or (c) the predawn quiet field level when lingering $F2$ region dynamo currents are gone. The midnight selection presents some problems at polar latitudes because disturbances often occur in

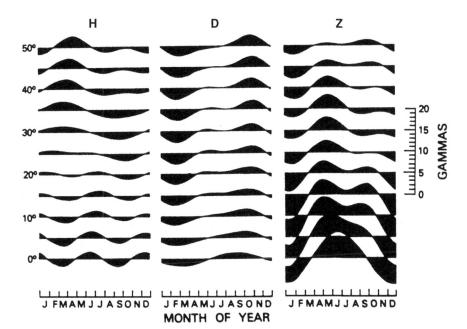

Figure 2.21. Annual and semiannual variation in the midnight field at middle and low latitudes, reconstructed from SHA of data obtained at North American observatories on quiet days. The three columns of field levels, in gammas, represent (left to right) the *H*, *D*, and *Z* components. Geomagnetic latitudes of the field representations are shown at the left side.

that region at night hours. In addition, the midnight values of the *H* and *Z* components at middle and low latitudes have been shown to track a night-side magnetospheric distortion of the Earth's dipole field that stretches seasonally into an antisunward direction, providing a magnetospheric component to the quiet-day midnight field level (Figures 2.21 and 1.29).

In the last chapter we learned that the number of SHA terms needed to represent a field is minimized when the selected coordinate system is arranged close to the symmetry of the source current. Should we consider the atmospheric properties most important and arrange the data by geographic latitude? Then, outside the polar latitudes, the insolation, thermotidal forces, winds, ionization, and conductivity would be well organized in a geographic arrangement.

Should the main field, from the interior dynamo process, dominate the mapping of Sq for a geomagnetic latitude arrangement of data? The Earth's geomagnetic field has large nondipole components, and the geomagnetic and geographic pole separations are noticeably nonsymmetric in the two hemispheres. The total field is particularly high in regions of central Canada, Siberia, and south of Australia; it is quite low near southern Brazil. In addition, the dynamo-generating *E* region of the ionosphere is just about 100 km from the Earth's surface; in this short distance the local field strength and direction (not the geomagnetic model field) determine the conductivity character. The distances from the Earth's spin pole to polar points where the field is perpendicular to the Earth's surface differ dramatically in the Northern and Southern Hemispheres. The dip equator is considerably offset from the geomagnetic equator in the African sector.

No analysis coordinate arrangement is ideal for Sq representation. However, most researchers find that Sq is best organized in a geomagnetic latitude arrangement. Dip coordinates would be preferable were it not for the utility of the SHA in application to Earth conductivity computations; the dip-latitude system is nonlinear and would therefore lead to irregularities in the interpretation of the induction-depth determinations. Although early researchers used a dip-latitude configuration, the geomagnetic-latitude coordinates, with an equatorial adjustment, are presently preferred. Geometry problems near the pole, created by the geomagnetic latitude versus time plotting of the data, usually have been ignored.

Three different techniques have been used to analyse Sq variations. One, called a *snapshot* method, fits the global field observations at a preselected fixed universal time (UT). Geomagnetic latitude and local time (geographic longitude) become the coordinates. Time selections are made to diagram the dynamo current through the UT day and seasons at world locations.

The second technique, called a *slice* method, makes use of the fact that the Sq source current system is relatively stationary with respect to the position of the Sun (Figure 2.22). Consider a longitude (slice) selection of observatories; as the Earth turns through the day, the surface observatories respond to 360° of the dynamo current pattern. Because the quiet field changes very slightly from day to day, the 24 hours of time at the observatories can be equated to longitude in the analysis. When the computation is completed, only the primary region is considered to be represented.

Two slice techniques can be used. The simplest type is the full pole-to-pole observatory arrangement. A second type is the use of a single-hemisphere set of observatories; the opposite hemisphere is modeled using a 6-month shifted copy of the original hemisphere records at equivalent locations in which the negatives of D and Z field vectors are assumed.

A third technique, called a *mirror* method, for one-hemisphere analysis by either the snapshot or slice method, is occasionally used when only one sample time is available. This method uses only the symmetric even values of $n-m$ (Figure 1.11); obtains the polynomial coefficients by integrating (summing), as in Equations (1.48) to (1.53), only through the latitudes from $\theta = 0°$ to 90°; and finally doubles the computed magnitudes to make up for the $\theta = 90°$ to 180° part. The odd values of $n-m$ that are omitted contribute mainly to low-latitude asymmetry in Sq (summer

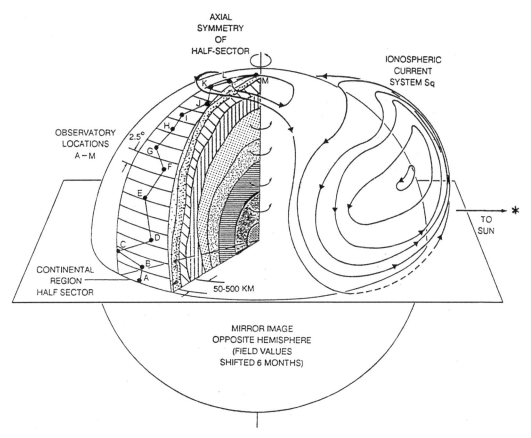

Figure 2.22. Representation of continental region observatories (A to M) used for a slice-type analysis in which Earth turns under an ionospheric current system (counterclockwise vortex of contour lines) that is fixed with respect to the Sun direction.

hemisphere intrusion of Sq into the opposite winter hemisphere), so this analysis technique has been used only for high-latitude studies.

Once the technique has been selected, the analysis proceeds in much the same way as main-field analysis (Section 1.4). Various methods are devised to fill in values for all latitudes and longitudes where there are no observatories. Then the SHA coefficients are obtained from the Fourier components of field at each latitude step. However, there are three small differences for the quiet field SHA: (1) there are no $m = 0$ terms because the Sq is cyclic in 24 hours, (2) both the external (source) and internal (induced) series of polynomials are important, and (3) significant seasonal changes that occur throughout the year must be modeled by repeated analyses.

Our spectral information (Figure 2.2) tells us that we need not extend the computations shorter than the 1- or 3-hr period variation where the unique quiet-time spectral features are lost. Typical analyses are carried no further than $m = 6$ and $n = 18$ because of noise evidence at higher degree and order. With such SHA analyses for each representative month,

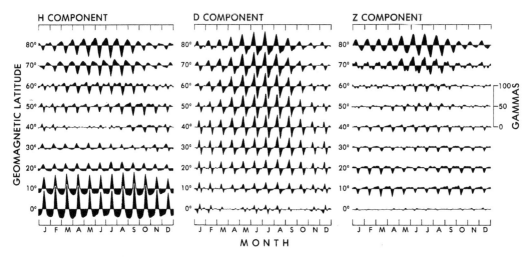

Figure 2.23. Annual picture of the midmonth daily variations of Sq (in local time) for the North American region from the equator to 80° latitude (bottom to top rows) displayed for H, D, and Z at left, center, and right sections, respectively. The scale size between baselines is 50 γ. Months of the year are indicated at the bottom of each section.

the Sq at all latitudes can be recreated. Figures 2.23 and 2.24 illustrate stack plots of the fields obtained for a slice analysis of North America and Australia region data representing the extremely quiet year 1965. Note the semiannual amplitude enhancements at the equator, the annual changes at higher latitudes, and how the patterns fit our expectations of an overhead source current vortex. Figure 2.25 shows how the SHA allows us to separate the external and internal contributions to the three components of field at the Earth's surface. It is the sum of these two fields that is observed.

The equivalent current function $J(\phi)$, in amperes, for an hour of the day equal to $\phi/15$ (the longitude divided by 15°) is obtained from

$$J(\phi) = \sum_{m=1}^{4} \sum_{n=m}^{12} \left(U_n^m \cos(m\phi) + V_n^m \sin(m\phi) \right) P_n^m, \tag{2.12}$$

with 4 for the maximum value of m and 12 for the maximum value of n. For the external current representation

$$U_n^m = -\left(\frac{5R}{2\pi}\right)\left(\frac{2n+1}{n+1}\right) a_n^{me} \left(\frac{a}{R}\right)^n, \tag{2.13}$$

$$V_n^m = -\left(\frac{5R}{2\pi}\right)\left(\frac{2n+1}{n+1}\right) b_n^{me} \left(\frac{a}{R}\right)^n, \tag{2.14}$$

or for the internal current representation

$$U_n^m = -\left(\frac{5R}{2\pi}\right)\left(\frac{2n+1}{n}\right) a_n^{mi} \left(\frac{R}{a}\right)^{n+1}, \tag{2.15}$$

$$V_n^m = -\left(\frac{5R}{2\pi}\right)\left(\frac{2n+1}{n}\right) b_n^{mi} \left(\frac{R}{a}\right)^{n+1}. \tag{2.16}$$

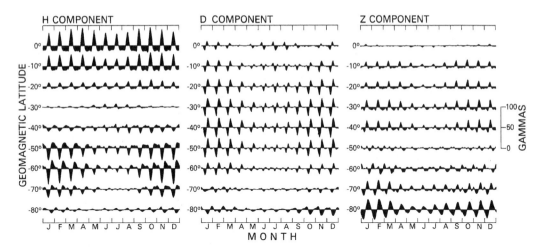

Figure 2.24. Annual picture of the midmonth daily variations of Sq (in local time) for Australian region from the equator to –80° latitude (top to bottom rows) displayed for H, D, and Z at left, center, and right sections, respectively. The scale size between baselines is 50 γ. Months of the year are indicated at the bottom of each section.

where R is the Earth radius in kilometers, P_n^m is the Legendre polynomial value for a given θ latitude, and the external or internal coefficients were determined as in Equations (1.54) through (1.57). The value of a is the radius of a sphere whose surface is located where a current could flow to give the fields described at the Earth's surface by the SHA; hence the name "equivalent current." Because we have other evidence that the dynamo current source is in the E-region ionosphere near 100-km altitude, the value of a is approximately equal to R, so their ratio ($a/R = 1$) may be omitted from the current computations.

The equivalent external current intensity I of latitudinal component θ and longitudinal component ϕ can be determined (in amperes) from J by

$$I_\theta = \frac{1}{r \sin (\theta)} \frac{\delta J}{\delta \phi} \tag{2.17}$$

and

$$I_\phi = -\frac{1}{r} \frac{\delta J}{\delta \theta}. \tag{2.18}$$

Figure 2.26 illustrates the derived external equivalent currents in a rectilinear display for a slice model that utilized stations along a longitude line through India and Siberia. The contour values are given in 10^4 ampere units such that 5 kiloamperes flow between the contours in the direction of the contour arrows. This interpretation means that where the contours are close together the current is intense. Note how the intense summertime current vortex of the Southern Hemisphere intrudes into the wintertime Northern Hemisphere. The occasional appearance of high-latitude current vortices near the midnight sector is really not dynamo current; soon we will see how to separate these polar disturbances (Chapter 3) from the data. A similar current pattern of smaller amplitude and with oppositely directed flow is induced in the conducting

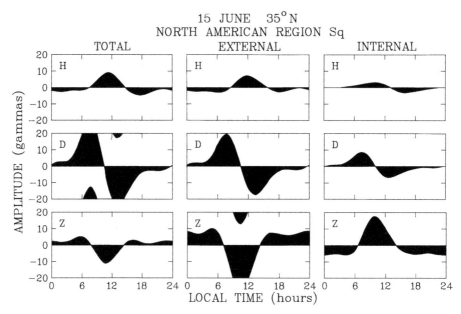

Figure 2.25. Example of the daily amplitude variation (total) and separated external and internal Sq field components, *H*, *D*, and *Z*, for the North American region at 35° latitude on 15 June 1965, a very quiet year.

Earth. The summation of the field effects from these external and internal currents determines the magnitude and direction of the observed quiet-time magnetic field at the Earth's surface.

To visualize the seasonal changes of Sq in one figure requires a special presentation. We identify the daily external current variation range as the difference in maximum and minimum current levels at a given latitude through a given analysis day. This range is a single value for a given date and latitude. Then a contour plot of these daily ranges on a latitude–date grid shows the full year's change in the external Sq current system. The trace of the midlatitude location of the maximum range on each date gives the year's variation of the Sq vortex focus location. Figure 2.27 illustrates the year's Sq behavior over India–Siberia in 1976. Note the seasonal change in focus track and the small activity contributions at the high latitudes.

The longitudinal wavelength property of SHA can be used to isolate the high-latitude current systems. Recall, in Figure 1.11, that the number of waves along a longitude circle was $n - m + 1$ (when m was not 0). The small, high-latitude vortices of Figure 2.26 can be isolated by a selection of the short wavelength (along a longitude) polynomials with $n - m > 1$. The remaining polynomials, with $(n - m) \leq 1$, describe the quiet-time Sq vortex completely. Figure 2.28 shows the effect of this separation and the absence of equatorial field contributions on the separated currents.

The high-latitude vortices that appear on our selection of quiet times are simply an equivalent current representation of low-amplitude,

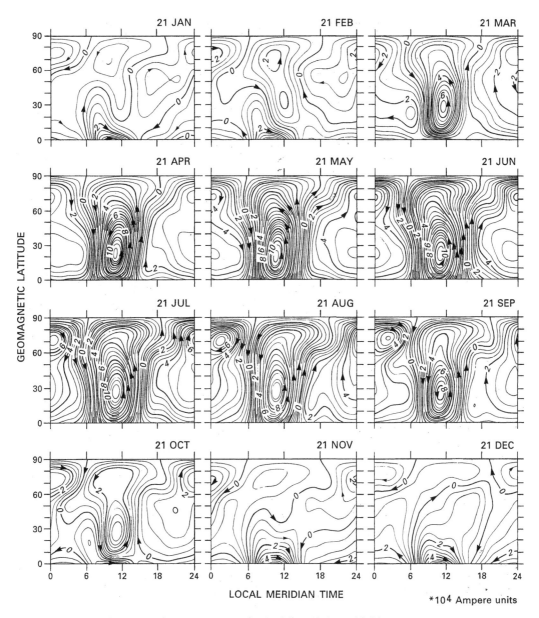

Figure 2.26. Equivalent external current contours for Sq daily variations of field in the India–Siberia region computed from a SHA on the 21st of each month in 1976. Each pattern, in local time versus geomagnetic latitude, shows the equivalent external current in 10^4 A units such that 5 kA flow between contour lines in the direction of the arrows. A midnight zero current level was assumed for the contour identification.

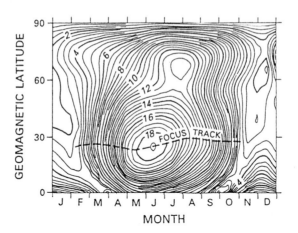

Figure 2.27. Day-of-year versus latitude contour representation of the daily range, at each latitude, for the external Sq equivalent currents of Figure 2.26. Contour units are 10^4 A. The seasonal track of the Sq vortex focus (represented by a heavy dashed trace) is determined by the latitude of maximum range for each day.

polar-region disturbance events that have been included because there were not enough truly quiet days with our Kp selection. These polar-region currents are sensitive to the direction of the solar-wind field (toward or away from the Sun) upon its arrival at the subsolar magnetosphere (see Figure 3.42). Early reports of sector effects (see Section 3.11) in quiet-day variation records seem to have resulted from an inclusion of the polar activity in the Sq data set.

At times of solar–terrestrial disturbances the E-region ionization changes, the atmospheric temperatures are enhanced, there is a density change in the upper atmosphere, and the thermospheric wind systems are modified. Studies even show a clear enhancement of the Sq and modification of the hemispheric current vortex focus location on quiet days selected during active years. At times of geomagnetic disturbances the thermosphere is heated by field-aligned currents (discussed in Section 3.10) and a pattern of global neutral winds develops. These winds can drive dynamo currents in the lower F region that modify the more usual Sq current system.

As will be seen in Chapter 3, the Earth's dipole field in space is distorted by the arrival of a wind of particles and fields from the Sun. An equatorial satellite, circling the Earth at about 6.6 R_e (the distance at which its orbit, in the direction of the Earth's spin, will be geostationary) will sense this compression of the main field on the sunward side of the magnetosphere and an expansion of the field into a tail formation on the dark side. This magnetospheric daily field change will appear to behave, at the Earth's surface, somewhat like an Sq field variation of ionospheric origin. The principal difference will be that the magnetospheric source would exhibit a major dominance of the sinusoidal diurnal term, no sharp cutoff of the variation at sunrise and sunset times, and no

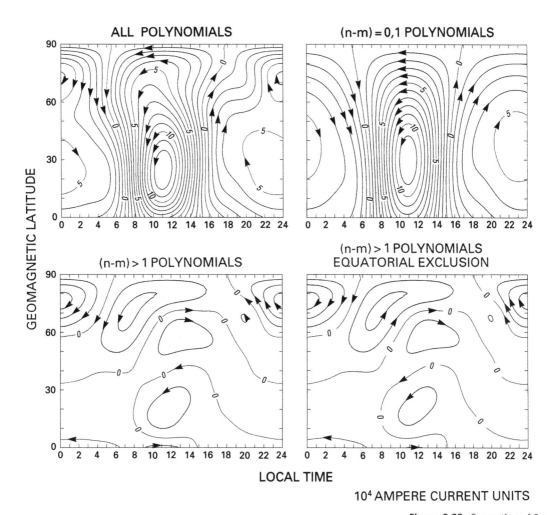

LOCAL TIME

10^4 AMPERE CURRENT UNITS

equatorial region enhancement. Figure 2.29 shows the magnetospheric field observed at a geostationary orbit satellite. Note the slight change from sinusoidal form as the satellite enters the influence of the (nightside) magnetospheric tail.

The daily change in the main field, as seen at the Earth's surface, has been estimated by some researchers to contribute as much as 10% to the local Sq variation, although the exact amount remains in dispute. The SHA modeling of field includes this contribution as part of its external equivalent current. If the magnetospheric contribution to Sq is the major source of the local nighttime hour Sq variation, then a first estimate of its daytime value might be obtained therefrom. However, some ionospheric specialists consider the nighttime lower F region to be sufficiently conducting to carry some nighttime dynamo currents.

Figure 2.28. Separation of Sq vortex from polar activity vortex using a selection of the longitude wavelength polynomials $(n - m)$. The exclusion of equatorial data (lower right panel) had a negligible effect on the identification of the polar activity. The India–Siberia data set for 21 June 1976 is represented. Currents are labeled for 10^4 Amperes flowing between contour lines in the direction of the arrows.

Figure 2.29. Median (and upper and lower quartiles) of the quiet field at GOES-5 geostationary satellite at 75° W for 24 hours of a spring equinox day during the 1983/1984 operation; the components are (a) *H*-northward, (b) *V*-vertical, and (c) *D*-eastward. Figure from Rufenach, McPherron, and Schaper (1992).

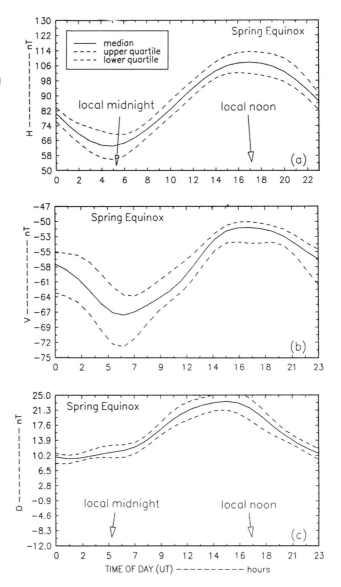

2.7 Lunar, Flare, Eclipse, and Special Effects

Two forces are in balance to keep a satellite of mass, m, such as our Moon, orbiting a much larger mass, M, such as the Earth. The masses are measured in kilograms. Directed outward is a centrifugal force

$$F_c = m\omega^2 r, \tag{2.19}$$

where ω is the angular velocity in radians/second ($\omega = 2\pi/T$ with T as the time for one orbit in seconds) and r is the radial distance of the small satellite in meters from the gravitational center of mass. Directed inward is the gravitational force

$$F_g = mMG/r^2, \qquad (2.20)$$

where G is Newton's gravitational constant. When m is much smaller than M, the radial distance of the satellite from the center of the larger mass is then obtained from equating F_c to F_g:

$$m\omega^2 r = (mMG)/r^2 \quad \text{so} \quad r = (MG/\omega^2)^{1/3}, \qquad (2.21)$$

a distance independent of the mass of the smaller satellite.

Now let us consider our permanent satellite about the Earth, the Moon. First let us define a *lunar time*, which is described in a way similar to our more familiar solar time. Lunar noon is when we see the Moon at its highest point above the horizon (upper transit); lunar midnight is fixed when the Moon is 180° (angle about the Earth) away from the lunar noon position. The Moon location can be computed in the **SUN-MOON** program of Section C.5. Viewed from above the North Pole, the Moon rotates about the Earth counterclockwise in the same direction that the Earth spins on its axis. The Moon appears to the Earthbound observer to rise in the east, later each day by about 50.47 minutes. Tides behave similarly.

On the side of the Earth toward the Moon (near lunar noon) the lunar gravitational force on the atmosphere dominates; on the side of the Earth opposite the Moon (near lunar midnight), the lunar centrifugal force on the atmosphere exceeds the lunar gravitational force. These *semidiurnal* (twice a lunar day) unbalances of forces give rise to a semidiurnal tide in the atmosphere (as well as the ocean). The resulting tidal motions transport the E-region ionization within the Earth's main field in such a way that dynamo currents are created that are of sufficient magnitude to be detected at the Earth's surface as a lunar-geomagnetic effect of about 1 to 10 gammas. Because the ionization changes in step with the 24-hr solar day and the twice-daily tides are in step with the lunar day, there is consequently not only a semidiurnal lunar-tidal component but also a solar component of the lunar field variations. The Sun, although much larger than the Moon, is at such a great distance that it produces only a small added tidal component to the computations.

Lunar variations can be extracted from magnetic records using a selection of close-by days in which there were no major magnetic disturbances. The speedy analysis method (called the "Fourier Analysis of Residuals" (Matsushita and Campbell, 1972)) involves the removal of the anticipated Sq variations from the data and the rearrangement of the daily samples into lunar-time increments for a *superposed epoch* fitting. From an average (or a locally weighted regression determination) of the values in each time block, a semidiurnal *lunar time variation, L2*, can be extracted. Figure 2.30 shows how the 12-hr Fourier component is the only significant lunar-time component. The solar modulation, buried by the lunar-time ordering, is then reestablished by appropriate shaping with the $\sqrt{\cos(\chi)}$ Chapman function (Figure 2.31). The **SUN-MOON** computer program, described in Appendix C, gives the moon's position and the Chapman function for any location, date, and time. A detailed lunar-analysis method of Chapman and Miller (1940) requires many more sample days but exactly establishes the lunar–solar variations. Figure 2.32 shows the global external equivalent current contours of $L2$ for the June solstice, December solstice, equinox, and yearly average periods.

Three ocean and island effects can be mentioned at this point: (1) The daily tidal motions of the ocean through the Earth's main field generate electric dynamo currents in the conducting salt water. The fields from such currents can contribute to observations at magnetic observatories that are uniquely located near major tidal basins. (2) The induced component of Sq depends upon the Earth's conductivity profile to depths of at least 500 km. The magnetic observatory in Hawaii is on the side of a volcanic mountain that rises from a deep seafloor and has an Earth conductivity that differs greatly from the deep surrounding ocean. Such observatories exhibit an unusual induced field that modifies the field components expected for their latitude location. At continental coast regions, where induced currents in the ocean are constrained to flow parallel to the shoreline, there is another modification of the induced contribution to the observations. (3) Ocean-flow electric currents (from the horizontal flow of the conducting salt water in the Earth's field) with periods longer than a day have been discovered, generating electric dynamo-current fields as large as 30 gammas.

On occasion, particularly during the peak of the Sun's activity, flares occur on the solar surface (Section 3.4), causing an intense emission of ultraviolet and X-ray energy that travels to the Earth at light speed. As a result, the atmospheric ionization in the subsolar region is momentarily enhanced, particularly at the ionospheric D and E regions. The conductivity modification causes a temporary increase in the local dynamo currents that can be recorded by surface magnetometers (Figure 2.33). The flare effect signature was called a *crochet* in the early literature (using the French word for the spurlike appearance on the recordings). Accompanying the ionization increase is an attenuation, called "short-wave fadeout" or SWF, of the high-frequency radio-wave signals going through the lower ionosphere.

From an ionospheric point of view, a solar eclipse is the inverse of a flare. When the ionizing radiation from the Sun is obstructed by the Moon, the E-region electron–ion recombination decreases the

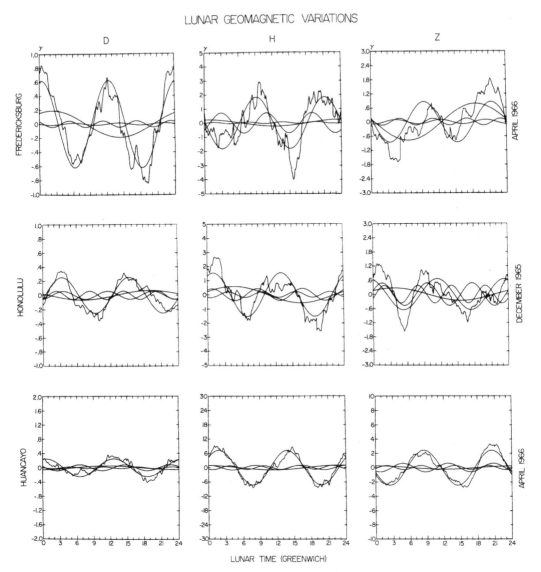

LUNAR GEOMAGNETIC VARIATIONS

Figure 2.30. Superposed epoch (in Greenwich lunar time) geomagnetic variations (irregular curves). The lunar $L2$ variation is shown as the large amplitude, semidiurnal variation of the Fourier field components (smooth curves) for D (left), H (middle), and Z (right) at Fredericksburg (top) in April 1966, Honolulu (center) for December 1965, and Huancayo (bottom) for April 1966. The declination, D, is in minutes; the field strengths of H and Z are in gammas. The variations' Fourier field components with periods of 24, 8, and 6 hours are shown for comparison to the 12-hr lunar effect.

Figure 2.31. Lunar contribution to the *D* field variation at Honolulu on 20 March 1965, determined by the Fourier Analysis of Residuals method and shaped by the Chapman function.

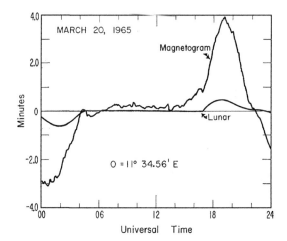

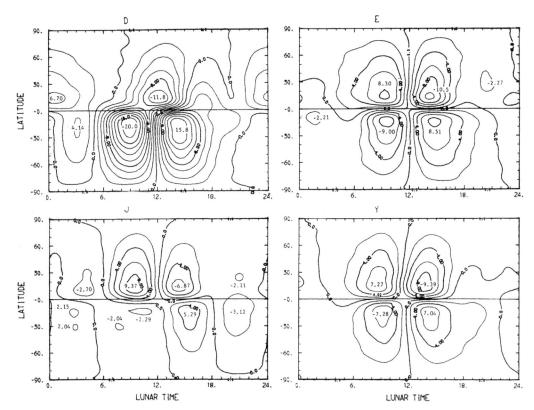

Figure 2.32. External lunar current systems for all four harmonic terms ($m = 1$ to 4) with respect to geomagnetic latitude and lunar time. Contours for (D) December solstice, (E) equinox, (J) June solstice, and (Y) yearly average months are depicted with contour intervals of 2 Ka. Positive and negative signs indicate counterclockwise and clockwise current directions, respectively. Figure from Matsushita and Xu (1984).

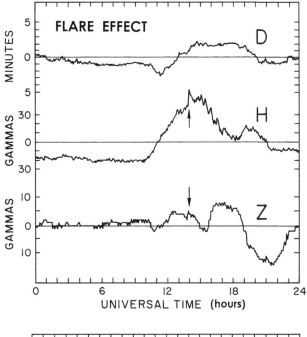

Figure 2.33. Flare effect near midday on 3 December 1965, recorded at Huancayo, Peru. The three H, D, Z field components are indicated. Baseline is average daily value. Data from World Data Center, Boulder.

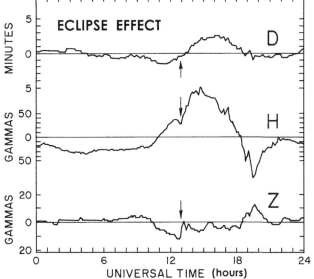

Figure 2.34. Eclipse effect near midday on 12 November 1966, recorded at Huancayo, Peru. The three H, D, Z, field components are indicated. Baseline is average daily value. Data from World Data Center, Boulder.

conductivity along the eclipse path. The pre-existing dynamo currents are then constrained to flow in the regions of higher ionospheric conductivity outside the eclipse path (*eclipse effect*). Ionospheric recombination characteristics can then be determined from the form of the temporary decrease in the Sq field (Figure 2.34) at the surface magnetic observatories in the eclipse path.

Figure 2.35. Ft. Yukon, Alaska, arrival of an atmospheric infrasonic pressure wave (top trace) that generated ionospheric dynamo current pulsations (bottom trace) with periods of 5 to 30 seconds in August, 1962. The source of this event was an atmospheric nuclear explosion at Novaya Zemlya, USSR, that caused a pressure wave to travel around the world with sonic speed (Campbell and Young, 1963).

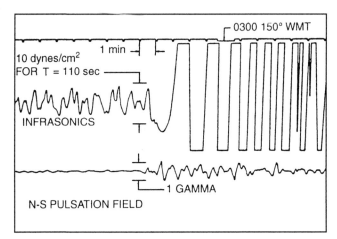

The ion displacement in the main field that occurs when a pressure wave arrives in the dynamo current region of the ionosphere can be seen as small field changes at the Earth's surface. The source of such pressure waves can be the sudden atmospheric joule heating by field-aligned currents during a major geomagnetic storm (Section 3.10) or a large atmospheric explosion, as from a major volcano or nuclear device. Figure 2.35 shows the result of a distant nuclear explosion that started an Earth-circling ionospheric pressure wave which could be measured by a specially designed, supersensitive barometer. The disturbance gave rise to a geomagnetic field pulsation measured with a sensitive induction loop antenna (Section 4.6). For this event, concurrent ionospheric absorption measurements confirmed a sonic propagation speed between Ft. Yukon and Fairbanks, Alaska (Campbell and Young, 1963).

2.8 Quiet-Field Summary

Although it was apparent that the geomagnetic field variations were unique at quiet times, conditions for quietness had to be assumed pragmatically in order to track seasonal and year-to-year changes. The principal source of the quiet-time field was discovered in the E-region of the ionosphere, where the ion density and collision frequency permitted the flow of an electron current. This current was driven by the daily thermal-tidal motions and thermospheric winds moving the ionization through the Earth's main field as in a dynamo. We used the term "tensor conductivity" to indicate the fact that the ionospheric plasma had different conducting properties depending on the direction of the forced motion with respect to the main field in the ionosphere.

The dynamo currents of Sq form two vortices on the dayside of the Earth, counterclockwise in the Northern Hemisphere, clockwise in the

Southern Hemisphere, and large in the summertime regions. A special conductivity property, in the region of the dip equator where the Earth's main field is horizontal, causes a concentration of eastward current called the equatorial electrojet. In the polar regions, disturbance-like currents can be separated from the Sq by wavelength discrimination in the spherical harmonic analysis.

Tidal forces acting upon the Earth's atmosphere can generate lunartidal dynamo currents in the daytime ionosphere as semidiurnal field variations in lunar time. These currents, usually several gamma in size, are modulated by the local ionization to show location and seasonal changes.

Any process that momentarily disrupts the lower regions of the ionosphere can be seen as a modification of the local dynamo current system. A sudden increase in E-region conductivity by intense solar flare radiation or a sudden decrease in conductivity along the path of a solar eclipse excites a detectable change in the quiet-day surface field. A sudden arrival of pressure waves in the ionosphere causes dynamo current effects.

Three computer programs have been provided so the reader can explore the Sq quiet daily field variations. The **SQ1MODEL** program (Section C.3) recreates the expected Sq at a given time and location for extremely quiet conditions. The **FOURSQ1** program (Section C.8) computes the baseline level, slope, and Fourier components of daily field variation values provided as an input. The **SUN-MOON** program (Section C.5) gives the positions for the Sun and Moon and the E-region ionization function of Chapman, $\sqrt{(\cos \chi)}$ at any given time.

In this chapter we have also introduced the annual and semiannual variations of geomagnetic fields. The seasonal changes in ionospheric conductivity, thermospheric winds, and tidal forces account for most of the seasonal changes in dynamo currents. The magnetosphere was also shown to have a seasonal deformation on magnetically quiet days, easily detected on the midnight side of the Earth but also manifested at all local times. In addition, we shall see in Chapter 3 that the geomagnetic disturbances have a semiannual variation. In all but the very quietest conditions, such activity is detected globally and contaminates most analyses of the not-completely-quiet records.

2.9 Exercises

1. Using program SQ1MODEL in Appendix C, determine the expected values of the 10-minute field components of the quiet daily variation, Sq, for the location of your home city on 21 June 2003.
2. Using program SUN-MOON in Appendix C, determine the solar zenith angles and values of the Chapman function for 12:00 hours Local Time for the date and location used in Exercise 1.

3. With the value of total field strength (that was found in Chapter 1, Exercise 7, using Equation (1.2)) for your home city determine the Larmor frequency (Equation (2.7)) of electrons at 110 km altitude moving perpendicular to that total field. Assume the total field at that altitude to be approximately the same as that at the Earth's surface below.

4. Use Equations (2.10) and (2.11) to show that the Longitudinal conductivity is approximately equal to a combination of Pedersen and Hall conductivities $(\sigma_0 \approx \sigma_2^2/\sigma_0)$.

5. Using Figure 2.10 read off the maximum Hall Conductivity (σ_2) at 110 km. Use the gyrofrequency you determined in Exercise 3 and the approximate Equation (2.1) to estimate the collision frequency at 110 km. Discuss the limitations of this computation.

6. Using Figure 2.19 draw an expected pattern of the H, D, and Z field variations for a June solstice Sq current measured at the surface near your home city latitude and longitude. Use program SQ1MODEL (in Appendix C) to produce the 30-minute values of field H, D, and Z for 21 June 2003 and compare the result with your drawing.

7. Use the 30-minute values of the H field that you obtained in Exercise 6 as input values to program FOURSQ1 (in Appendix C) to determine the harmonic spectral components. Make a graphical display of the spectral information.

8. Use 1-hour values of the H field that you obtained in Exercise 6 to determine the median Sq (H) level for that day with program SORTVAL at C.9 in Appendix C. Then use the same values to determine the sample mean and standard deviation with program ANALYZ at C.10 in Appendix C. Compare the mean and median values.

9. Using Figure 2.33 estimate the size of the field change in H that was caused by the flare at the equatorial location. Next, assume that a December flare on the Sun (having a comparable flare effect size at the equator) is recorded near midday at the latitude of the magnetic observatory near your home city. Estimate the size of that flare effect in the H component.

10. An aeromagnetic company wants to know what are the expected values of the quiet daily field variation (Sq) on the solstices and equinoxes at the new observatory in your country. What are your instructions to those who scale the magnetograms from the observatory to insure that a pure estimate of the Sq values are determined?

Chapter 3
Solar–terrestrial activity

3.1 Introduction

When the phrase *solar–terrestrial activity* is used, the intent is to describe those changes of energetic particles and electromagnetic fields that originate at the Sun, travel to the Earth's magnetosphere, and have drastic effects upon the Earth's atmosphere and geomagnetic field. The activity is on time scales that are short in the human perception of events. The Sun is said to be "active" when the magnitude of such changes is distinguishably large with respect to the average behavior over tens of years. A specific region or process on the Sun is said to be an active source region when a particle or field disturbance in the Earth's magnetosphere can be traced to some special change in that region of the Sun. The vagueness in these definitions should disappear as we become more specific in the description of such phenomena as sunspots, flares, coronal holes, coronal mass ejections, solar wind, geomagnetic storms, ionospheric disturbances, auroras, and substorm processes.

We call the moving plasma of ionized particles and associated magnetic fields that are expanding outward from the Sun the *solar wind*. Its associated field is the *interplanetary magnetic field* (IMF). The wind exists out past 150 times the Sun–Earth distance because the pressure of the interstellar medium is insufficient to confine the energetic particles coming from the hot solar corona. We call this solar-wind dominated region the *heliosphere*.

Outer space is filled with particles and fields originating from the formation of the universe and from stars. The name *cosmic rays* is given to the corpuscular radiation reaching the Earth from all directions of outer space. Low-energy cosmic ray particles that penetrate the region between the Sun and the Earth are swept away by the energetic solar wind streams (*Forbush decrease*) that attend occasions of solar–terrestrial activity. Radioactive carbon isotope ^{14}C is created in the Earth's atmosphere during the bombardment of common nitrogen ^{14}N by low-energy cosmic rays. The resulting radioactive carbon eventually gets absorbed, along with the common ^{12}C, by all living (organic) matter. ^{14}C has a half-life of 5,730 years (the time for half of it to decay to carbon-12). The $^{14}C/^{12}C$ ratio rises at times of low solar wind, in years of quiet Sun conditions, and thereby becomes an indicator of cyclic solar variation that can be dated back several thousand years using fossil-carbon samples.

As the Earth travels about the Sun in this dynamic solar-wind environment, the Earth's field reacts to the solar flux. The blast of solar wind envelops and distorts the dipole geomagnetic field into an elongated teardrop shape. On average, this wind compresses the magnetosphere to about eleven Earth radii (R_e) on the sunward side and stretches the magnetosphere outward, far past the Moon's orbit (at about 60 R_e) on the "downwind" side. At times the wind squeezes the subsolar magnetosphere past the geostationary satellite orbits at about 6.6 R_e.

Because the multipole geomagnetic field components become less important with increasing distance from the Earth, the field in space is dipolar out to several Earth radii. Farther out, the would-be dipole field increasingly experiences severe distortion by the impinging solar wind and special currents generated by the disturbance-related processes. Satellite studies have shown the magnetosphere to be a dynamic region of particles, currents, and fields whose behavior is our primary concern in this chapter. To help the reader achieve an understanding of the activity-related Earth-surface field disturbances, I will start this chapter with a review of the Sun's normal characteristics and changeable features. Next, we will look at the disturbed solar wind and how it impinges upon the magnetosphere and creates an unstable condition called the *substorm*. Finally, we will look at the resulting Earth-surface field changes and the indices of geomagnetic activity.

As we proceed into this chapter it may be helpful to have available several relative dimensions. The mean Earth radius (R_e) is taken to be 6.371×10^3 km. The Sun's radius (6.960×10^5 km) is about 109 times that of the Earth. The mean Sun–Earth distance, called one *astronomical unit* (AU), is about 1.496×10^8 km (215 times the solar radius). The Sun's distance from the Earth is about $23,480R_e$, whereas the Moon is about $60.3R_e$ from the Earth. Although the Moon is only 0.27 of the Earth's diameter, its angular size, as viewed from the Earth, is about equal to that of the Sun's visible sphere, permitting our solar-eclipse-time

Table 3.1. *Coordinate systems for geomagnetism*

SYSTEM	ARRANGEMENT
solar-ecliptic	X to Sun; Z northward & normal to ecliptic
solar-magnetospheric	X to Sun; Z northward & dipole in X–Z plane
solar-magnetic	Z along dipole; X in plane of Z & Sun–Earth
geomagnetic	Z into Earth; H along dipole field

observations of a solar corona that extends far into space. Also, because the gravitational pull decreases with the square of the distance, it is the Moon (389 times closer than the larger Sun) that dominates the Earth's tidal changes. A computer program **SUN-MOON** for locating solar and lunar positions is described in Section C.5.

In discussions of the solar-terrestrial environment, four *coordinate systems* are used on special occasions. In Table 3.1 I have listed the coordinate title and the direction arrangement. They are all right-hand systems (first two fingers and thumb pointing in the X, Y, and Z or H, D, and Z orthogonal directions respectively); therefore, two given directions define the third. The "ecliptic plane" is determined by the path of the Earth about the Sun.

For the interplanetary magnetic field (IMF) region studies, the solar-ecliptic system is favored. At the boundary of the solar wind and magnetosphere, the solar-magnetospheric system is usually used. The solar-magnetic coordinate system is important inside the magnetosphere. At the Earth's surface, of course, the geomagnetic system is most valuable.

3.2 Quiet Sun

Atoms and molecules can both absorb and emit energy. At suitable wavelengths, the absorption and emission spectra are at discrete color wavelengths (lines) that can be used to distinguish the responsible substance and its physical state. The chemical composition, temperature, magnetic field strength, and general processes transpiring on the solar surface have been determined through the use of sensitive spectrometers (spectra measurement instruments) aimed at the Sun.

Our Sun is categorized by astronomers by the terms "G-2," "Yellow," and "Dwarf." This means that the Sun belongs to an abundant category of stars in our Milky Way galaxy whose spectra, a G-2 type, are stronger in calcium line emissions than in hydrogen and show a profusion of metal lines with iron conspicuously present. "Yellow," corresponds to the 5,000 to 6,000 K surface color temperature (recall the Kelvin temperature scale from Section 2.4). The "dwarf" term (not related to the Sun's large radius)

Figure 3.1. The Sun's position in our galaxy. Distances are given in *light years (ly)*, the distance traveled in one year at the speed of light (5.9×10^{12} miles or 9.5×10^{12} kilometers).

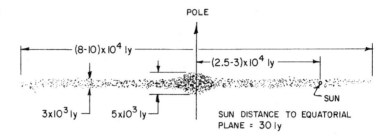

indicates a grouping within a central main sequence of stars on a special plot of photovisual brightness versus color temperature.

The Sun is located about 3×10^{17} km from the center of our galaxy on a spiral arm moving with an orbital velocity of about 250 km/sec (Figure 3.1). Galactic position viewed from the Earth (Figure 3.2) varies through the *sidereal day* (rotation period of the Earth with respect to fixed stars). This day is 23 hr 56 min 4.091 sec long, about 4 min shorter than the solar day (sidereal time is computed in the **SUN-MOON** program, Section C.5). What we call the Milky Way is our edge-on view of our galaxy, and the time the Milky Way takes to circle our zenith is the sidereal day. Selected stars, which emit strong ionizing X rays, seem to be congregated toward the galactic center. The overhead transit of such stars has been found to contribute to a detectable but small sidereal-time increase in nighttime E-region ionization. A corresponding increase in nighttime dynamo-current Sq (Chapter 2) may have been observed.

The Sun is the only star whose surface characteristics are observable. The Sun has a significant dipole magnetic field that can be tilted as much as 30° to its rotational axis. Except for the Sun's mass (1.99×10^{30} kg), volume (1.41×10^{27} m^3), and energy output (3.86×10^{26} J/sec) we have no direct information about the Sun beneath its opaque surface. By volume, the Sun is thought to be 81.76% hydrogen, 18.17% helium, and the remaining 0.07% is thought to be made up of carbon, nitrogen, oxygen, and various metals. Fusion processes form helium (^{4}He) from four protons (^{1}H) in the Sun's interior and produce approximately 4×10^{-12} joules of energy for each helium atom. The Sun radiates

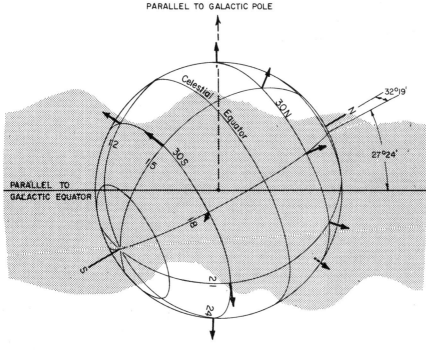

PARALLEL TO GALACTIC POLE

View From Center of Galaxy

Figure 3.2. The Earth viewed from the galactic center. Heavy arrows show the zenith direction from 30° N and S latitude Earth stations to illustrate how our view of the Milky Way varies throughout the night. Three-hour sidereal time longitudes 12, 15, ..., 24 are indicated.

its energy in all wavelengths, much like the radiation from an object heated to the temperature of the solar surface. This emission is called "black-body radiation." Energy, flowing outward from the core-fusion process as radiation, reaches a point at about 0.86 solar radii where turbulent hydrogen convection occurs. Above this region is a *photosphere* layer of about 400-km thickness that extends to the boundary of the visible disk. The temperature in the photosphere decreases rapidly outward from about 6,000 to 4,000 K. The hydrogen ion density in this layer diminishes so rapidly with altitude that the region seems to be just a narrow 100 km thick; it gives off light at all wavelengths. Spectral absorption lines of the higher solar atmosphere identify the outer composition. The *chromosphere* is the irregular transition region extending as much as 10,000 km from the relatively cool photosphere to the relatively hot *corona*. This corona has temperatures that start at 20,000 K and grow, with distance outward, to 100 times that value. The coronal region extends from the chromosphere deep into space to merge with the interstellar medium. Because of the high temperatures, the molecules of this outer atmosphere of the Sun are completely broken apart (ionized) into electrons and charged atoms stripped of electrons (ions).

3.3 Active Sun, Sunspots, Fields, and Coronal Holes

Active changes on the solar surface are more easily recognized when measured at selected light wavelengths of relatively strong emissions. Light wavelengths are conventionally expressed in Ångströms (Å), although the SI system preference is nanometers (10 Å = 1 nanometer). The lines that are selected for solar observations of hydrogen are the Lyman-alpha line (Lα, 1216 Å) and the H-alpha line (Hα, 6563 Å line). For calcium, the singly ionized calcium-two lines (CaII 3968, 3934, and 8542 Å) are selected; and for iron, the atomic lines (FeI near 5250.2 Å) are selected. X-ray radiation from the Sun is monitored at wavelengths between about 0.5 and 8.0 Å. A broad band of solar radio-wave emissions with frequencies between 10 and 30,000 MHz is also tracked. (Note: Hz stands for Hertz, a unit of one cycle per second; MHz is a frequency of one million cycles per second.)

Solar electromagnetic radiations from the short-wavelength, high-energy X rays, through the visible spectra to the long wavelengths of the infrared, illuminate our planet and its atmosphere (Figure 3.3). Generally, the shorter the wavelength of a disturbance in the spectrum, the greater the solar emission response to the variable solar activity. The Earth's atmospheric ionization reacts to the changing energetic solar emissions after the 500 seconds (8.3 minutes) it takes light to travel to the Earth from the Sun. Larger, more delayed ionospheric changes at the Earth occur after the time it takes for particles and associated fields to make the trip from the Sun to the Earth's magnetosphere (about 1 to 5 days) plus a reaction time within the magnetosphere.

Photographs of the solar surface regularly show dark spots. These are small, relatively cooler (4,600 K) regions in the photosphere appearing as dark, more or less bowl-shaped, depressed areas, often occurring as groups. Sunspot activity is tracked using a *sunspot number, R,* which takes into account the total number of individual spots, the number of groups, and a visibility adjustment factor. One unit of R is roughly equal to an area of 6×10^{-8} of the visible solar surface hemisphere. The

Figure 3.3. Names of wavelengths for electromagnetic spectra. Scale is given in Ångströms (Å) (10 Å = 1 nanometer). Note the short visible range of about 4000 to 7000 Å.

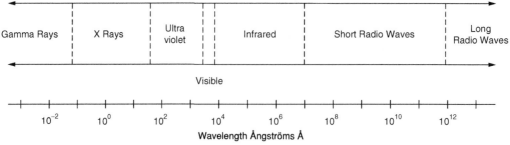

average of the maximum sunspot numbers is near 108; the average of minimum numbers is about 5. The sunspot numbers are not always exactly balanced between the north and south solar hemispheres. There are periods of many years when one hemisphere is more active than the other.

Long-lived large spots grow rapidly in about 10 days to their maximum size, often 37,000 km in diameter. These then decay slowly over the next 50 or more days; some fully developed spots persist for as many as 10 of the 27-day solar rotations. Magnetic fields of a singular polarity grow in association with the sunspot area; in large spots these fields can reach 4×10^8 gamma. Spot groups are often concentrated into a preceding and a following part, with opposite magnetic field polarity. Most preceding spots have a similar polarity in one hemisphere and the opposite polarity in the opposite hemisphere for one activity cycle. The spot polarity order reverses in each succeeding activity cycle, with a corresponding reversal of the Sun's dipole field. Both the sunspots and the regions free of activity appear more frequently in narrow-longitude intervals. These intervals differ in the two hemispheres, occasionally showing 180° displacement. The first spots of a new cycle start at about 30° solar latitude; the last spots of a cycle reach about 8° (Figure 3.4). There seems to be a poleward shift of these latitudes during the more active cycles.

The long-persisting spots are used to determine the Sun's surface rotation. Looking down from above the Sun's north pole, the surface moves counterclockwise, just like the Earth's orbit about the Sun and the Moon's orbit about the Earth. Unlike the motion of a solid body, the Sun's surface rotation seen from the Earth (*synodic period*) varies with

Figure 3.4. Example of variation in sunspot number (top graph curves) and location (lower graph dots in "butterfly diagram") for five solar activity cycles. N and S (north and south) correspond to the field polarity of the preceding spots and the alternation of solar field direction.

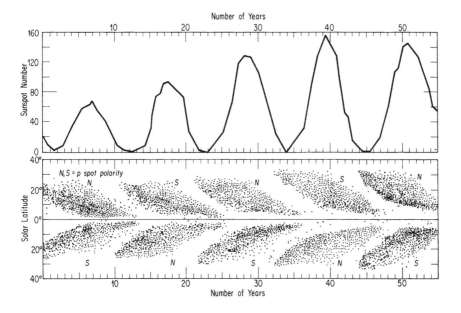

solar latitude from about 26.6 days at the equator to 27.2 days at 20° latitude, 28.1 days at 30°, and 31 days near the pole. For the activity influencing the Earth, a 27-day rotation is usually assumed for the principal period. For identification purposes, each 27.2753-day period has been assigned a *Carrington rotation number*, starting from 9 November 1853, the beginning of Carrington's study. That amounts to approximately 13.391 rotations per year so that the rotation number will reach 2035 in 2005. The rotation is said to begin at the *central meridian* of the Sun viewed from the Earth; positions within the rotation are assigned Carrington longitudes that are measured westward (west direction for an earth-bound viewer of the Sun) from the central meridian.

Various other periodicities have been recognized in sunspots. There is a slight indication of; 25- to 26-month variation in recurrence. Figure 3.4 indicates the Maunder "butterfly" diagram to show the location of sunspots on the solar surface over many years. It can be seen that at a mean activity minimum in R, the spots are associated with two different cycles. The average overlap is 2.9 years. The average duration of one sunspot family is about 13.6 years so that the major *solar activity cycle* duration becomes 10.7 years (called the "11-year solar cycle"). The alternation of field polarity for every new group of sunspots means there is a basic 22-year cycle. The maxima and minima in the display of sunspot numbers (Figures 3.5 and 3.6) indicate a possible superposed

Figure 3.5. Annual sunspot number from about 1610 to 1850. Only a few values have been determined prior to 1650. Figure adapted from Eddy (1977).

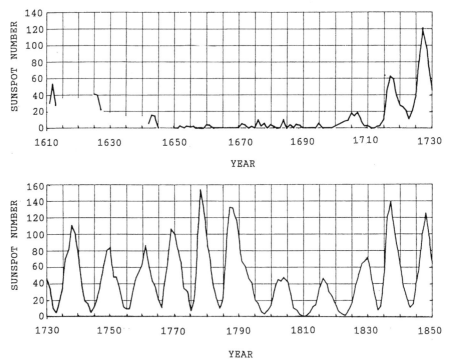

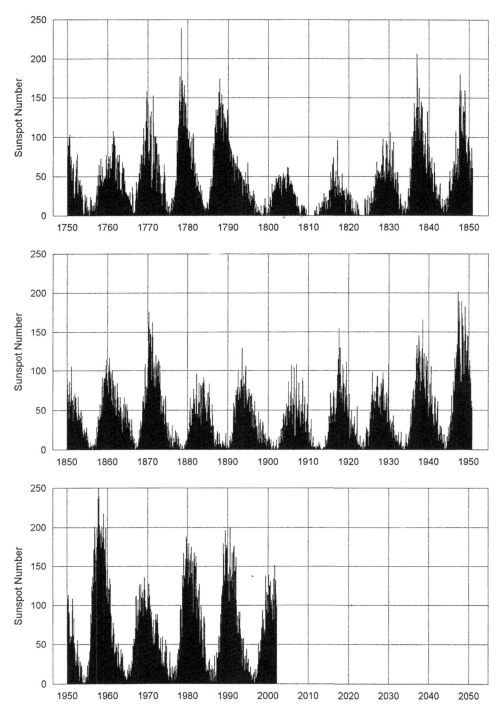

Figure 3.6. Monthly sunspot number from January 1750 to March 2002. Figure prepared by E. Erwin of NGDC/NOAA.

Figure 3.7. Solar fields measured with the Mt. Wilson solar magnetometer employing the strong field Zeeman effect (Section 4.11) upon spectral emissions. Np and Sp are the North and South Poles, respectively. Note: east (E) and west (W) directions on the Sun correspond to the same Earth surface directions from the point of view of an earthbound observer.

80- to 90-year cycle. The average zones of activity move poleward for the more active cycle years. There was a period from about 1645 to 1715, called the *Maunder minimum*, when the sunspots and solar activity seemed to have almost disappeared (see *EOS*, 22 Oct. 2002, Eddy interview). A significant increase in the $^{14}C/^{12}C$ ratio (related to the change in cosmic rays) has been identified with this extensive solar activity minimum. Each cycle of solar activity has been assigned a number counting from the sunspot maximum of the year 1760. When the cycle identification was initiated, that year was the best-defined cycle following the Maunder minimum. Solar cycle number 24 will begin near 2006.

In strong magnetic fields the iron spectral lines near 5250.2 Å (525.02 nanometers) shift their wavelength by about 10^{-5} Å/Gauss (see Zeeman-effect magnetometer in Section 4.11). Detailed monitoring of these photospheric fields shows a correspondence to regions of activity.

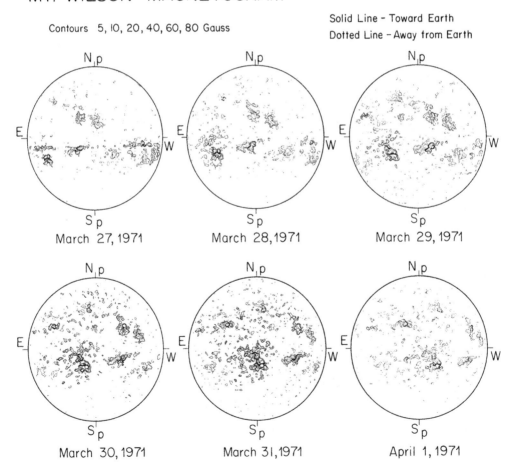

MT. WILSON MAGNETOGRAM

Contours 5, 10, 20, 40, 60, 80 Gauss

Solid Line – Toward Earth
Dotted Line – Away from Earth

March 27, 1971   March 28, 1971   March 29, 1971

March 30, 1971   March 31, 1971   April 1, 1971

The more prominent irregular field forms persist from day to day with noticeable changes in extent and intensity. Such fields seem to rotate along with their associated unique local visual features across the solar surface (Figure 3.7). Special techniques have been devised to identify "computed background photospheric field regions" for each Carrington rotation. Such patterns show an alternation of outward (+) and inward (−) direction of field polarity.

Usually, solar regions of one magnetic polarity have a nearby companion region of opposite polarity. However, on occasion large regions of a single polarity appear. These unipolar regions have been shown to match low-density coronal areas that appear as large dark regions in pictures of the Sun, representing the emission of longer-wave ("soft") X rays (Figure 3.8). Field lines from these *coronal holes*, which are usually found at high latitudes, stretch far from the solar surface into the interplanetary medium. The holes are often asymmetrically located with respect to the Sun's rotational axis. Polar holes sometimes last for many years; the holes at a more equatorial location may persist for as many as 18 solar rotations. It is now thought that these coronal-hole open-field lines provide a major highway for a continuous outflow of

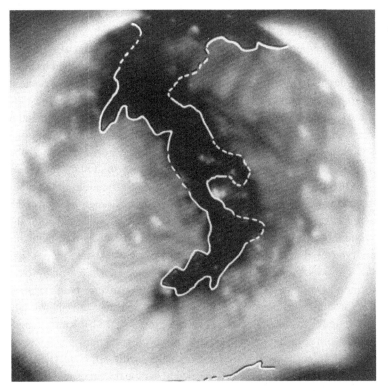

Figure 3.8. Image of X-ray emissions from Sun taken by NASA Skylab spacecraft. The coronal hole area, outlined by solid and dashed lines, extends from the polar region to the equator and was found to lie in the middle of a large positive unipolar magnetic region. Figure from Eddy (1979).

solar particles and recurrent solar–terrestrial disturbance events. In the early studies of geomagnetic storms the 27-day recurring activity was attributed to mysterious "solar *M* regions" that are now known to be the coronal holes. Multiple coronal holes cause corotating streams of particles in the solar wind. Long-period magnetic waves appear in the flow from the polar corona and cause continuous variation in the associated field directions. The occurrence of coronal holes follows the solar cycle but they are more prevalent in the year or two following sunspot maximum.

3.4 Plages, Prominences, Filaments, and Flares

Typically, the solar active areas arise near the sunspot regions. For some unknown reason, for a number of sunspot cycles, the radiative output increase due to the active areas predominates over the cooling effect of the sunspots. Then for the following several cycles, the dominance reverses with the sunspot cooling of radiative output more than the active area contribution. The change in such radiative balance affects the world's climate (Hoyt and Schatten, 1997).

There are a number of transitory phenomena associated with solar activity that follow the sunspot cycle in their intensity and occurrence number. *Plages*, the most common of these, are small but relatively bright areas seen on the solar photosphere. Plage is a French word meaning a beachlike region. Solar plages are best observed using selective color filters for hydrogen and calcium. Typically, a local magnetic field increase precedes plage formation and, in developed plage areas, this field remains above the surrounding values. Some radiowave emissions are traced to these regions. The X-ray photographs of the solar surface show 20 to 60 Å emissions that seem to come from the coronal areas above the chromospheric plages. Plage area, brightness, and location are monitored as solar activity indicators. Sunspots often start their growth in plage regions.

On the limb of the Sun, large chromospheric formations of luminous gas arise as arches extending into the darker coronal region (Figures 3.9 and 3.10). These are called *prominences*. Viewed on the solar disk, the prominences appear as long dark ribbons called *filaments*. The filaments are typically born in the active solar regions of preexisting plages near the sunspots. Some change form slowly and persist for many days to months; others show great variations within several hours or less. The prominences begin as archlike structures, often with intermittent appearances in the first week They can grow to 2×10^5 km in length and extend to a height of 4×10^4 km above the photosphere. Some quiet prominences suddenly become active, with sufficient energy to exceed the 618 km/sec velocity needed to escape the Sun's gravitational

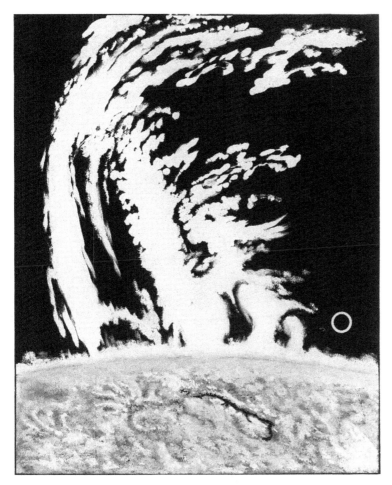

Figure 3.9. Solar prominence and filament appearance compared to the size of the Earth (small circle).

field. Most are stable structures, lasting more than 2 to 3 months. Many prominences are seen along the neutral line that separates plage regions of opposite magnetic polarity.

The earliest clear connection between solar activity and a disturbance in the Earth's magnetic field occurred in 1859 when Carrington first observed a flare (Figure 3.11), which was followed by a severe magnetic storm about one day later. A *solar flare* is now defined as the transient brightening of a small region on the solar surface observed in the Hα emission line. A typical flare would probably begin as a bright speck near sunspots that, in their growing phase during the week or two after formation, show rather complex, rapidly increasing magnetic fields. Flares often recur in almost the same general region. At times of high activity an average flare occurrence rate of one every two hours is typically observed. A flare usually shows a major impulsive increase in

Figure 3.10. Photograph of Sun showing hydrogen-alpha emission regions at 1358 UT on 31 July 1988, from solar telescope at Boulder, Colorado. Figure provided by Space Environment Center, NOAA.

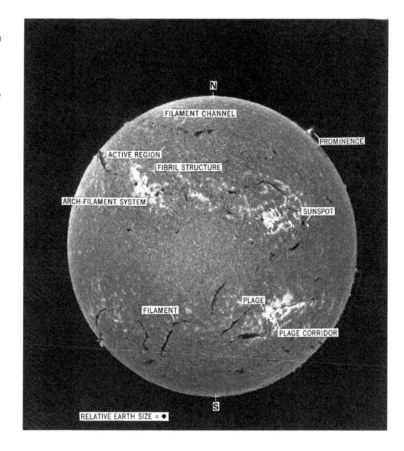

Hα and CaII emission in the first minutes. An area increase follows the brightness; large flares reach over 10^9 km² in area. The decay is always slow: Large flares may persist for 2 to 3 hours. With the most active flares, there occur microwave and broadband radio-wave bursts, ultraviolet emissions, X rays, and particle ejections. Although primarily electron and proton fluxes are measured, alpha particles and heavier nuclei have been detected. Typically, 30-keV to 10-MeV electrons are thrown into space. (Note: an electron volt, eV, represents the gain in energy of an electron when it moves through an electric field created by a one-volt

Figure 3.11. (Opposite) (Top) Carrington's sketch of a solar flare on 1 September 1859. The dark regions were sunspots. The bright regions marked A and B were flares that seemed to migrate to positions C and D before disappearing. (Bottom) Kew observatory records on which the time of Carrington's flare at 11 hr 15 min is indicated along with the subsequent storm onset in the early hours of 2 September. The D, H, and Z components of field are indicated (top to bottom) in the three magnetograms. Records for three days are offset top to bottom in each field component diagram. Figure from Bartels (1937).

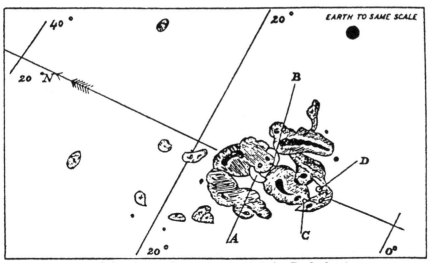

Solar sketch, September 1, 1859, by R. C. Carrington

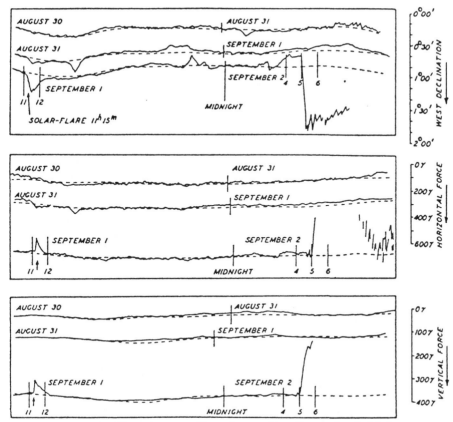

Magnetograms, Kew, August 30 to September 2, 1859

electric potential difference.) On special, but rare occasions, active regions produce large flares for which particularly high proton fluxes have been detected in the MeV to GeV energy range; these solar proton events are rated by both their energy level and sustained duration at that level.

Solar-activity observers categorize most flares by area and X-ray output. The flare areas, in square degrees (an area with the Earth's circumference would be about 1 square degree), are classified as importance 0 for *subflares* with less than 2 square degrees, to importance 4 for greater than 24.7 square degrees. The order of magnitude of the peak burst intensity, in watts per square meter for a band of X-ray emissions from 1 to 8 Å, separates the flare classifications (in increasing output) into B, C, M, and X flares; an attached number indicates a level within the order of magnitude. A single-letter brightness scale, F, N, or B, may also be added to indicate a faint, normal, or brilliant optical characteristic, respectively.

Radio-wave bursts from 10 MHz to 30,000 MHz produced by flares are identified as Types I to V depending upon their unique spectral intensity signatures on a time-versus-frequency plot (Figure 3.12). The *Type II radio-wave bursts* of the more important flares have been associated with solar atmosphere shock-waves. The emission frequency, at any one time in the event, is roughly equal to nine times the square root of the plasma electron density at the altitude of the shock in the solar atmosphere. By comparison with model solar atmospheres, the radial speed of the expanding explosion can be derived from the drift in frequency of the Type II burst. This shock speed (of the order of 1000 km/sec) is utilized in estimating the travel time for the arrival of solar–terrestrial disturbances at the Earth. Type IV emissions, lasting for hours, accompany

Figure 3.12. The frequency-time diagram and names of dynamic electromagnetic emission spectra types associated with flares. Not all types occur with a particular flare. Figure provided by S. Mangis, Space Environment Center, NOAA.

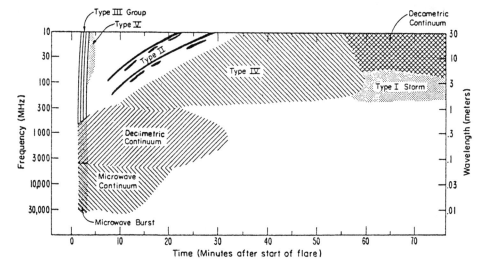

some solar particle emission events. Radio-wave emissions at 10.7 cm (near 2.8 GHz), are routinely monitored; their close relationship to the sunspot numbers provides a continuous solar-activity index. The monthly average level of these emissions, called F10, the *ten-centimeter flux*, is

$$F10 \approx 50.0 + 0.968 \text{ SSN}, \qquad (3.1)$$

where SSN is the monthly mean sunspot number above 90.

3.5 Mass Ejections and Energetic Particle Events

During a solar eclipse the Moon exactly covers the bright solar disk, allowing observers to record, with a *coronagraph*, the luminous coronal emissions blasting far into space (Figure 3.13). In recent years, specially constructed coronagraphs on satellites artificially "eclipse" the solar disk and record the continuous progress of the luminous ejections of matter. With this and other new observing techniques the term *coronal mass ejection*, or CME event, began to be used to describe a special category of particles leaving the Sun. The CMEs were not necessarily flare related but included the complete range of disturbance processes associated with magnetic field line regions of the corona. The CMEs occur at a broader extent of latitudes than the flares, which seem to be restricted to lower latitudes. When the CMEs lift off from a closed field line of the solar

Figure 3.13. Luminous ejection of matter from the Sun during the solar eclipse of 7 March 1970. This is a composite figure with a superposed picture of the Sun's surface taken on the same day. Figure from the National Center for Atmospheric Research.

corona, the topology causes superthermal electrons to move outward from both footprints, causing a counter-streaming flux as a signature. Scientists have found three to four CMEs per day in active years and about one every five days in quiet years. The CMEs are responsible for populating the solar wind with principally protons and electrons (0.5 to $50 \, cm^{-3}$) and with about three to five percent helium. Leading-edge wind plasma speeds vary from 50 to 1200 km/sec, with an average between 350 and 400 km/sec.

About 10^{15} to 10^{16} grams of the Sun's matter are irregularly blasted into interplanetary space as solar energetic particles (SEP). From the monitored temporal changes in characteristics of the ejected protons and electrons, SEP events are subdivided into impulsive and gradual types. For impulsive events, flares themselves are thought to account for the particle acceleration, while for gradual events, acceleration of the CME particles are believed to arise in the coronal and interplanetary shocks. At solar maximum, about 1000 events per year are of the impulsive SEP type that reach their peak output in minutes and typically follow solar flares. They are rich in electrons, are associated with Type III and IV radiowave bursts (Figure 3.12), last only hours, have a higher than average amount of ^3He (an abnormal form of helium) and iron, and are of limited angular extent (i.e., usually they are detected only when the observing location can be traced to a flare site along a solar field line).

Only about ten SEP events per year (at solar maximum) are of the gradual type that reach their peak output in hours to days and can originate from disturbances anywhere on the solar disk. They are rich in protons, are associated with Type II and IV radiowave bursts (Figure 3.12), last for days, have helium and iron composition typical of the Sun–Earth environment, and are of broad angular extent. Most of the large SEP events are of the gradual (or a composite) type. The gradual SEP are now thought to arise from an acceleration of the particles in the solar–terrestrial environment by a shock propagating outward from the Sun's surface. Table 3.2 from Cliver and Cane (2002) illustrates how the impulsive and gradual X rays from the Sun can distinguish the particle composition and other features of the SEP events.

3.6 Interplanetary Field and Solar Wind Shocks

There is an *Archimedes spiral*, or "garden hose" spiral (in reference to water from a rotating garden water sprinkler) direction of the solar wind in the heliosphere (Figure 3.14). Although the individual solar wind particles themselves are moving radially from the Sun, the rotation of the emitting solar surface provides a spiral flow pattern for the

Table 3.2. *Composition of Solar Energetic Particle Events*

Particles	IMPULSIVE electron rich	GRADUAL proton rich
^{3}He/^{4}He	~1	~0.0005
Fe/O	~1	~0.1
H/He	~10	~100
Duration	hours	days
Longitude Cone	<30°	~180°
Radio Type	III, V, (II)	II, IV
X-rays	impulsive	gradual
Events/year	~1000	~10

particle density and associated fields. The spiral angle depends upon the ratio of the solar wind particle velocity to the solar surface angular velocity, taking into consideration the solar longitude of the departing particle. At Earth distance, the average spiral angle is close to 45°.

The arrival of solar wind disturbances at the Earth's magnetosphere is traced to and from solar source regions using the spiral information. The central meridian longitude of the visible solar disk is often used as a reference line for the position of active areas. As the solar disk rotates, the times of *central meridian passing* (i.e., crossing the apparent longitude at the center of the visual solar disk) of the solar phenomenon are sometimes noted to describe the delay in the start of a corresponding magnetosphere disturbance and to estimate the solar wind particle velocity (along the spiral) for the event.

For distances of about 1 AU and beyond, the solar field, **B**, components parallel and perpendicular to the radius vector are given by

$$B_\parallel = \frac{B_0 s^2}{r^2} \tag{3.2}$$

and

$$B_\perp = \frac{B_0 s^2 \alpha}{v r}, \tag{3.3}$$

where B_0 is the solar magnetic field strength measured at distance s, v is the solar wind particle velocity, and α is the solar surface angular velocity. Note that as the distance, r, increases, the parallel B component decreases much more rapidly than the perpendicular component.

Beyond a few solar radii from the Sun, the average distance between encounters (*mean free path*) of the solar wind particles is considerably

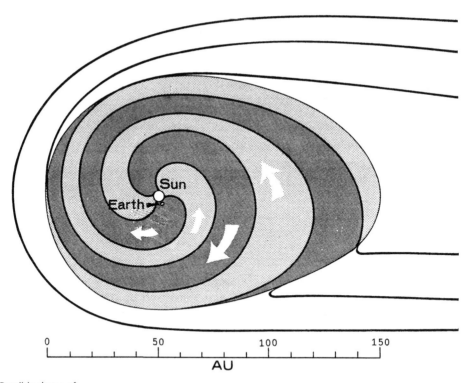

0   50   100   150

AU

Figure 3.14. Possible shape of solar wind Archimedes spiral and boundary of heliosphere (see *EOS*, 29 Oct. 2002). Distance scale is in astronomical units (AU). Arrows indicate sector direction when four regions are active in the solar ecliptic plane.

larger than the scale of most disturbances measured within the region. This means that the wind can be treated as a *collisionless plasma* and the important interaction of the charged particles is with the local magnetic field. The appropriate scale for measurement of disturbance phenomena is the gyroradius (*Larmor radius*), r, described by the charged particle's natural circular motion about a magnetic field line:

$$r = \frac{mv}{Be} \text{ meters,} \qquad (3.4)$$

where e is the charge (coulombs), m is the mass (kilograms), and v the velocity (meters/second) perpendicular to the field B (Teslas) as in Equation (2.8). At 1 AU the Larmor radius is of the order of several kilometers for electrons and several hundred kilometers for ions.

In the solar wind, the magnetic field is weak so the combined motion of the solar wind particles in the magnetic field drags along the field itself. The solar-wind electrical conductivity is about proportional to the mean free path, near 10^4 Siemens/meter. For computational purposes, this value is so very large that the wind can be considered a perfect conductor. A perfect conductor moving in a magnetic field generates currents that hold constant the internal magnetic field within the conductor. Thus a selected sample of the solar wind, although it may change shape with time, will continue to hold the original magnetic flux through its contour. For this reason we say that the solar wind plasma has a *frozen-in* field.

Satellite measurements of the solar wind magnetic field spiral angle (in a coordinate system aligned with the spiral direction) show alternating sectors of the field to be directed principally *toward* or *away* from the Sun. These sectors, often two or four in number, are separated by their *sector boundary*. Dynamic solar processes modify the solar–coronal dipole magnetic field into a unique pattern that opens the field lines at high latitudes and forms an equatorial *heliospheric current sheet* (Figure 3.15), separating currents directed toward and away from the Sun. The field polarity along the open field lines is identified with the polar region in which the field line originates. Active processes on the solar surface modify the systematic pattern of the equatorial current sheet into a "ballerina skirt" pattern such that the Earth's environment is awash in the solar fields, sometimes directed from the solar north polar region and sometimes toward the south. These fields form the observed toward and away sectors.

Figure 3.16 shows the relationship between solar wind speed observed at the low solar latitudes and the brightness contours of the corona

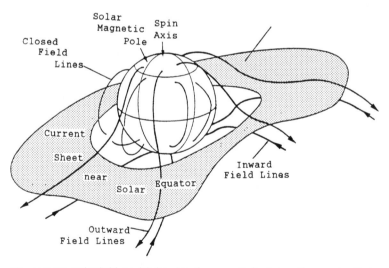

Figure 3.15. Solar field model showing the current sheet near the equator that separates open field lines directed outward above the sheet and inward below the sheet. Closed field lines, beginning and ending in the solar corona, are also depicted. Solar activity and the offset of the solar magnetic pole from the spin axis create a "ballerina skirt" effect of the current sheet at the Earth's distance in the solar-ecliptic plane so that the outward- and inward-directed fields identify toward and away sectors encountering the Earth. Figure adapted from Smith, Tsurutani, and Rosenberg (1978).

Figure 3.16. Correspondence of polar coronal holes to fast solar wind speed and to solar magnetic polarity. (Top) Wind speed measured near the solar equator (shaded to show the associated frozen-in + and − field polarity) identified with Carrington (rotation 1616) longitude positions. (Bottom) The corresponding Carrington longitude versus solar latitude display shows coronal brightness contours and + or − shading to identify two opposite hemisphere polar coronal holes that have positive or negative field polarity. Figure adapted from Hundhausen (1977).

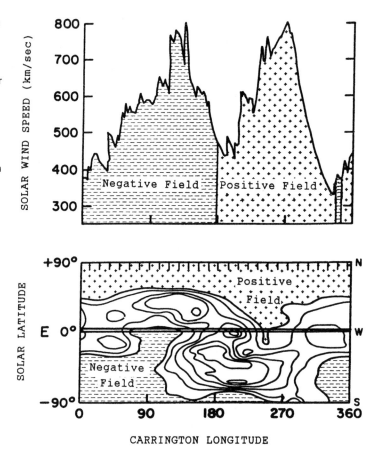

upon which the solar magnetic field polarity is plotted to show the polar coronal holes.

As the streams of coronal-hole particles from the Sun encounter the low-velocity (300–350 km/sec) plasma of the heliospheric current sheet, a field compression results. At this interaction region the field direction can be highly variable, with large southward components lasting less than a few hours. We shall see that, on arrival at the Earth's magnetosphere, these corotating stream disturbances can initiate recurring geomagnetic storms.

Shocks that occur in the solar wind environment are similar in many ways to those familiar to us on Earth. We have heard the boom of an aerodynamic shock as a jet aircraft moves faster than the speed of sound and causes a shock front of air particle collisions. The *Mach number, Ma,* is the ratio of the speed, V_x, of the disturbance to the speed of sound in the medium, V_s. We observe the bow shock of waves piling up ahead of a speeding boat as it causes water-particle collisions while plowing onward, faster than the ocean waves. The solar-wind plasma of electrons and charged nuclei have such a large mean free path that the collision-type shocks cannot occur. In place of collisional constraints, the solar wind magnetic field acts on the plasma, forcing the charged particles, which are moving perpendicular to the field, to circle the field (Equation (3.3)); whereas particles moving parallel to the field continue in that direction. Together the two motions cause the particles to spiral around the field line. In this way the collisionless plasma exhibits motion constraints that can be considered analogous to the elastic collision properties of an atmospheric gas.

Because the solar wind is a plasma, some terminology originally developed for fluid and gas dynamics is used in the literature. One of these terms, the *Reynolds number*, is a measure of the ratio of internal forces to viscous forces. The number is defined in a way such that a very large value indicates that viscous properties can be neglected, that mixing occurs, and that the mathematical description of the gas behavior is simplified. When charged particle interactions must be taken into consideration, a special magnetic Reynolds number is defined. For the highly conducting solar wind plasma this Reynolds number is exceedingly large so viscous forces may be neglected in the analysis of shock-wave interactions.

Collisionless shock fronts of compressed solar wind develop as fast-moving plasma overtakes slower-moving plasma. During active times, swift streams of energetic particles are ejected from the location of a flare, a coronal hole, or other irregular solar-surface CME region. The fast ejecta from the Sun describe a velocity-dependent trace away from the Sun and overtake slower particles along an expanding front. The faster the particle speeds, the less tightly wound is the Archimedes spiral. If the difference between the two speeds of the interacting plasmas exceeds the local characteristic wave speeds, a *shock front* (Figure 3.17) will be

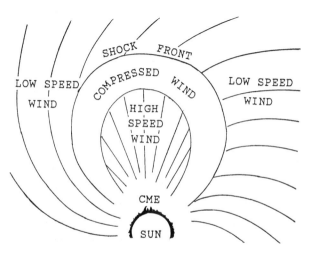

Figure 3.17. Shock front formation and region of compressed solar wind resulting from the interaction of high-speed wind with ambient low-speed wind following a flare or coronal mass ejection.

formed at the leading edge, the plasma density and azimuthal field will be increased (by a factor of 4 at 1 AU), and a tangential velocity discontinuity will be formed. Because the shock front moves faster than the local characteristic waves, no disturbance is communicated ahead of the front. The shocks are sometimes separated into driven or blast types depending on whether the front is supported by a continuing flow of fast plasma or is considered an explosive-type event. Electromagnetic noise bursts at radiowave frequencies are believed to be generated in the shock front.

Both electromagnetic field equations and fluid-flow hydrodynamic equations are needed to follow the solar wind properties of a completely ionized collisionless gas in a magnetic field. The combination brought about the terms *hydromagnetic* (hm) and *magneto-hydrodynamic* (mhd) waves. The turbulence of shocks in the solar wind gives rise to hm waves, which are called transverse and longitudinal hydromagnetic waves when they travel in the respective directions along, or perpendicular to, the ambient magnetic field, **B**. *Alfven waves* (transverse hm) have a speed

$$V_a = \frac{B}{\sqrt{\mu \delta}},$$ (3.5)

where μ is a constant of the region called the permeability and δ is the ion density. A fast *magnetosonic* wave travels at a transverse speed V_a and at a longitudinal speed

$$V = \sqrt{V_a^2 + V_s^2},$$ (3.6)

where V_s is the sound speed (less than V_a). The slow magnetosonic wave has only a transverse velocity V_s. With distance from the Sun, the magnitude of V_a decreases as the inverse of the radial distance, whereas V_s decreases as the inverse fourth power of the radial distance. Near 1 AU the V_a and V_s magnitudes are about 30 km/sec and 100 km/sec, respectively. In plasmas, hydromagnetic waves can carry information of disturbance out from a source in a manner analogous to sound waves. If the hm waves travel in all directions with equal Alfven velocity (i.e., when $V_s \ll V_a$), an Alfven Mach number, M_{aA}, can be appropriately used:

$$M_{aA} = \frac{V_x}{V_a},$$ (3.7)

where V_x is the velocity of the source of disturbance. This Mach number describes the conditions for the shock-front plasma flow. Ions move back upstream from the shock front but are limited in their travel by interactions with Alfven waves generated at the shock.

3.7 Solar Wind–Magnetosphere Interaction

A shock is continuously formed as the solar-wind particles and their frozen-in *interplanetary magnetic fields*, IMF, encounter the Earth's magnetosphere, a bubblelike region defined by the presence of the Earth's main field in space. In solar–magnetospheric coordinates (Table 3.1) the arriving solar-wind fields are described in IMF B_x, B_y, and B_z directions of which the last two are particularly significant for special interactions. Figure 3.18 shows the distribution of these field strengths for a 3-year sampling. The region of the magnetosphere just inside the envelope that contains the characteristic fields and particles is called the *magnetopause*. Figure 3.19 is a diagram of the magnetosphere illustrating the names assigned to the various significant regions. The *magnetopause*

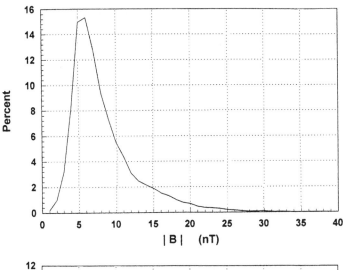

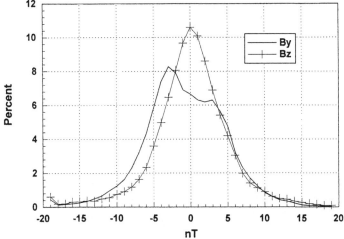

Figure 3.18. Distribution of interplanetary magnetic field for 1988–1990. (Top) Percentages of 5-min averages of the IMF magnitude that are in the magnetic field amplitude range of 0.99 to 0.0 nT of the value shown. (Bottom) Percentages of 5-min averages of the IMF in which B_y or B_z is ±0.5 nT of the field value shown. Figure from Rich and Hairston (1994).

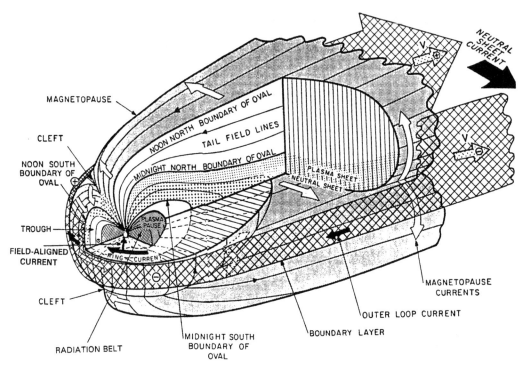

Figure 3.19. Diagram of the magnetosphere with regional names and average currents indicated. Figure adapted from NGDC/NOAA drawing.

stand-off position (closest approach of the compressed boundary), R_{mso}, is usually between 5 and 20 R_e with a typical location of 11 to 12 R_e. The shock front stand-off position, R_{sso} is usually 3 to 4 R_e farther from the Earth. Positions of the dayside bowshock and magnetopause follow a parabolic shape about the stand-off locations (Figure 3.20). A very good approximation of the shape of the sunward boundary of this parabolic front in the Earth's geomagnetic equatorial plane is obtained by using only the balance between the Earth's magnetospheric magnetic pressure and the pressure of bombarding solar wind particles:

$$\frac{B^2}{8\pi} = 2nmv^2 \cos^2(\beta), \tag{3.8}$$

where B is the Earth's field strength at the magnetospheric stand-off position; and for the solar wind, n is the average particle density, m is the average particle mass, v is the average wind velocity, and β is the angle between a line perpendicular to the boundary and the incident wind particles.

The well-defined magnetopause outer boundary layer of the magnetosphere separates the geomagnetic field from the solar wind (c.f. Lakhina et al., 2000; Tsurutani et al., 2001). The geomagnetic field compression by the arrival of strong solar wind can be represented by a magnetopause current (sometimes called the Chapman–Ferraro current).

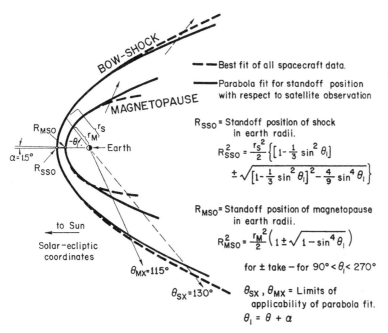

Figure 3.20. **Figure 3.20.** Parabolic shape of bow shock and magnetopause boundary in solar-ecliptic plane, derived from spacecraft data. The equations to the right (identified with symbols shown on the diagram) allow the satellite magnetopause and bow-shock boundary crossings, at various locations, to be converted to equivalent stand-off positions for use in storm-effect computations.

Looking at the momentary increase of field about the Earth at the onset of a storm, we can apply our old friend, spherical harmonic analysis, to compute the equivalent external magnetopause currents that could sustain such a distorted dipole shape and tie the surface field to the stand-off location. Using only the significant g_1^0 term from such a SHA, the added horizontal component, H_b (nanoteslas), due to the compressed boundary is approximately

$$H_b \approx \frac{250 \sin(\theta)}{R_{ms}}, \tag{3.9}$$

where R_{ms} is the magnetopause stand-off distance in Earth radii and θ is the Earth surface location geomagnetic colatitude of the field measurement. For example, for a stand-off position of 10 R_e the H field at the equator will increase 25 nanoteslas. The energy, E, required to confine the magnetic dipole field at the equator is approximated as

$$E = 4.1 \times 10^{13} H_b \text{ joules} \tag{3.10}$$

For the 10 R_e compression (above) the energy would be about 10^{15} joules (energy units). The increase in the H-field component is usually detected at the onset of a magnetic storm at middle- and low-latitude observatories.

Magnetic field directions are important to the interaction of the solar wind with the magnetosphere (Zhou and Tsurutani, 2001). Field lines

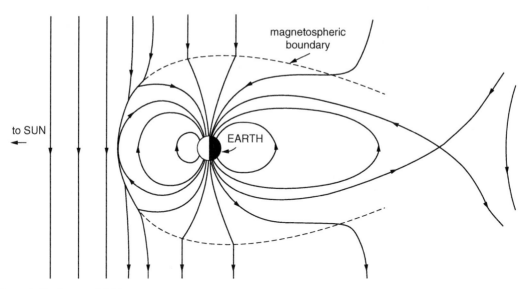

Figure 3.21. Expected field line interconnection when the B_z component of the interplanetary magnetic field is negative (directed southward). Figure adapted from Levy, Petschek, and Siscoe (1964).

appear to cross in the region where the magnitude of the field goes to zero. It has been suggested that an *x-type neutral point* occurs when two regions have opposing field line directions; a *field line reconnection* is then initiated, and the particles flowing from the two regions toward the neutral point are subsequently ejected at right angles away from the neutral point. The presently accepted model supposes that a southward directed (solar-magnetospheric coordinate system) interplanetary magnetic field of the solar wind ($-B_z$ IMF) encounters the northward directed magnetic field of the Earth (recall that the Earth's dipole field is directed from the South Pole to the North Pole); then the field lines interconnect, distorting the Earth's dipole field and providing entry for the solar-wind particles into the magnetosphere (Figure 3.21). The B_z component of the arriving solar wind thus acts as an on/off switch: The *southward turning IMF* becomes the critical requirement for initiation of major magnetic disturbances at the Earth. This storm-initiation model has some difficulties: The full reconnection onset requirements are unknown; sometimes IMF conditions favor reconnection but it doesn't happen; a steady state of field-line interconnection is assumed but unproven.

Storms that recur for many solar rotations have been traced to the streams of particles from long-duration coronal holes and fluctuating IMF B_z arising from the stream interactions with the heliospheric current sheet. Such events often result in a long period of solar–terrestrial activity and have been given the name *high-intensity long duration continuous AE activity* (HILDCAA). The AE in the name is an activity index, described in Section 3.15 below.

The Archimedes spiral angle of the solar-wind flow arriving at 1 AU causes the $+B_y$ IMF (eastward) to correspond to the away-from-the-Sun sector and the $-B_y$ IMF (westward) to correspond to the toward-the-Sun sector of the solar wind. The direction of favored solar-wind convection around the magnetosphere is determined by the sign of the IMF B_y. We shall soon see that these sector field interactions become important for the magnetospheric polar cap field lines that have folded to the nightside of the magnetosphere by the solar wind.

The *polar cusps* (see "cleft" in Figure 3.19) occur where the magnetospheric field is folded downstream by the solar wind. Here the field strength is extremely low and the field lines, which have just merged with the IMF, create funnels for solar-wind particle entry into the Northern and Southern Hemisphere upper atmosphere of the auroral zones. Typically, the cusp region is near $77°$ geomagnetic latitude (about $1°$ spread in latitude) and is about 3 hours wide near noon magnetic time. The solar-wind characteristics, seasonal dipole orientation with respect to the IMF, and location of the principal storm impact region can all modify the cusp location. During a storm the flow of IMF particles into the cusp is essentially continuous. Associated with the equatorward side of the cusp is a plasma of particles and fields immediately inside the magnetopause that extends from the cusp tailward to form a *low-latitude boundary layer* (LLBL) (or sometimes just LBL) of hotter ions and lower densities than at the cusp.

Along the sides of the magnetospheric boundary, viscous interactions occur, transferring solar-wind momentum to the closed field lines of the magnetosphere, hauling them tailward. The field lines that merge with solar wind by the reconnection process are transported over the poles and dragged far behind the Earth to form a long magnetospheric tail (*magnetotail*) more than 200 R_e long. A neutral sheet current is created in the tail (Figure 3.19) whose cross section shows a two-loop tail lobe structure circulating within the boundary and across the interior as a *cross-tail plasma-sheet current* almost 3 to 7 R_e in width. The term *magnetospheric convection* is used to describe all the flows resulting from reconnection and viscous interaction.

3.8 Geomagnetic Storms

Dynamic processes on the Sun deliver a plasma of charged particles (principally protons and electrons) and associated fields to the Earth's environment, causing geomagnetic disturbances at the Earth's surface that have been named *geomagnetic storms*. At midlatitudes on the Earth, for a long-term average, about one storm each year is larger than 250 gammas in H and about ten storms per year are over 50 gammas. The number

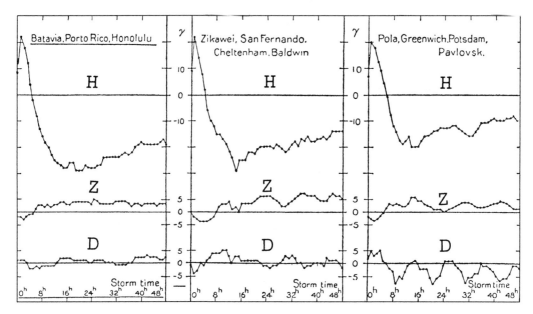

Figure 3.22. Averaged (on storm onset time) appearance of H, D, and Z field components for forty storms of moderate intensity from eleven middle- and low-latitude stations (names indicated). The declination is positive westward in this display. Figure from Chapman (1936).

and intensity of storms vary with the eleven-year solar activity cycle with about a one-year lag. At middle and low latitudes many storms display a similar general appearance for the H component of field (Figure 3.22), although some of the average features may be absent occasionally. On occasion, the storm starts with a *sudden commencement*, SC, (sometimes called SSC) that occurs almost simultaneously (within minutes) everywhere on Earth. The sudden commencement is caused by the shock wave at the magnetosphere, formed by the arrival of the fast solar-wind plasma beginning the storm. The SC may be followed by a general increase in the magnetic northward field as an *initial phase*, a compression effect (Equation (3.8)) that can continue up to several hours. Many storms occur without this initial phase. The next (or sometimes first) appearance of the storm is called the *main phase* (or *growth phase*), in which this principal component of the field decreases and shows major fluctuations for a longer time, and with larger amplitude, than the initial phase. Typical structured auroras and intense electrojet currents are observed then. Finally, in a *recovery phase*, the storm spends its greatest time gradually returning to an undisturbed level over as much as several days. This main and recovery storm-field pattern, seen in the H field component at middle and low latitudes, results from the addition of many contributing disturbance fields. In general terms, the solar wind drives a convection of the magnetosphere that dissipates in geomagnetic storm-related processes. The storm energy is divided between the input to the auroral ionosphere, the

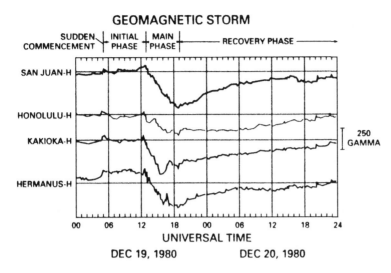

Figure 3.23. The variation in *H* component of field at the four observatories indicated to the left. Names of the four parts of a typical geomagnetic storm are given at the top of the figure. Scale size is indicated to the right.

creation of the magnetospheric currents, and down-tail (magnetospheric) processes.

Figure 3.23 illustrates the storm nomenclature for the *H* component of field during a 19–20 December 1980, storm seen at the low-latitude observatories that are used for preparation of an activity index (see Dst in Section 3.15). On rare occasions, unusually large storms occur as a major decrease in field but do not show the clear phases of Figure 3.23. Such storms are typically associated with intense polar-cap proton bombardment and with the uncommon, red-glow auroras.

The storm SC is related to the arrival of the solar-wind hydromagnetic shocks at the geomagnetic field interface (*stand-off position*); the SC size is roughly proportional to the square root of the solar-wind dynamic pressure. If the *interplanetary magnetic field* (IMF) attending the arrival of the solar–terrestrial disturbance wind is directed southward $(-B_z)$, then a geomagnetic storm usually follows the SC. The IMF $-B_z$ often occurs following a transition between toward and away sectors. If the IMF remains northward behind the arrival shock, then there is usually no subsequent storm; the shock effect stands alone and is named a *sudden impulse* (SI). It is believed that the SC and the SI seen at world observatories are produced by magnetopause currents, compressional waves propagating within the magnetosphere, and resulting ionospheric currents. Northern Hemisphere studies show that in most cases the *D* component of the SC or SI is eastward, the *H* component is northward, and the Z component is downward.

The storm SCs are largest within the summer hemisphere and at the dip equator owing to the enhanced conductivity that allows stronger ionospheric currents to flow in those regions. The main phase depends upon a sustained southward directed interplanetary magnetic field of the solar wind at the magnetospheric boundary. The SC size is independent of the main phase, and almost all SCs are followed by a storm main phase. During the storm main phase there are major increases in the plasma of trapped ions and electrons between $L = 2$ and $L = 9$ (see L-shell definition in Section 1.3). With the injection of plasma into the Earth's field environment, a charge separation arises across the dawn and dusk sides of the magnetosphere that drives some equatorial drift of groups of injected particles in partial *ring current* paths and feeds *field-aligned* current systems that transfer energy between the magnetosphere and ionosphere to cause auroras and ionospheric disturbances. Strong storm currents flow from the high-latitude ionosphere though the conducting lower-latitude E and lower-F regions. The recovery phase of the storm attends the switching off of the southward $-B_z$ IMF.

Figure 3.19 illustrates, with arrows, some of the current systems that arise during a storm. It is important to realize that many of the currents depicted in this figure are not continuously flowing but represent long-term average patterns for many storms in which bundles of particles temporarily follow parts of the defined paths. Before the extensive exploration of the magnetosphere by satellites it was believed that the main phase and recovery phase of storms (Figure 3.23) were simply a direct result of the growth and decay of a *ring current* encircling the Earth. However, it is now understood that many disturbance fields, in addition to partial ring currents, are contributing to the middle- and low-latitude surface field measurements.

3.9 Substorms

Only the middle- and low-latitude geomagnetic fields describe the gradual storm-time growth and decay. Satellite observations in the magnetosphere show no single mechanism that closely follows the main and recovery storm phases. The idea of *substorms* (sometimes called *auroral substorms* or *polar substorms*) arose out of the need to tie together the in situ observations of storm period bursts of activity that are linked together on time scales shorter than the main and recovery phases.

While the high speed (500 to 900 km/sec) solar-wind plasma compresses the magnetosphere (to as much as 5 R_e) the direction of the frozen-in IMF B_z determines further interaction. If the IMF B_z is strong and northward (called the *NBZ system*), then small complicated field-aligned currents flow into and away from the sunlit polar cap ionosphere; no substorm ensues. If the IMF B_z is fully southward ($-B_z$), allowing field-line connection between the solar wind and the northward directed magnetospheric field (Figure 3.21) and there is a significant

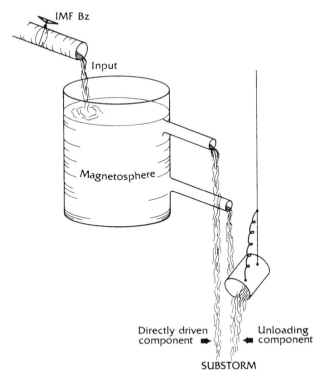

IMF Bz

Input

Magnetosphere

Directly driven   Unloading
component   ➡   ◀   component

SUBSTORM

Figure 3.24. Drawing depicts two types of magnetospheric storm sources. When the IMF B_z is turned southward, the solar–terrestrial particle input flows to the reservoir. From this magnetospheric region a directly driven component of the disturbance flows and a bucket is loaded which, when full, tips to deliver the unloading component of the disturbance. Figure from S.-I. Akasofu.

turning of B_z toward the north, then a major substorm can be triggered in which the arriving particles enter and modify the shape and composition of the magnetosphere. The substorm onset is typically accompanied by an enhancement of $Pi2$ pulsations (see Section 3.13).

Once the energetic solar-wind particles gain entry into the magnetosphere, two methods for substorm current creation are identified (Figure 3.24). In the first, called the *directly driven* process, the arriving particles immediately find their way into the cusp region of the dipole field and into the partial ring current path that causes field-aligned currents, auroras, and strong ionospheric currents (an eastward ionospheric electrojet current on the dusk-side auroral region and a westward current on the dawn-side of the Earth). The directly driven processes have a characteristic response time of about two hours.

In the second, called the *loading-unloading* process, there is a magnetospheric tail (*magnetotail*) particle flux buildup and reconfiguration in which particles are further energized before precipitating into the auroral-zone atmosphere via what has been modeled as a virtual *substorm current wedge* centered near the midnight sector (Figure 3.25). The variations in the loading-unloading and the field-aligned current system having characteristic lifetimes of about 10 to 20 minutes are more rapid compared to the directly driven system. About an hour after the auroras and the electrojet currents of a substorm reach their peak development, there is often a *poleward leap* of these phenomena to very high latitudes followed shortly by the end of the substorm. The intermediate substorm processes are complex and still under some debate among upper atmospheric physicists. Rather than list all mechanism viewpoints, I will describe some of the important observable substorm features.

3.10 Tail, Ring, and Field-Aligned Currents

The magnetotail structure has a cross section of a north and south lobe separated by an east–west, cross-tail plasma-sheet current (Figure 3.19) (see also Ohtani, 2001; Ohtani et al., 2002). During the usual loading-unloading substorm there is an intensification of the magnetic flux in the tail and the plasma sheet thins near the Earth (Figure 3.26). A strong cross-tail current forms inside 10 R_e. Reconnection of field lines occurs in the center of the tail current sheet, causing an explosive release of energy; the substorm expansive phase attends the sudden collapse of this cross-tail current. A new tail-current system forms that directs a portion of the cross-tail current to feed the growth of partial ring currents and flow, along field lines, into the auroral zone and then westward in the nighttime ionosphere. The precipitating particles excite auroras and a massive flow of currents into and away from the auroral region that can be represented as a virtual substorm current wedge in Figure 3.25

Figure 3.25. Virtual current model of the substorm current wedge arrangement. Field-aligned currents flow into and away from the auroral zone where auroras and a closing current called the auroral electrojet occur. Adapted from figure by Kamide and Fukushima as shown in Rostoker, 1974.

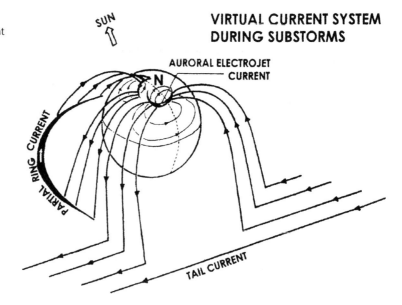

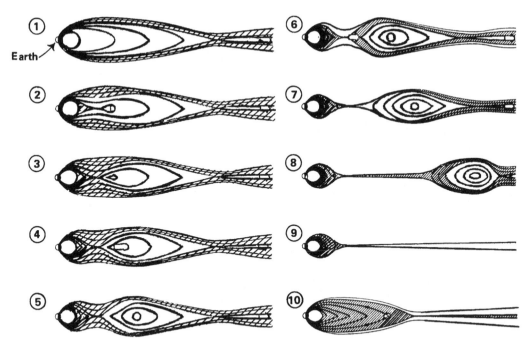

Figure 3.26. Magnetotail reconfiguration pattern during loading-unloading substorm event. The solar wind is directed left to right. Sequentially numbered changes illustrate the energization of particles, causing them to flow earthward, and the pinching-off of a downstream plasmoid before it returns to prestorm conditions. Figure from NASA International Solar–Terrestrial Physics Program Bulletin; the source was Hones (1985).

(c.f. Rostoker, 1996). A later decrease of auroral-region currents signals a magnetotail plasma sheet thickening and a downwind retreat of the tail-current sheet, ending the substorm process, and the magnetosphere returns to a more dipolelike form. The typical geomagnetic storm period is composed of many substorms.

During intense magnetic storms, a direct effect of the cross-tail plasma-sheet current can be observed at the Earth's surface. Midnight levels of field variation (with the main-field level and induced-field contributions removed) at low latitude and equatorial observatories have been shown to track the effects of the growth and motion of the cross-tail current's earthward edge (Figure 3.27; Campbell, 1973). With changing seasons, the average location of this current naturally shifts position to the downwind (winter hemisphere) location from the Sun, just as the dipole field of the Earth distorts seasonally to the antisunward position (Figure 1.29).

The average of many early satellite passes through the equatorial magnetospheric region of 2 to 9 R_e showed that the storm-period values of field, from about 2 to 3.5 R_e could represent an average eastward current system, and the field from about 3.5 to 9 R_e could represent a larger westward current system (Figure 3.28). This Saturnlike, *ring current* model became the accepted geomagnetic storm picture supporting

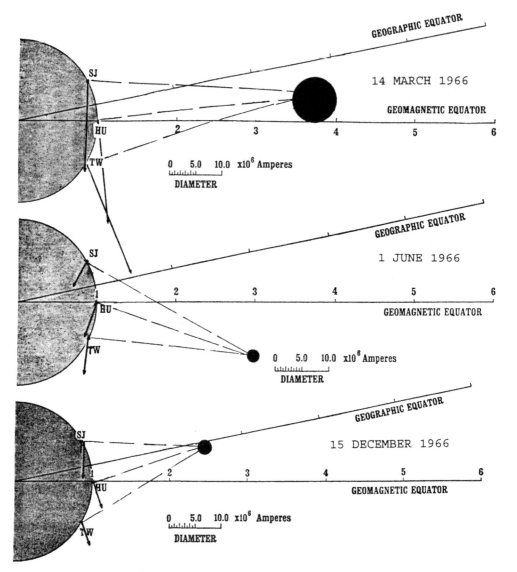

Figure 3.27. Effect of storm-time magnetospheric tail current upon Earth surface fields at midnight. Solid dark circles represent the magnetospheric current measured by the midnight field changes at Earth surface locations on three storm days near (top) the March equinox, (middle) the June solstice, and (bottom) the December solstice. Shaded half-circles at left depict the nightside of the Earth with station locations at San Juan (SJ) and Trelew (TW) in the midnight plane. Values of H and Z field components, with the main field level and induced field removed, then averaged for four hours near local midnight, are indicated by arrows at the stations. In the midnight plane these fields triangulate (dashed lines) to a storm-time, magnetotail current sheet directed toward the viewer. Radial distances are indicated in R_e. The equivalent current magnitude, computed from the fields, is indicated (with the scale) by the diameter of a solid circle at the triangulated location. Note the unequivocal surface response of the magnetospheric storm current. The location of this current shifts about the geomagnetic equator in agreement with the seasonal, downstream, solar wind direction.

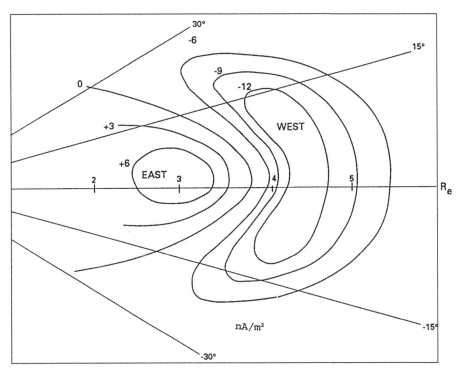

Figure 3.28. North-south cross section of long-term averaged partial ring currents in nanoamps per square meter. Figure adapted from Hoffman and Bracken (1967).

the notion that the main and recovery phases of the storm were due to the growth and decay of the large westward ring current.

Later, better-instrumented satellites discovered that the simple east-ward region represented the radiation belt (see Section 3.12) and the westward ring region had major temporal and spatial fine structures (Figure 3.29). There is not one simple global ring current or a simple superposed partial ring current. Rather, there are many large and small partial ring currents varying greatly in time and location; radial as well as azimuthal currents exist on a variety of scales. At best, less than half of the low latitude storm-time field behavior might be reconciled by the observed westward ring-current flow. Storm-time, field-aligned currents connect the main westward partial ring currents, at 3.5 to 9 R_e, to the high-latitude ionosphere where active auroras are observed. The lower-latitude, radiation belt (see Section 3.12) corresponds to the area of faint (subvisual), stable, subauroral, red (oxygen emission), *SAR arc* glow that often occurs at *F*-region altitudes during disturbance periods.

Figure 3.30 illustrates the averaged *field-aligned* (Birkeland) current (FAC) directional patterns into and away from the ionosphere during a typical auroral substorm. Interactions of the bombarding particles with the Earth's atmosphere define an *auroral zone*. Three regions are shown in the figure. Closest to the geomagnetic pole is the flow into and away

Figure 3.29. Radial profiles of ring-current region current densities during two storms, (top) 4–7 September and (bottom) 18–20 September 1984. Four passes of the AMPTE/CCE satellite are overplotted in each example. Current density units are given to the left; radial distance in *L*-shell units is given at the bottom. Many partial ring currents, of relatively short duration, were observed. Figure adapted from Lui, McEntire, and Krimigis (1987).

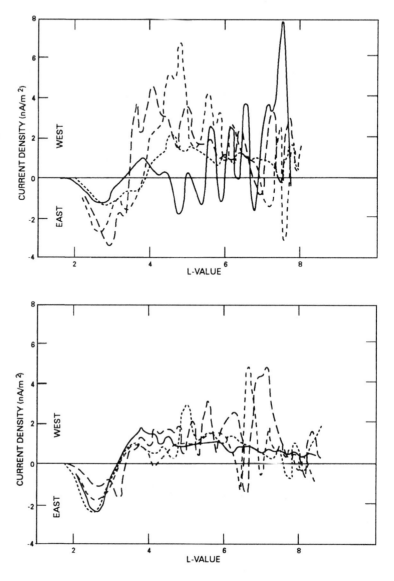

from the *magnetospheric cusp* region (cleft in Figure 3.19) of the midday magnetosphere at the highest auroral latitude near 80°. The cusp region is formed because the original dayside field lines, which were connected to the opposite hemisphere, have folded back to the downwind direction.

The next charged particle flow to define the auroral latitudes is called *Region 1* field-aligned current at the poleward position; that flow is into the ionosphere on the morning side of the Earth and out of the ionosphere on the evening side. The Region 1 current density is reported to be about 2–3 μA/m^2 when IMF $B_z = 0$ or positive.

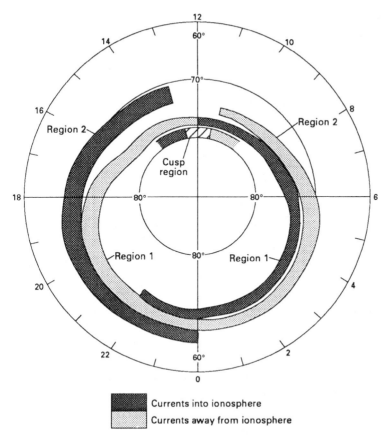

Figure 3.30. Averaged Region 1 and Region 2 field-aligned (Birkeland) currents in the polar region. Circles of invariant latitude and local time are indicated. Figure adapted from Iijima and Potemra (1976).

Currents into ionosphere

Currents away from ionosphere

As IMF B_z becomes more negative the Region 1 current density increases to a value near 5 μA/m² for IMF $B_z = -5$ nT. These currents are typically most intense between about 0700 and 0900 and between 1300 and 1500 geomagnetic time (Figure 1.22). The cusp and dayside Region 1 are identified with a dayside low-latitude boundary layer of the storm particle precipitation from about 0900 to 1500 geomagnetic time. Westward flowing, partial ring currents can be directed along Region 1 routes.

The *Region 2* currents, at the equatorward side of the polar region, flow opposite to the Region 1 currents at any given local time. These Region 2 currents might connect to the eastward ring current. The density of Region 2 and 1 currents is about equal on the nightside of the magnetosphere, but the Region 2 currents can be one-third to one-quarter the size of the Region 1 currents on the dayside.

Together the three regions define an oval of activity encircling the polar cap. The projection of Region 1 and 2 upon the ionosphere is approximately coincident with the locus of the auroral electrojet. A number of speculative routes have been proposed for the Birkeland currents for connection to the tail currents via the ring-current region, but there is no agreement on a unique pattern at this time. It is important to realize that Figure 3.30 is an averaged behavior diagram of Birkeland currents; throughout an individual storm there is not a continuous flow of particles in all the routes shown (see Ohtani et al., 2000; Higuchi and Ohtani, 2000). The field-aligned system is driven by multiple magnetospheric currents whose ionospheric projection averaged in time may look like a large continuous pattern.

Some scientists have computed the surface projections of the field-aligned currents at high and polar latitudes using magnetic data from an extensive array of surface observatories, from detailed ionospheric profiles, and from the occasional pass of satellites through the region. The Tsyganenko (1987, 1989) magnetospheric models are often used for such computations because he included the satellite observations of current systems. In the substorm analysis shown in Figure 3.31, note the increase of field-aligned current during conditions of IMF $B_z \leq 0$ and the shift, in time of maximum, of the afternoon field-aligned currents as the IMF B_y field changes from negative (toward) to positive (away) sector directions. A seasonal variation in field-aligned currents has been reported by Christiansen et al., 2002.

The modeling of the field-aligned currents provides not only a view of the expected surface-field response but also of the thermospheric heating that attends these strong substorm Birkeland currents. This heating is a result of current forced to flow through a nonperfect conductor (as in home electric heaters) and is given the name *Joule heating*. Current, I, in a wire of resistance (reciprocal of conductivity), R, radiates heat energy at a rate $I^2 R$ units per second. A "Joule" is defined as the amount of heat generated by a one-ampere current flowing for one second through a resistance of one ohm. This Joule heating represents the energy expended to keep the current flowing. Similarly, the field-aligned and ionospheric currents, driven by a substorm in the conducting upper atmosphere, provide a considerable heat source in the upper atmosphere at high latitudes.

Figure 3.31. (opposite) Contours of the distribution of field-aligned currents in the north polar region above 60° latitude computed from 70 days of magnetic records taken at a longitudinal chain of ten Greenland stations operated during the summers of 1972 and 1973. Ten-degree, north polar latitude circles are shown. Meridian hours are indicated around the circumference, with the midnight meridian at the bottom of each diagram. The distributions are divided into groups corresponding to magnitude and directions of the IMF field components B_z (top to bottom rows) and B_y (left to right). Shaded contour areas represent upward current regions and unshaded areas represent downward current regions. Figure from Friis-Christensen et al. (1985).

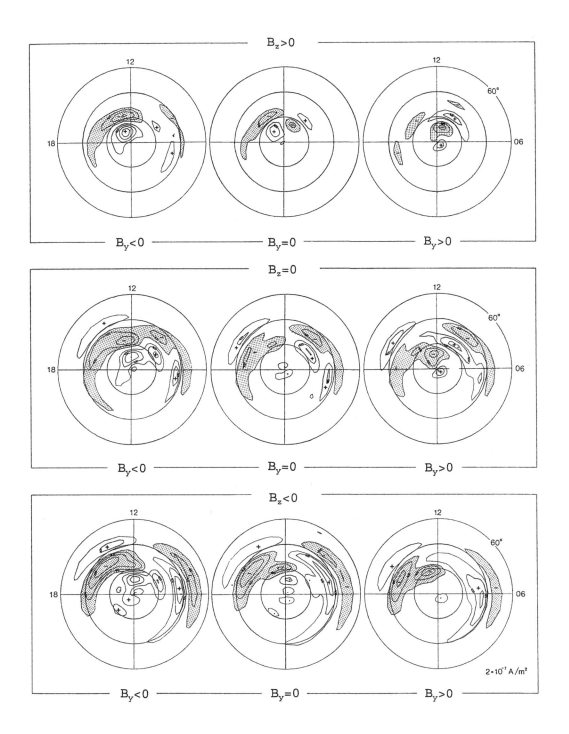

During a substorm, such intense thermospheric heating causes a high-altitude global wind surge in the conjugate auroral zones. Propagating equatorward into the opposite hemispheres at velocities up to 700 m/sec, the surges are strongest in the night sector of the auroral zones near the longitudes of the geomagnetic poles and begin to reach the equator in about 3.5 hours after the onset of the storm (Fuller-Rowell, personal communication). Together, the conjugate hemisphere-heating surges drive a change in the global thermospheric wind circulation system at middle latitudes and the development of zonal wind flows. A large part of the substorm energy is dissipated by the Joule heating. Such energy is not only responsible for major changes in the global thermospheric wind system but also for alterations of the ionospheric F region (see Section 5.6) and some dynamo currents located at the E region and bottom of the F region. A "flywheel" effect has been ascribed to the thermospheric heating because the slow buildup, as heating is accumulated, and slow recovery, as the thermospheric winds dissipate, resemble the inertial response of a spinning heavy wheel.

3.11 Auroras and Ionospheric Currents

Substorm *auroras* often appear as a single arc or parallel system of arcs with bright lower borders; they extend across the sky from horizon to horizon, generally parallel to geomagnetic latitude circles. Arcs often develop folds, move rapidly, and show intensity variations. Some are broken into individual rays, which may seem quiet or rapidly moving, giving the appearance of heavenly draperies. The draperies manifest the coronal form when they occur in the magnetic zenith. Flaming auroras, near the peak of an intense display, consist of strong waves of light rapidly moving upward in regular succession. At times one sees only diffuse luminous surfaces or auroral glows that occasionally pulsate.

The arcs are often a hundred to several thousand kilometers in length with widths of merely a few hundred meters. The usual maximum emission heights have been measured at 90 to 130 km; those with red lower borders extend down to about 70 km, whereas the diffuse red arcs typically exhibit their maximum intensity at 250 to 300 km. Sunlight on the upper atmosphere stimulates the excitation processes, expanding the upper limit of measured aurora to over 1000 km.

The auroral occurrence distribution is organized with a geomagnetic coordinate system. The maximum incidence of auroras is near 64° to 70° geomagnetic north or south latitude. These regions are called the *boreal auroral zone* in the Northern Hemisphere and *austral auroral zone* in the Southern Hemisphere. The auroral forms are aligned near geomagnetic east–west at latitudes from 55° to 68° (north or south) and in the sunward direction at latitudes above ±80°. In the northern auroral zone the apparent horizontal motions are westward in the evening and eastward in the morning; the meridional motions are generally southward throughout the night. Mirror-image behavior of auroras is observed in the two hemispheres when darkness permits viewing at conjugate locations.

The annual auroral occurrence maximum follows the well-established 11-year solar activity cycle. The 27-day recurrence tendency of auroras is related to the Sun's surface rotation period. There are two seasonal maxima roughly coinciding with the equinoxes. In the auroral zones the daily occurrence is greatest near the local midnight hours.

Auroras are visual evidence of energetic particles that bombard the upper atmosphere. When incoming electrons, spiraling along a field line, enter the high-altitude atmosphere, collisions occur with two consequences: (1) For higher-energy particles the impact can strip an electron away from a molecule of air thereby ionizing that molecule, doubling the number of electrons (incoming original plus that one newly removed), and thereby increasing the electrical conductivity. (2) For lower-energy particles, but above a limiting energy level, the atmospheric molecules receive the energy, become excited, and (in a time determined by the molecule and energy state) dissipate the energy as light emissions called photons (unless a subsequent collision with another atom or molecule takes away the energy).

In the world of atoms and molecules, excitation and emission can only occur at prescribed steps; such *quantized levels* have characteristic patterns for each atom or molecule and have prescribed energy unloading requirements such that only particular light wavelengths can be emitted after a specified delay time. In a way, auroras are similar to the glow in commercial neon signs; the sign-tube gas is bombarded by an electric current of particles that delivers the correct energy for the gas to glow at its characteristic color.

The heights of intense auroras are determined by four factors: the energy of the precipitating particles, the composition (and density) of the atmosphere, the permitted emission levels of the atmospheric particles, and the excited state half-life for the characteristic emission. The *half-life* is the time for half of the excited-level photons to be emitted. Between 60 and 100 km altitude, oxygen and nitrogen molecules dominate the spectral sources; from 100 to 200 km we find mainly oxygen atoms and nitrogen molecules; and above 200 km, oxygen atoms dominate. Above about 150 km the atmospheric density is so low that there are very few collisions to remove energy from excited atoms and molecules. Incoming particle collisions with atmospheric hydrogen and helium occur at subvisual levels.

The primary contribution to visual aurora comes from the intense green 5577 Å oxygen line and from the ionized molecular nitrogen band of blue-violet emissions near 3914 Å and 4278 Å in the atmospheric region from about 100 to 300 km. Although the nitrogen emissions are about as strong as the oxygen below 200 km, our eyes favor the green. The oxygen emissions have a three-quarter second half-life, whereas the nitrogen emissions are relatively instantaneous. A red 6300 Å atomic oxygen line is limited to the higher altitudes because, with its 130 second half-life, the relatively short time for more collisions at the lower altitude would remove the energy before emission. The high-altitude red-glow auroras are an impressive but relatively rare phenomenon. Other red-color band emissions from nitrogen just below 100 km occasionally form a lower red border to the auroras.

Electrons with energy of about 10 keV (kiloelectron volts) or protons of several hundred kiloelectron volts can penetrate to the 100-km level. It takes electrons of about 2 to 5 eV to excite the oxygen emissions and about 2 to 4 eV to excite the nitrogen; at higher energies the incoming electrons can remove an electron from the air atom. The processes can be illustrated by the following energetic electron (e) impact upon a nitrogen molecule (N_2) that first strips off an electron and leaves a singly charged ($+$) excited ($*$) nitrogen ion:

$$e + N_2 \rightarrow N_2^{+*} + 2e \tag{3.11}$$

and then excites molecular band emission of the ionized nitrogen:

$$N_2^{+*} \rightarrow N_2^{+} + h\nu(3914 \text{ Å}, 4278 \text{ Å, etc.}), \tag{3.12}$$

where $h\nu$ indicates photon emission. The electron density $[n_e]$ is recoverable from the photon emission rate, p, and for the above reactions can be approximated by

$$[n_e] \approx (1.6)^4 \sqrt{p}. \tag{3.13}$$

The auroras often appear as drapery-like structures, with folds following the nearly vertical high-latitude field lines (Figure 3.32). The gyroradius of the precipitating electrons limits the width of any luminous element. The *auroral zone* is a global region of auroral occurrence maximum near 65° geomagnetic latitude that forms an *auroral oval* circling the eccentric axis pole (Figure 3.33). During some magnetic storms, a light polar rain of energetic electrons has been detected inside the auroral oval. This special phenomenon, occuring when the interplanetary field is properly aligned, is thought to have its origin in the unique field conditions at the distant magnetospheric boundary whose field lines connect to the polar cap. As a disturbance event grows, the region expands both poleward and equatorward. At times of low geomagnetic

Figure 3.32. Drapery-like auroral arcs with almost vertical field-aligned elements as they appeared in the sky above the Geophysical Institute, University of Alaska. Photograph by V.P. Hessler. Other auroral pictures can be found in Eather (1980) and at Hutchinson website (www.ptialaska.net/~hutch/aurora.html).

activity the auroral region (usually subvisual at such times) shrinks to a location near 70° to 75° geomagnetic latitude. The auroral displays have been tracked to field-aligned currents and auroral electrojets. Auroral intensity and position changes exhibit a close relationship with geomagnetic field measurements made at the surface below the aurora.

On occasion, soon after the IMF B_z field turns southward and following a prolonged period of northward direction, there occurs a transpolar bar of luminosity in the noon-midnight direction. This *theta aurora* seems to represent magnetotail plasma that is surrounded by open field lines. The event can persist for several hours and, in the north polar region, appears to move from the dawn side toward the dusk side for positive IMF B_y (and oppositely in the southern polar region).

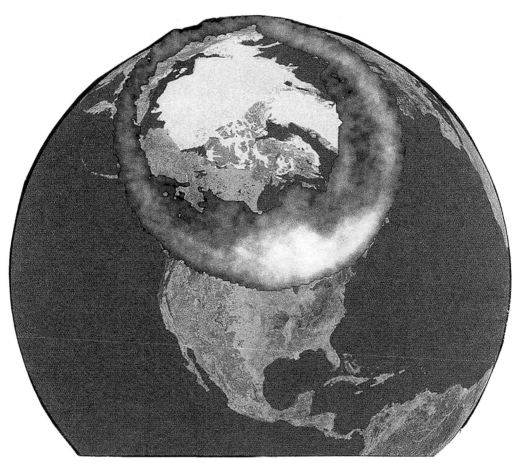

Figure 3.33. NASA picture of the Earth's northern auroral oval centered near the eccentric-axis dipole axial pole location.

Figure 3.34 is a full polar-cap presentation of the time sequence of an auroral form change during a substorm that would be seen as an overhead display to a surface observer. The *auroral oval* expands and shrinks with the input of solar wind energy. At the *expansion phase*, more auroral arcs are formed on the poleward side of the oval, spreading the display to higher latitudes. Each arc drifts to lower latitudes. As the substorm continues, disturbed forms in the arcs spread to the east and west. Identifiable *westward traveling surges* move at an average speed of 1 km/sec. Simultaneously, forms drift eastward to become *pulsating patches* on the eastward (dawn) side. The nightside (usually just before midnight) region between westward and eastward drifting motions is called the *Harang discontinuity*. This was called the auroral *breakup* in early literature and was identified on magnetograms as the demarcation line between regions of positive and negative H-component field perturbations (see Zhou et al., 2000, for breakup relationship to solar wind). The *recovery phase* of the auroral substorm

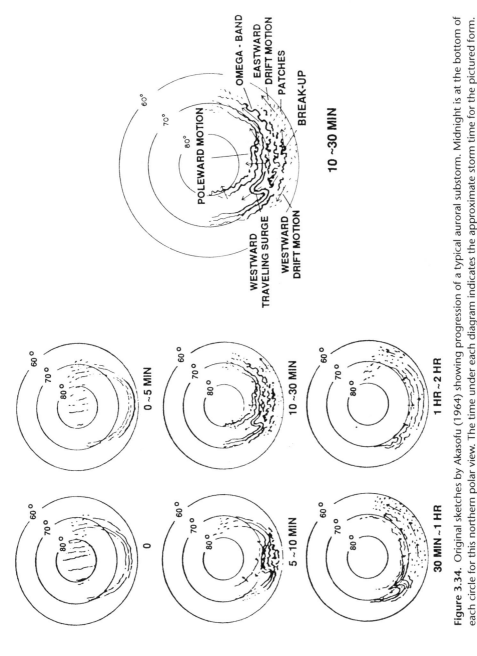

Figure 3.34. Original sketches by Akasofu (1964) showing progression of a typical auroral substorm. Midnight is at the bottom of each circle for this northern polar view. The time under each diagram indicates the approximate storm time for the pictured form. The larger diagram to the right identifies auroral forms for the principal activity at 10–30 minutes into the substorm.

is signaled by the poleward auroras' return to lower latitudes in the midnight sector.

In the auroral nightside ionosphere an intense westward ionospheric current, the *auroral electrojet*, driven by the Birkeland currents, flows in the region of enhanced conductivity created along with the auroras. This strong westward electrojet current depresses the northward geomagnetic field at those latitudes with violent variations in synchronization with the overhead auroras. Very short-period (seconds to minutes) field pulsations (Section 3.13) and associated auroral luminosity fluctuations are measured at this time. The auroral region of maximum field disturbance is generally restricted to less than 5° in latitude and 100° in longitude on the nightside of the Earth. The ratio of high- to low-frequency components of the geomagnetic disturbance decreases rapidly with distance from this region and with level of activity.

The conducting lower F-region and E-region ionosphere allows a closure of the strong substorm westward electrojets to be communicated to other longitude and lower-latitude ionospheric locations. Compared to the magnetospheric current systems, ionospheric currents are a short distance from the surface magnetic observatories, so these currents have a major influence upon the H, D, and Z variation fields. At the auroral zone, field measurements triangulate to the regions of intense visual auroras and track all their form variations, such as long auroral arcs, spatial twists, and rapid pulsations (Figures 3.35, 3.36, and 3.37). The substorm ionospheric currents travel all the way to the equatorial region where the unique ionospheric high-conductivity condition

Figure 3.35. The directed field, **B**, obtained from two surface magnetometers (or just one magnetometer and the assumed 100-km current altitude) is used, after adjustment for induction contributions, to triangulate to the equivalent auroral electrojet current, I, flowing in the E region of the ionosphere at the location of the maximum auroral arc intensity.

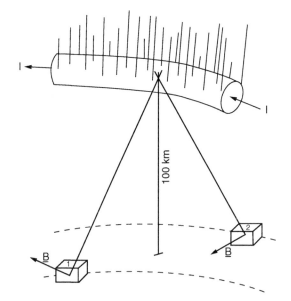

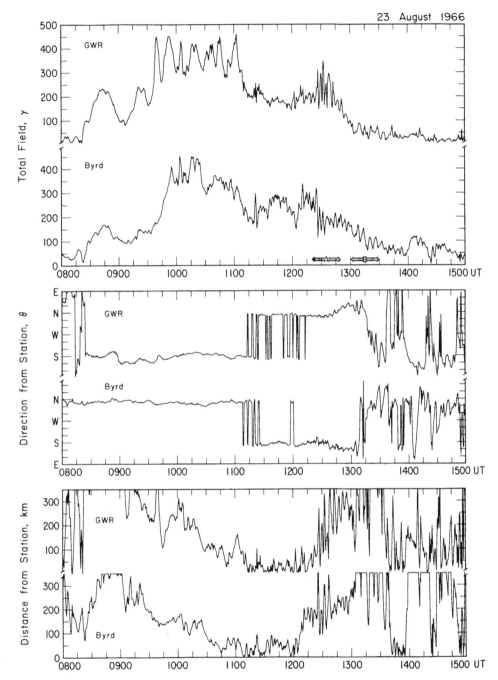

Figure 3.36. Behavior of an equivalent line current representing the auroral electrojet (see Figure 3.35) above the conjugate stations of Great Whale River (GWR), Canada, and Byrd, Antarctica, for the days and hours indicated. The top panel displays the total variation field, $\sqrt{(\Delta H)^2 + (\Delta D)^2 + (\Delta Z)^2}$, at the ground (corrected for induced effects); the middle panel shows the angular direction, θ, for a surface projection of a line toward the nearest approach of the equivalent source current (indicated in cardinal geomagnetic directions NSEW); and the bottom panel shows the surface-projected distance in kilometers from the station to that current. Note the conjugate behavior in amplitude and location of the electrojet in the north and south auroral regions.

BYRD 24 APRIL 1967

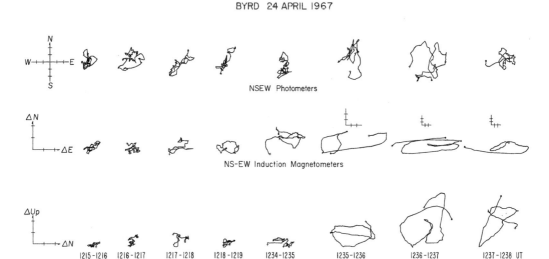

NSEW Photometers

NS-EW Induction Magnetometers

NS–V Induction Magnetometers

Figure 3.37. Comparison of polarization patterns of light and field for sample data when motion of aurora is indicated. Successive 1-min intervals of 0.5 sec data at Byrd, Antarctica, on 24 April 1967, are shown. A trace begins with a dot and ends with a plus sign. The top patterns show the locus of 4278 Å auroral fluctuations whose coordinates are the difference in output of photometers that were directed upward (at 45° angles to the horizontal) in the cardinal directions indicated to left. The middle and bottom patterns show the corresponding field fluctuations in the directions indicated to left; scale changes are indicated by scale marker separations. Note the similar variation in location and intensity that identifies the rapid field pulsation behavior with the number and location of bombarding auroral-zone electrons.

in the daytime E region causes an equatorial (dip latitude) enhancement (Figures 3.38 and 3.39). The narrow latitudinal scale of this equatorial enhancement proves it is of ionospheric origin (because that region is much nearer than magnetospheric sources). The current must reach the equator through the middle latitude ionosphere, a feature that verifies the global extent of the current system originating in the auroral electrojet.

The location of auroral-zone currents corresponds to regions of increased E-region ionization (extra electrons stripped from atmospheric molecules by precipitating particles). The auroral-zone instrumentation for measuring this ionization is called a *riometer* (relative ionospheric opacity meter). A minimum signal level is recorded as the instrument sweeps through a band of frequencies near 30 MHz (the minimum is used to exclude man-made transmissions). Near this frequency the monitored natural cosmic radiowave noise from the galaxy has a diurnal pattern (in sidereal time). When the ionosphere becomes more conducting, a

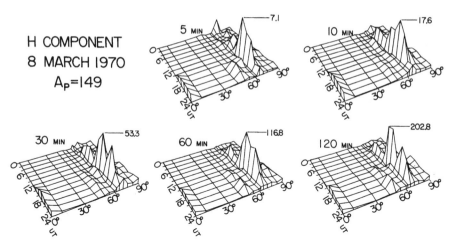

Figure 3.38. Spectral amplitudes of 5-, 10-, 30-, 60-, and 120-min period Fourier components from a 4-hr analysis sample of the *H* component of field variation for every 2 hr of universal time (UT) covering the storm-time disturbances over North and South American locations during a day of very high activity index (Ap = 149). In these projections the geomagnetic latitude is given in degrees; constant latitude lines indicate the location of the American sector observatories contributing to the analysis. The maximum amplitude in each period sample is normalized to about the same vertical height for the projections; the amplitude scale can be determined from the peak values (gammas) indicated on each display. Note the auroral region activity maxima through most of the UT day as well as the subauroral and equatorial electrojet region enhancements during the sunup hours of significant $\sqrt{\cos(\chi)}$.

decrease in the riometer quiet-day cosmic noise pattern occurs, from which a measure of the increase in ionospheric electron density, n_e, within the antenna window is determined. Approximately, the absorption (measured in decibel units), A_{dB}, is given by

$$A_{dB} \approx 1.6 \times 10^{-17} \int_0^\infty n_e v \, ds, \qquad (3.14)$$

where v is the collision frequency, and the $\int$ summation (see Section A.7) of the incremental steps (ds) of altitude has contributions only within the ionosphere. The correspondence of fluctuations in the riometer-measured absorption, the auroral luminosity, and the disturbed magnetic field have been explained in terms of the electron density and physical processes relating the phenomena.

When incoming electrons are turned from their forward motion by encounters with atmospheric molecules, the braking of forward motion causes an emission of X rays called *bremsstrahlung radiation* (using the German word for braking). This radiation is too weak to penetrate to the Earth's surface. However, at the stratospheric altitudes attained

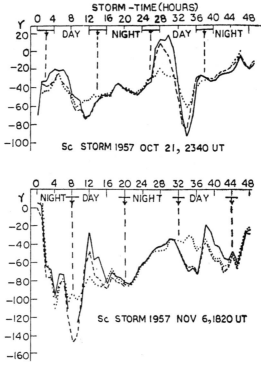

Figure 3.39. Storm-time hourly values of *H* (with quiet-day variations removed) in India for storms on (top) 21 October 1957, and (bottom) 6 November 1957, for the equatorial stations Trivandrum at 0.3°S dip latitude (solid line) and Annamalainagar at 2.7°N dip latitude (dashed line) compared with the low-latitude station Alibag at 12.9°N dip latitude (dotted line). Time is given in hours from sudden commencement (Sc) with day and night periods indicated. Amplitude scale is in gamma (γ). Records show that disturbances are of similar magnitude at night but are enhanced at equatorial latitudes during the day because of the *E*-region intensification of ionospheric currents near the dip equator. Figure from Yacob (1966).

by research balloons the radiation can be detected and used to count the energy and number of precipitating electrons during auroras. Figure 3.40 illustrates the substorm relationship between the bremsstrahlung X-ray count, the riometer-measured ionospheric absorption, the geomagnetic pulsations, and the field of the westward auroral electrojet current.

On rare occasions, when the entire polar ionosphere *D* and *E* region becomes absorbing, completely blocking out high-frequency HF and VHF (3–300 MHz) radiowaves and phase-shifting LF and VLF (3–300 kHz) frequencies, the name *polar-cap absorption* (PCA) is used. This disturbance phenomenon is the result of high-energy (1–400 MeV) proton bombardment and reaches a maximum before the associated

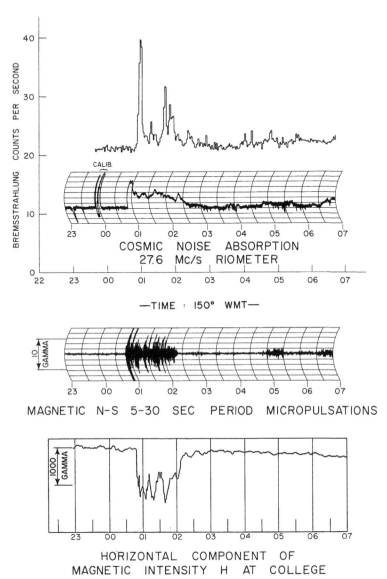

Figure 3.40. Simultaneous records of geomagnetic substorm commencing about 0050 local time at College, Alaska. Traces, top to bottom, are bremsstrahlung counts from a balloon-borne sensor at about 20-km altitude, riometer-measured cosmic noise absorption (increase positive upward), north–south axis geomagnetic micropulsations in the 5- to 30-sec period range, and the H-component of field at the magnetic observatory.

major geomagnetic disturbance commencement. PCAs have been traced to extremely high-energy, proton-rich flares and CMEs from the Sun that are sometimes called *proton events* or $H\beta$ photon emissions.

The simultaneous H-, D-, and Z-component surface field changes can be represented as an overhead shift in magnitude and location of an equivalent current system. For example, a field that would be downward for an overhead current to the west of zenith would shift to upward for an overhead current to the east of zenith. With a chain of polar-cap observatories, it was possible to draw equivalent two-dimensional currents (i.e., ionospheric currents that would have the observed surface-field effect). However, we realize that such current models obtained from the magnetic fields measured in the polar regions are largely the surface projections of three-dimensional magnetospheric current systems.

Independent of the storm, geomagnetic field variations in the polar regions change in prescribed ways that reflect the toward (T) and away (A) sector changes of the interplanetary magnetic field upon its arrival at the Earth's magnetosphere. In particular, the surface observatory Z component of field, after removal of the baseline level and averaged-quiet-day change, exhibits a diurnal phase shift that is B_y IMF dependent (Figure 3.41). Such behavior is oppositely directed at the two geomagnetic pole locations. The sector shift can be represented as a change in the daily mean level variation in Z. The discovery of the sector effect allowed researchers to recover, from old magnetic records, evidence of IMF changes that predated the advent of satellites (see Vennerstroem et al., 2001).

A spherical harmonic analysis of geomagnetic data at relatively quiet times allows the removal of the quiet-day Sq current behavior by simply removing the long-wave Legendre polynomial for indices with $(n - m) = 0$ or 1. The shorter-wavelength currents describe the low-level polar region activity. Typically, two vortices appear; these shift in position by a few hours for toward ($-B_y$ IMF) or away ($+B_y$ IMF) direction of the arriving solar wind (Figure 3.42). At the polar cap, in very quiet times, the Sq ionospheric dynamo current would be a simple continuation of the lower-latitude system. However, polar-region activity often occurs, even when a relatively quiet period persists at most other latitudes. Then the currents, called Sq^p, are a mixture of the dynamo current (that doesn't shift with sector change) and a sector-sensitive disturbance current located poleward of the normal auroral zone.

The ionosphere at middle and low latitudes can be significantly disturbed during a geomagnetic storm. This disturbance takes on various forms that change with the thermospheric wind system (Section 3.10, above), the season, and the time of day. Often, during storm conditions, the F region and total electron content are enhanced in the morning hours and depleted in the evening. However, the behavior varies with onset time of the storm and location of the most intense substorm fields with respect to local time. Although the hours of duration of a disturbed ionosphere generally correspond to the period of magnetic storm conditions, there is usually no close correlation between the two phenomena. That is to be expected from the fact that there are numerous local contributing processes for the ionospheric changes that are unmatched to the contributing processes for magnetic field variation. Figure 3.43 illustrates one very rare

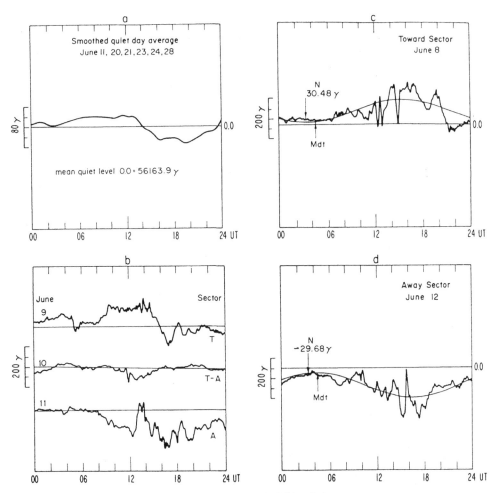

Figure 3.41. Example of daily polar station geomagnetic field variations associated with the interplanetary magnetic field sectors for Thule *Z* component in 1965: (a) the quiet variation obtained from the mean of the six days indicated; (b) example of a three-day sector change sequence with toward (T), transition (T-A), and away (A); (c) toward sector day with horizontal line indicating the quiet secular baseline, the diurnal sine curve of the Fourier component analysis, the local midnight location (Mdt), and the nighttime value at the quiet period (N); (d) similar to (c) but for an away-sector day. Note that the scale in part (a) is 2.5 times larger than the scales in the rest of the figure.

storm event in which a change in the ionospheric foF2 corresponded to an oppositely directed change in the local magnetic field.

3.12 Radiation Belts

Charged particles collect in a region around the Earth called the "Inner Radiation Belt" which starts at about 400 to 1200 km altitude and extends

POLAR QUIET EXTERNAL CURRENTS

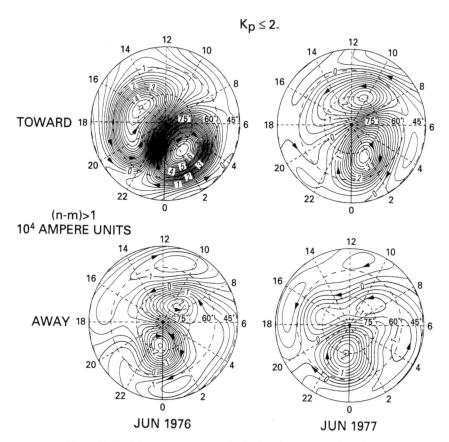

$K_p \leq 2-$

TOWARD

$(n-m) > 1$
10^4 AMPERE UNITS

AWAY

JUN 1976   JUN 1977

Figure 3.42. Polar current patterns in the Northern Hemisphere for toward- (top row) and away- (bottom row) sector separated averaged data representative of the June solstice periods of 1976 (left column) and 1977 (right column) on relatively quiet days when the geomagnetic activity indices Kp $\leq$ 2−. Contour levels are in units of 10^4 Amperes and the display covers from 45° geomagnetic latitude to the pole. Local times are indicated around the circles. Data from an Indian–Siberian longitudinal observatory chain were analyzed for the external components having short Legendre wavelengths with $(n - m) > 1$ to separate these polar contributions from the Sq. Note the shift to earlier hours for the nightside current foci as the sectors change from toward (IMF $B_y < 0$) to away (IMF $B_y > 0$). Figure from Campbell et al., 1994.

to about 3,900 km ($1.6\,R_e$). This region arises because the storm-time particles that find their way into the magnetosphere are required by physical laws to spiral around field lines (Equation (3.4)) and drift inward along the dipole field, gradually circling the Earth (negative particles one way, positives the other). Because the dipole field lines converge closer to the

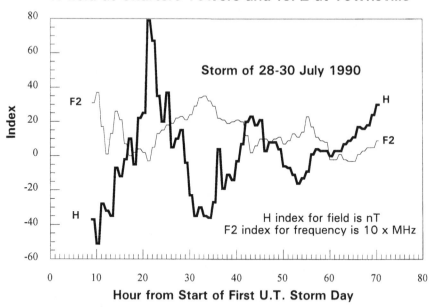

Figure 3.43. The Australian ionospheric station at Townsville is close to the magnetic observatory at Charters Towers. For the storm period shown in this figure, the changes in ionospheric foF2 about the quiet-day frequency levels are compared to the *H* component of field (from which the main field level, quiet-time field variation, and a lognormal representation of the Dst index have been removed). The *H* and *F*2 variations here show a significant value of −0.6 for the linear correlation coefficient (Section A.10).

Earth (dipole field increases) the forces on the charged particles cause them to reach a point where they turn around to spiral and drift outward and then inward to the opposite hemisphere, only to turn around again (bounce). The typical bounce time on the field line is estimated to be about 1/10 second while spiraling at about several thousand times a second. The particles are largely trapped in a donut-shaped shell about the Earth, forming a belt of radiation that is sometimes called, for its discoverer, the "Van Allen Belt" (Van Allen, 1969).

In the region of South Atlantic Ocean and South America there exists an anomalous low field (Figure 1.23d) resulting from the eccentric axis field offset (see Section 1.5). There the bouncing particles need to go deeper into the atmosphere to find the field strength usual for their reversing direction requirement. At such depths they are more likely to have collisions with the increased density of atmospheric atoms (Figures 2.6 and 2.13) slowing down the charged particle bouncers to increase their accumulation in that field region of the radiation belt and forming what has been named the "South-Atlantic/South-American Anomaly".

A second region, sometimes called the "Outer Radiation Belt" had been reported at about 4 to 8 R_e, although this feature has subsequently been identified as the partial ring current region (Figure 3.29).

3.13 Geomagnetic Spectra and Pulsations

We have learned that the geomagnetic field has variations that cover period ranges from fractions of a second to millions of years. Some people have unwisely talked about the "D.C. level" of field as if there were a "direct current" or steady level of reference. In reality, there just isn't such a behavior; there is only a wide range of slow and fast changes. In this chapter we have discussed active times, with respect to the previous chapter that focused upon quiet behavior. Here we have concerned ourselves with the changes that are more rapid than a day, excluding the spectral peaks that occur regularly at 24, 12, 8, and 6 hours which are primarily the result of ionospheric dynamo currents.

Figure 3.44 illustrates the typical quiet and disturbed geomagnetic spectra with periods from five minutes to several hours. Note the clear pattern of increasing amplitude with increasing period and contrast the active and quiet times. Of course, such relative spectral behavior depends upon latitude. Figure 3.45 shows some samples of the time-latitude-amplitude patterns for representative spectral slices. There are large amplitudes at the auroral zone and a daytime enhancement at equatorial latitudes (Figures 3.38 and 3.39). Figure 3.45 also indicates the change in field amplitude for quiet, moderate, and active geomagnetic conditions.

At shorter periods, below a few minutes, we enter the domain commonly called by the equivalent names *geomagnetic pulsations, micropulsations*, or *ultra-low frequency* (ULF). The field pulsations shown in the central part of Figure 3.40 are clearly part of the substorm process. Pulsations have been observed simultaneously in space and on the ground (Mursula et al., 2001; Vaivads et al., 2001; Thompson and Kivelson, 2001). Geomagnetic pulsations have been subdivided into the groups called *continuous and irregular pulsations, Pc* and *Pi*, corresponding to their appearance on a time-versus-amplitude trace, then numbered in bands from short to long period (Figure 3.46). Some scientists believe that pulsations arise as magnetospheric hydromagnetic (Alfven) waves that are generated by the *Kelvin–Helmholtz instability* process, which results when one magnetospheric plasma streams over another one. Other scientists assign importance to the shock arrival of the solar wind disturbance. Veroe, 1996 (and papers in JGR 1994 Monograph 81 on ULF waves) considers the observed field line resonance characteristics of the pulsations. Resonance periods seem to be dependent on field line length (Equation (1.30)). An enhancement of the *Pi*2 pulsations is taken to be

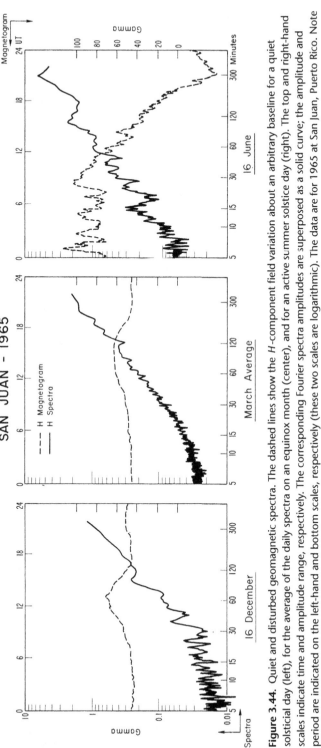

Figure 3.44. Quiet and disturbed geomagnetic spectra. The dashed lines show the *H*-component field variation about an arbitrary baseline for a quiet solsticial day (left), for the average of the daily spectra on an equinox month (center), and for an active summer solstice day (right). The top and right-hand scales indicate time and amplitude range, respectively. The corresponding Fourier spectra amplitudes are superposed as a solid curve; the amplitude and period are indicated on the left-hand and bottom scales, respectively (these two scales are logarithmic). The data are for 1965 at San Juan, Puerto Rico. Note that with increasing activity there is a change in both the amplitude and the relationship of the spectral contributions.

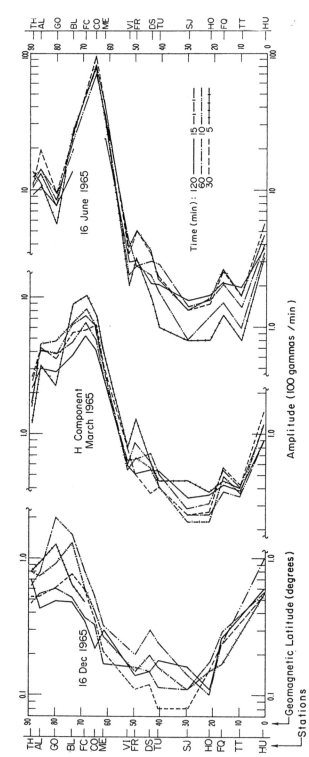

Figure 3.45. Geomagnetic latitude display of the spectral amplitudes at 5-, 10-, 15-, 30-, 60-, and 120-min periods for 16 December (a quiet winter day), average of all days in March, and 16 June (an active summer day) 1965. The logarithmic amplitude scale represents 100 times the amplitude in gammas divided by the period in minutes. Only the *H* component of field is shown. Two-letter codes of the magnetic observatories providing data are indicated at the sides. Note the auroral zone maxima and equatorial region enhancements.

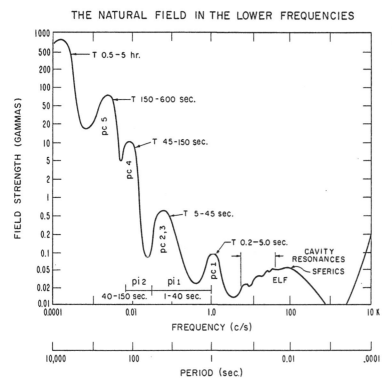

THE NATURAL FIELD IN THE LOWER FREQUENCIES

Figure 3.46. Nomenclature for the natural field fluctuations. The geomagnetic domain is considered to be at frequencies below about 3 cycles/second (Hz). The natural signals above this frequency arise mostly from lightning sources (called "atmospherics" or "spherics"). Amplitudes depicted in the figure are the typical sizes to be expected for moderate activity at midlatitude locations. The "continuous" pulsation, Pc 1–5, nomenclatures are assigned the period ranges indicated near the peak amplitude positions. The "irregular" pulsation, Pi 1 and 2, nomenclatures are identified with their period ranges at the bottom.

an indicator of the current wedge formation at the onset of a substorm (Figure 3.25).

The pulsations are transported to the auroral latitudes along the magnetospheric field lines, and therefore they have equivalent behaviors in conjugate regions (see Figure 1.8). The longer-period oscillations have been interpreted as resonant oscillations within a magnetospheric field shell. Almost all but the $Pc1$ pulsations seem to display auroral luminosity (Figures 3.37 and 3.47) and ionospheric current modification relationships. Their average amplitude change with latitude is similar to that of the longer-period geomagnetic disturbances (Figures 3.38 and 3.45) with maxima in the region of the auroral electrojet and a secondary enhancement at the dip equator. Some geomagnetic pulsations can arise from dynamo currents driven by irregular thermospheric winds. Such pulsations can occur with the substorm onset of joule heating, a pressure wave from a sudden atmospheric nuclear blast (Figure 2.35), or the atmospheric disruption from a volcanic explosion.

The $Pc1$ pulsations are unique in many ways. They have an interesting, rising-frequency structural appearance and travel away from their high-latitude arrival location in guided ionospheric ducts (Figure 3.48).

Figure 3.47. Correspondence of geomagnetic and auroral pulsations. The Byrd, Antarctica, north-viewing photometer (4278 Å at 45° angle) and north–south axis pulsation magnetometer are compared with the magnetic pulsations from a similar magnetometer at Great Whale River, Canada, a conjugate location. Top section shows records from about 1148 to 1201 UT. Bottom section is an expanded time scale from about 1157 to 1200 UT.

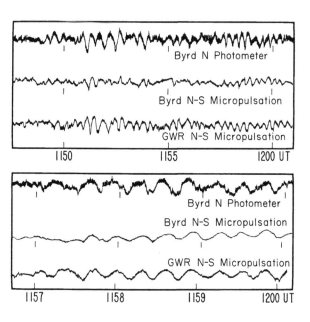

Figure 3.48. Spectral display of simultaneous *Pc*1 micropulsations on 16 February 1969 recorded at Maui, Hawaii; Boulder, Colorado; and Newport, Washington. Darkness indicates the spectral amplitude of signals at the frequency (given to right) for the times (given at the bottom). Horizontal markers indicate where timed amplitude samples were taken, together with other observations, to establish (by triangulation) an apparent signal source location near $L = 4$ and to find propagation velocities of about 2.3×10^3 km/sec.

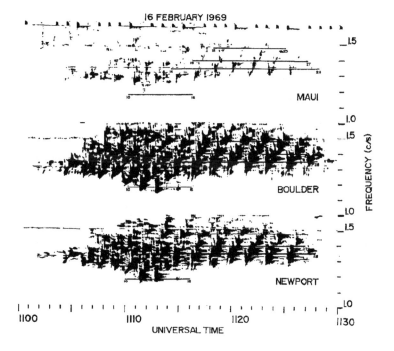

They are not enhanced at the Equator. An amplitude–time display of these pulsations shows a beating structure, appearing on a slow-moving analog chart recording like beads on a string; this gave rise to their original name *pearls*. Conjugate Earth locations show alternation of the *Pc*1 fine

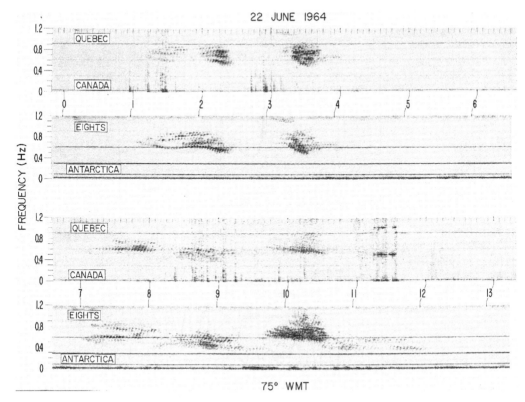

Figure 3.49. Pulsation spectral presentation similar to that in Figure 3.48. Conjugate *Pc*1 pulsation events recorded at Baie St. Paul (Quebec), Canada, and Eights, Antarctica, on 22 June 1964. Local time is given (in hour marks between the paired records) for the Quebec location, 75° WMT. The long horizontal lines and some vertical lines are noise. The short dark horizontal bars near 1130 on Quebec data are the daily calibrations within a noisy period. The comb markers at the top of records are the magnetometer timing marks. It is local nighttime at Eights throughout the period. Note the relative fading of the *Pc*1 event at Quebec as the day develops and the lower ionosphere becomes more opaque to the signals after 0900.

structure and an attenuation of the signal by increased daytime ionization (Figure 3.49). Because the pulsation frequency structure, polarization, location, and occurrence are rather uniquely defined, numerous attempts have been made to unravel magnetospheric source-region characteristics from the hydromagnetic waves observed at the Earth's surface and their propagation in the ionosphere (Campbell and Thornberry, 1972).

3.14 High Frequency Natural Fields

We have moved to ever higher frequencies in our geomagnetic disturbance survey; we might ask "How long does this continue?" The short

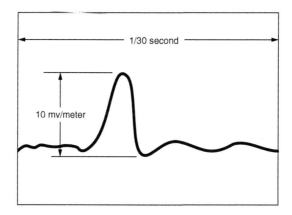

Figure 3.50. ELF "slow tail" signal from lightning in which the filtering has suppressed the major high frequency energy.

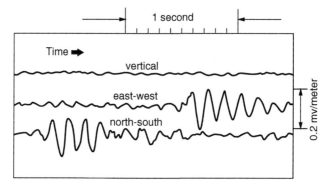

Figure 3.51. The spherical Earth–ionosphere cavity allows the global excitation of the Schumann oscillations initiated by lightning.

wavelength side of the geomagnetic spectra is assumed to end below a few cycles/second. It is not that natural electromagnetic signals are absent past the Pc1s but rather that the fields called spherics (atmospheric signals born of lightning), begin to dominate the scene at the higher frequencies (Figure 3.46). Radiowave researchers have provided the naming in this region: the Extra Low Frequency (ELF) of about 3 to 3,000 c/s. The lowest frequency group of oscillations in this range are the "slow tails" shown in Figure 3.50. These pulses of field travel, with low attenuation, away from the transient lightning stroke in the Earth–ionosphere electromagnetic wave guide. A signal energy maximum appears between 30 c/s to several hundred c/s.

At 100 c/s the ELF wavelengths are of the order of 300 km. At longer wavelengths, nearly the Earth's diameter in size, cavity resonance frequencies from the lightning, called Schumann Resonances, are excited at about 8 c/s (Figure 3.51) along with some of their harmonics (see Satori and Zieger, 1996). Higher than the ELF range other spherics phenomena dominate, such as frequencies in the lightning stroke itself and "whistlers" in which the electromagnetic lightning energy travels along geomagnetic field lines to bounce between hemispheres. All ELF phenomena are sensitive to solar–terrestrial disturbances.

3.15 Geomagnetic Indices

Over the years, three geomagnetic indices have become useful for providing a global picture of degree of disturbance level. All three indices were simply derived from a limited number of observatories and were initiated at a time before modern digital recording and processing techniques were widely available. As scientists began to discover the sources of many geomagnetic phenomena, a more complete understanding of the applicability of the indices unfolded. No one index was found to isolate a particular part of the storm phenomenon, although each index had its own unique representation of the activity. Revision of the indices has traditionally been validly resisted on the grounds of the need for historical continuity.

The principal geomagnetic disturbance index is called the *K index* obtained from the *H* component of field (or the *D* component if it is more disturbed than *H*) and divides activity into ten levels. The letter K was selected by Julius Bartels to stand for the word "kennziffer," meaning the index of the logarithm of a number (see Section A.4). At individual observatories participating in the preparation, an estimated value of the local quiet daily variation is subtracted from the daily records and the range (largest minus smallest) value of geomagnetic activity is determined for each 3-hr UT period. Using a predetermined adjustment to match the statistical occurrences of activity levels among the contributing stations (and some small correction for diurnal and seasonal effects), a 0-to-9 scale value is assigned to the disturbance level.

Starting in 1932, a "planetary" indication of activity, called *Kp*, has been derived using an average of 11 selected observatory K values (Figure 3.52). The resulting index is specified in one-third unit sublevels indicated by the −, 0, or + added symbol. The Kp scale of 28 levels is pseudologarithmic in field amplitude; a convenient conversion back to the "adjusted" field strengths for an equivalent amplitude (gamma) at midlatitudes, called "ap" is shown in Table 3.3. A daily activity index, called "the equivalent daily amplitude," *Ap*, is an arithmetic mean of the eight daily ap values. The monthly, seasonal, or yearly Ap values can be formed from appropriate averages because such indices are linear measures. At low and middle latitudes, storm-time magnetic field variations are generally similar in form (linear regression correlation coefficients of 0.9 or better) over distances often greater than 30° in longitude and more than 20° in latitude. Correlations between

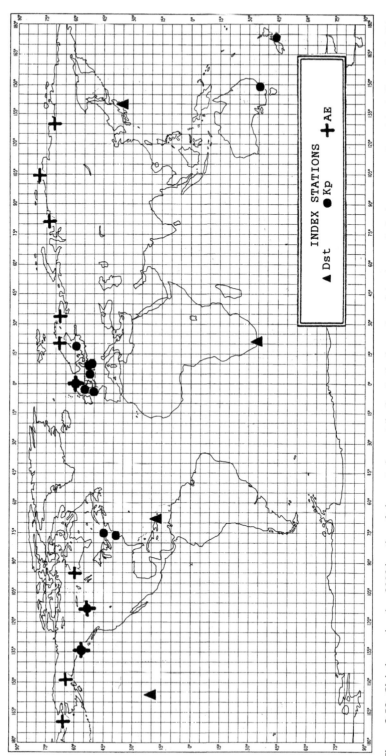

Figure 3.52. Global map showing 2002 location of observatories contributing to the formation of the three principal geomagnetic activity indices, Dst (solid triangle), Kp (solid circle), and AE (plus).

Table 3.3. *Equivalent ap values for given Kp*

Kp index	ap index	Kp index	ap index
0_0	0	5–	39
0+	2	5_0	48
1–	3	5+	56
1_0	4	6–	67
1+	5	6_0	80
2–	6	6+	94
2_0	7	7–	111
2+	9	7_0	132
3–	12	7+	154
3_0	15	8–	179
3+	18	8_0	207
4–	22	8+	236
4_0	27	9–	300
4+	32	9_0	400

ap indices and the local magnetic-storm fields in these regions are typically poor, mainly because the ap and Kp are derived from the range of disturbance within each hour, thereby ignoring the longer period field changes.

Figure 3.53 shows the *musical Kp index diagram* employed to illustrate the index changes through a month laid out in 27-day solar rotations indicated at the left of each row. This rotation number (differing from the Carrington rotation that is not exactly 27 days) was started by Bartels for an 1831 date such that rotation number 2000 began on 12 November 1979.

There are two difficulties with Kp as a global magnetic activity index: (1) many of the Kp stations are at subauroral latitudes and the locations favor the Northern Hemisphere and European continent (Figure 3.52); (2) by using a three-hour-range selection, a frequency dependence is introduced. Kp discriminates against slowly changing, major departures of the field (e.g., the storm recovery phase) and, because magnetic field disturbance amplitudes increase with increasing period (Figure 3.44), the high-frequency portion of the disturbance spectra is essentially ignored; Kp favors the irregular variations near 3 hrs in period. Nevertheless, the index has considerable value in the preliminary selection of disturbed and quiet days (Chapter 2) and is used for the selection of days in main-field modeling. Automated procedures for deriving Kp exist (c.f. Takashi et al., 2001).

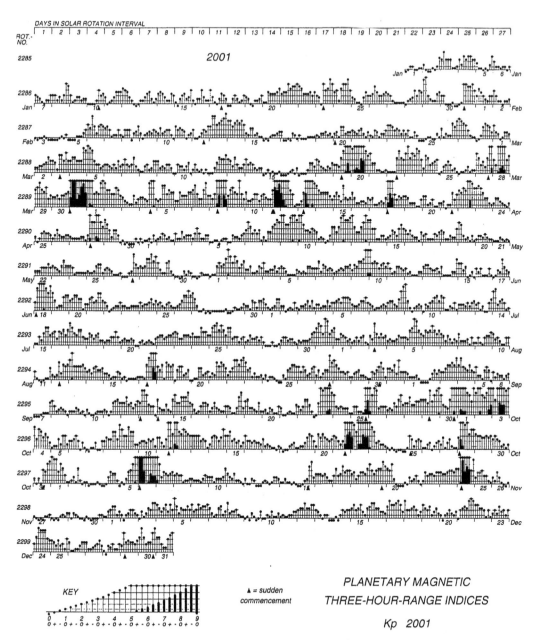

Figure 3.53. Diagram of Kp index values for 2001 prepared by the University of Göttingen. Scale symbols are given at bottom. Each row of records represents a 27-day interval, close to the solar rotation period, that is identified with a count started by Bartels (numbers shown at left of each row). Recurrent periods of quiet or activity are emphasized by the arrangement. Sudden commencements of geomagnetic storms are marked with small triangles. Figure provided by H.E. Coffey, World Data Center, Boulder.

Table 3.4. *Occurrence of Ap index levels*

===

Days per year with high activity $\geq$ Ap

$\quad$ N(high) $= 4.14 \times 10^4\ \text{Ap}^{-2.25}$

$\quad$ ($\pm 6\%$ for Ap $= 15$ to 150)

Days per year with low activity $\leq$ Ap

$\quad$ N(low) $= -52.3 + 267.4 \log(\text{Ap})$

$\quad$ ($\pm 4\%$ for Ap $= 2$ to 20)

===

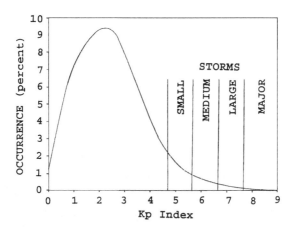

Figure 3.54. A smoothed curve indicating the expected occurrence distribution of Kp indices with peak occurrence near Kp $= 2+$; values at this level and lower are usually considered to be quiet conditions. The geomagnetic storm sizes and assigned ranges are indicated.

$\quad$ Figure 3.54 gives the percent occurrence of Kp indices and the storm designation levels. Table 3.4 shows mathematical representations of the Ap geomagnetic activity levels obtained from a 40-year sample.

$\quad$ A geomagnetic activity enhancement at the equinoxes (Figure 3.55) has recently been associated with the favored alignment of the magneto-spheric boundary with respect to the solar wind interaction region at these times. Earlier explanations considered the favored equinoctial alignment of the Earth with respect to a radial solar disturbance particle flow.

$\quad$ Figure 3.56 illustrates the variation of Ap with sunspot number through the years. The magnetic activity seems to lag behind the sunspot maxima by one to three years. This delay may be due to the tendency for high-velocity solar wind streams to occur during the sunspot declining phase or correspond to the latitude change of the Sun's active regions during the cycle. Thompson (1993) has related the number of occurrences of high Ap index in the current sunspot cycle (N_C is the count of days with Ap ≥ 25 from the preceding to following sunspot minima) and the maximum sunspot number in the current cycle (R_C) to a prediction

Figure 3.55. Comparison of 30-day running means of all Ap indices (dots) and the absolute solar declination, δ, (solid curve) for a 30-year sample beginning in 1932. Figure adapted from Rosen (1966).

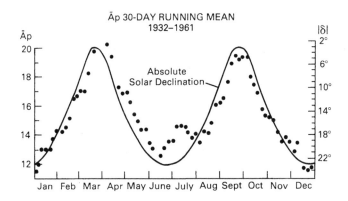

Figure 3.56. Comparison of annual number of disturbed days with magnetic index Ap ≥ 40 and the annual average sunspot number (SSN) from 1930 through 2001. The Carrington sunspot cycle number is indicated above the corresponding maximum. Figure provided by E. Erwin of NGDC/NOAA.

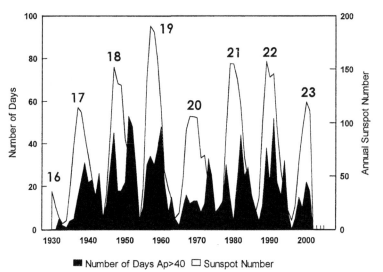

of the maximum sunspot number in the next cycle (R_N):

$$R_N = \frac{(N_C + 47.7)}{2.04} - R_C \pm 17 \qquad (3.15)$$

Variations of the Kp index are the Kn, Ks, and Km (and corresponding an, as, and am) field values. The derivation of these indices represents an attempt to alleviate some of the Kp location problems by using five Northern (n) Hemisphere and four Southern (s) Hemisphere observatories representing separated longitude sectors and adjusted to a uniform latitude location. Hemispherical (Kn and Ks) indices are formed and a mean value (Km) is computed from them. With just two stations (Hartland, UK, and Canberra, Australia) a simplified aa index is derived.

The hourly (and sometimes a 2.5-minute) *AE index* is the result of an attempt to indicate the aurora-related geomagnetic activity around the auroral oval of the Northern Hemisphere. The index values are thought

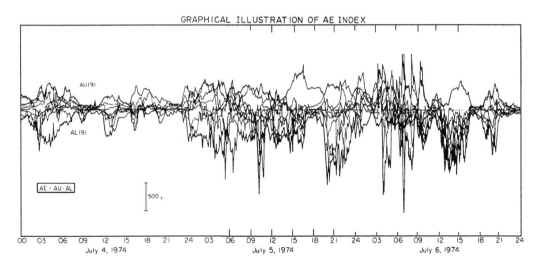

GRAPHICAL ILLUSTRATION OF AE INDEX

Figure 3.57. Superposition of nine auroral zone geomagnetic observatory *H*-component records from which the hourly range between the upper, AU, and lower, AL, becomes the auroral electrojet, AE, index. Figure from World Data Center files.

to measure the two current systems that represent the directly driven and loading-unloading process described above. AE is formed from the over-plot envelope of *H*-component field variations about the quiet level of observatories (5 to 11 in number) selected to have locations near the auroral zone maximum and widely distributed in longitude (Figure 3.52). The largest values in the sampling interval become the AU (U for "upper") and the smallest values become the AL (L for "lower"). The difference between AU and AL becomes the *auroral electrojet index*, AE (Figure 3.57).

The AE index has been prepared irregularly over the years because of difficulties in obtaining contributing observatory data and in finding organizations to handle the computational duties. Despite the obvious problem that the latitude of maximum auroral activity may move equatorward or polarward from the fixed AE observatory location, causing a distance rather than intensity change in the auroral electrojet, the index has found great use in relating the magnetic activity to magnetospheric observations. As might be expected, because of the subauroral latitude preference of the Kp observatories, there is a linear correlation (coefficient 0.89) between the AE and Ap indices (averaged for similar periods of activity):

$$AE \approx 1.94 + 11.2 \text{ Ap} \tag{3.16}$$

Figure 3.58 shows how the distribution of AE index values changes between years of major quiet (1965) and extremely active (1958) solar disturbance levels.

The exact formulation of an hourly *Dst index* (Dst means storm-time disturbance) was firmly settled in 1964. At that time the view of

Figure 3.58. Comparison of the number of hourly values of indices AE (top), Dst+ (bottom left), and Dst– (bottom right) for an extremely active year (1958) and the most quiet year (1965). The Dst are separated into positive and negative groups because a single magnetospheric compression process seems to dominate the + values; whereas many magnetospheric and ionospheric processes contribute to most negative values. The number of hourly values in 1-, 2-, or 10-gamma bins are indicated for increasing index field levels (gammas).

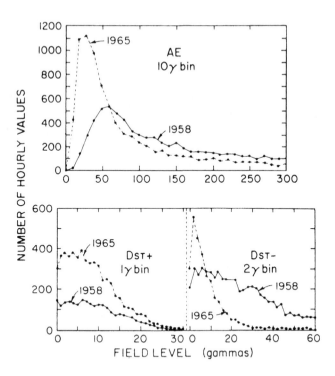

a geomagnetic storm was simply this: following a compression of the magnetosphere, with the arrival of the solar wind disturbance, there was a rapid growth and slow decay of a Saturnlike ring of westward current that formed about the Earth. It was believed that the current encircling the Earth was part of the Van Allen radiation belt discovered just before that time. Such a westward current would cause an H-component field depression worldwide. For that reason, the index has carried the alternative name *ring-current index*. Originally, a large number of observatories were employed in the preparation; now, only four remain (Figure 3.52). The Dst observatories were selected to be located at lower latitudes than the subauroral zone and away from the field-enhancement region of the dip equator.

The basic idea for the Dst is to indicate the global part of a geomagnetic disturbance as an average of field strengths that remains after the typical quiet-day variation features and a baseline of main-field level are removed from low-latitude (nonequatorial) station data. The assumption of a ring source-current (so the resulting field should be parallel to the geomagnetic axis) led the index designers to use only the H component of field and to adjust for different station geomagnetic latitudes through division of the averaged values by the cosine of the average of station geomagnetic latitudes. The (supposed) adjustment to equatorial latitude amplitudes leads to the use of the name "Equatorial Dst" and

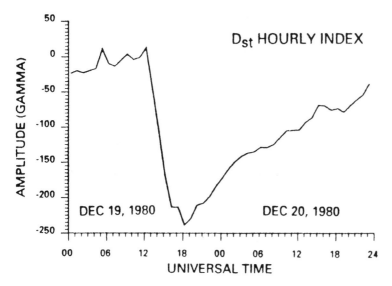

Figure 3.59. Dst hourly index for 19–20 December 1980, derived from values of the H-component field shown in Figure 3.23.

does not imply that the index is derived from equatorial records. Figure 3.59 shows the resulting index obtained from the station records in Figure 3.23. For geomagnetic storms such as this one, the positive values of the index initially occur as a result of the magnetospheric compression and disappear with the increasing arrival of overriding negative H-field contributions. A form of the Dst index derived from the H-component field variation at several widely-spaced middle or low latitude observatories is usually highly correlated (coefficient greater than 0.95) with the formally published index value.

A primary problem for the Dst index derivation has been the limited distribution of contributing observatories. The four locations are unevenly spaced in longitude about the Earth, with a large gap in the central-Asia sector; three of the observatories are located in the Northern Hemisphere. A secondary problem is the removal of the quiet daily variations, Sq, from a projection of the quietest five days of the month. The ionospheric dynamo currents (dependent upon thermospheric winds and ionospheric conductivity; Sections 3.10 and 3.11) are known to shift both in amplitude and location during active times. Therefore at storm time, Sq variations are not adequately removed by using such quiet-day field levels. Effects of induced currents on Dst are also a problem (Haekkinen et al., 2002). Figure 3.58 shows the changing distribution of Dst index values (separated for + and − value groups) for extremely quiet (1965) and active (1958) years. Attempts to find a linear relationship between the Dst indices and Ap show an absence of significant correlation (coefficient = 0.017) for Dst positive values and a moderate correlation (coefficient = 0.69) for Dst negative values.

Satellite measurements in the magnetosphere in recent years have failed to find any process that paralleled the storm-time growth- and decay-phase shape shown by the Dst index; rather, many partial ring currents of relatively short duration were discovered (Figure 3.29). Determinations within the partial ring current region showed insufficient field to generate the low-latitude surface observations of Dst. Further studies indicated that the field-aligned currents were contributing significantly to the low-latitude surface measurements of magnetic disturbance. Satellite drag (an indicator of thermospheric winds) is better correlated with Dst than with any other geomagnetic disturbance index. A seasonal activity change (Figure 3.55) appears in Dst (Cliver et al., 2001). At low latitudes, field measurements on the nightside of the Earth track nighttime cross-tail currents to show location changes responding to activity level and season (Figure 3.27, Campbell, 1973 and Ohtani et al., 2001). Also, there is clear evidence that the substorm auroral electrojet currents can close, through the middle-latitude ionosphere (affecting the Dst observatories), all the way to the dip equator and appear as an equatorial enhancement (Figure 3.39). The Dst does not monitor simple ring-current behavior, but rather it represents an ensemble of magnetospheric and ionospheric fields detected at middle and low geomagnetic latitudes on the Earth.

It was recently noticed that the negative values of the typical storm-time Dst index followed a lognormal form (see Section A.9, Campbell, 1996, and Figure 3.60). The correspondence between the Dst storm and lognormal shapes may be accidental because the Dst is a time series, not a statistical count of occurrences in each amplitude bin. However, the Dst measurements over the time of the storm are values obtained from a summation of simultaneous sources arising at a number of locations, each of which represents both the summation of elementary processes and the sequential series of processes (Aitchison and Brown, 1957).

Program **DSTDEMO** in Section C.4 illustrates the lognormal fit and Dst prediction.

3.16 Solar–Terrestrial Activity Summary

The Sun is the source of geomagnetic field disturbances. Active regions of the solar surface, such as flares and coronal holes, are responsible for coronal mass ejections (CME) of energetic particles. Although relatively random in occurrence, CME events are organized by an 11-year solar cycle and a 27-day surface rotation period. The solar blasts of CME cause hydromagnetic shocks that define the solar-wind disturbed features. Upon intercepting the Earth, the solar wind blows the Earth's

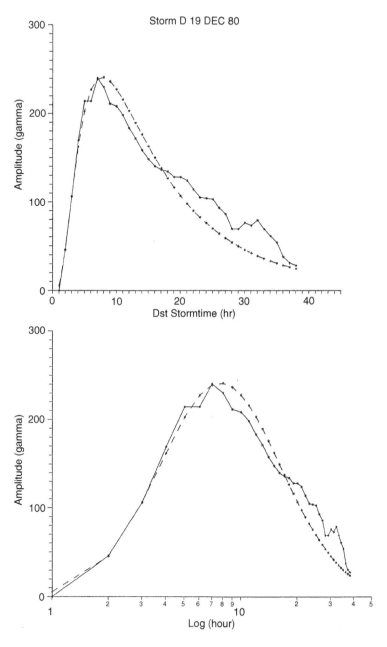

Figure 3.60. Geomagnetic storm of 19 December 1980. (Top) Amplitude of Dst–hourly values (connected with solid line segments) from storm onset time. (Bottom) The Dst– amplitudes are plotted with a logarithm of time scale. The best-fitting normal curve to the Dst distribution is displayed as a dashed curve. The appearance of this normal distribution in the linear time scale is shown as a dashed lognormal curve at top.

dipole field into a cometlike structure, compressing the dayside geomagnetic field from 25 to as little as 6 R_e at its stand-off location and extending the downwind tail far past the Moon's orbit.

The interplanetary magnetic field (IMF), whose component directions are defined on leaving the Sun, travels along with the solar wind, frozen-in with the solar-ejected particles. When the IMF arriving at the

Sun-side magnetosphere is southward, field-line interconnection occurs and a major geomagnetic storm ensues. Particles collect in a radiation belt close to the Earth forming a South-Atlantic/South-American Anomaly. The east–west IMF at the magnetospheric boundary defines polar-region current pattern shifts in direction called sector effects. A geomagnetic storm is a period of surface field disturbance as noted by averaged low-latitude and midlatitude observatories. The program **DST-DEMO**, in Appendix C, illustrates the general behavior of geomagnetic storms and the predictability of the storm recovery from the storm main phase.

Individual disturbance-connected magnetospheric and ionospheric processes that occur during a geomagnetic storm period are called substorms. Substorms involve severe changes in the magnetotail, partial ring currents, and field-aligned currents that precipitate particles (mainly electrons and protons) into the high-latitude auroral region. Auroras occur as this precipitation excites the oxygen and nitrogen of the upper atmosphere to glow at the prescribed wavelengths that are allowed by their atomic and molecular structure. The ionospheric conductivity is enhanced in the aurorally active region. A major portion of the substorm energy is dissipated by global thermospheric heating.

Geomagnetic pulsations are classified into oscillatory period groups by their amplitude-time trace appearance as continuous (narrow frequency content) *Pc* or irregular (broad frequency content) *Pi*. The pulsations are magnetospheric disturbance phenomena whose characteristics, when fully explained, may provide surface-monitored diagnostics of our space environment.

Fields of the magnetospheric and ionospheric currents at selected observatories are grouped to form geomagnetic activity indices. Three indices commonly used are Kp, AE, and Dst; all differ in their response to substorm currents because of their selected-station locations and the prescribed data processing methods.

3.17 Exercises

1. For a disturbance (shock front plasma flow) source in space traveling at 300 to 400 km/sec, what is the Alfven mach number when the Alfven wave velocity (transverse hydromagnetic velocity) is 30 km/sec?

2. A satellite magnetometer indicates passage through the magnetopause boundary at $r_M = 6\ R_e$ when the satellite is located at a latitude of 30° (Figure 3.20). What is the magnetospheric standoff (R_{MSO}) location in Earth radii at this time?

3. Compute the field increase (nT) resulting at the Earth's surface geomagnetic latitude (program **GMCORD** at Appendix C.1) of your home city

when the above described satellite measures the compression by the solar wind.

4. Find the energy (joules) required to confine the Earth's magnetic dipole field for the above compression.

5. Go to the website listed in the Figure 3.32 caption to view the auroral displays photographed by Dick Hutchinson. For two of these auroras, write a complete description that would allow a non-scientist, who has never seen an aurora, to understand what an aurora is like.

6. Select two Ap magnetic activity index levels: (a) in the interval between 50 and 70, and (b) in interval between 2 and 10. Determine the number of days in a year that one should except (on average) for Ap ≥ (a) and Ap ≤ (b) to occur (with the estimated error given).

7. Determine the expected values of the AE indices on your highest selected Ap level of Exercise 6. Then find the most recent time when those two levels of Ap occurred, by using website http://www.ngdc.noaa.gov and selecting "Geomagnetism" from the list on the left. Next select "Geomagnetic Data" at the bottom right. Then select "Space Physics Data" on the right and choose the Kp and Ap indices. (Read the description before selecting the highlighted FTP at the bottom.) If you select "kp-ap.fmt" near the list bottom to view the format to be used in the data files, you will see that the data line starts with yymmdd for the year, month, and day in the first 6 positions and the Ap values are at positions 56 to 58. Go back to the previous screen and select a year for your data reading.

8. Go to website http://www.ngdc.noaa.gov and select "Space Physics Data", as in the previous exercise. Then select "Disturbance Storm-Time (Dst) Indices". In the description text select highlighted FTP and select the 2000 data. Display the values for January 2000 and copy the hourly values for Dst index from the geomagnetic storm that occurred between 22 and 25 January. The 24 hourly UT values of Dst appear in 24 groups of four spaces each, following the date and lead 000; ignore the last group of four numbers.

9. Select program **DSTDEMO** (in Appendix C) and enter the hourly values of the Dst event of Exercise 8 above starting with 15 UT on 22 January 2000 and ending at 24 UT on 25 January 2000. Determine how well the values match a lognormal distribution and if the main phase predicts the recovery phase.

10. Go to website http://www.sec.noaa.gov and select "Space Weather Now" from the list on the left. Report the information on the X-ray flux, solar wind particles, auroral maps, present position in the solar cycle, and the prediction of today's activity forecast.

11. Suppose that SEC reports that there was an impulsive X-ray SEP event on the Sun (Table 3.2). What will be the likely composition of the ejected particles? Assuming a light speed of 3×10^5 km/sec and a mean solar wind speed of 300 to 500 km/sec, how much warning time will you have before the arrival of a possible geomagnetic storm?

12. Your nation's space agency, in conjunction with the national communications industry, has obtained funding for a satellite to be sent into geostationary orbit. There is space on the satellite for some scientific instrumentation. You are asked to attend a scientific advisory panel to propose the placement of a magnetometer on the satellite. What will such magnetic field measurements show and what will be their scientific value?

Chapter 4
Measurement methods

4.1 Introduction

Man has no obvious sensation of the presence or change of the Earth's magnetic field such as he does for the sensation of rain, wind, or earthquakes. He must rely upon the field's interaction in other physical processes to produce measurable effects. In this chapter we will look at some methods of providing such geomagnetic information.

From a physicist's viewpoint the geomagnetic field we wish to measure has some interesting singular characteristics. It is ever-present; we must take deliberate action to create any required field-free environments. Because of the great spatial extent of the field with respect to available sensor dimensions, only single-point measurements are typically obtained. The natural field is constantly changing and cannot be stopped at will by the experimenter. A conglomeration of Earth-core, magnetospheric, ionospheric, and induced currents all contribute to the simple measurement of a geomagnetic field magnitude and direction at each instant of time; occasionally, special frequency-analysis techniques allow us to identify some of these contributing sources.

The Earth's field changes are not easily stoppered in a bottle and brought to the laboratory for testing like a paleomagnetic rock sample. Those who want the measurement usually must move to a sampling spot that they have selected with care in order to minimize unwanted "noise" and to indicate special upper-atmosphere or deep-Earth characteristics. Everywhere at the Earth's surface, the "steady" field (i.e., slowly varying with respect to the spectral components at active times) is quite strong compared to the relatively infinitesimal fields of rapid (micro) pulsations.

Sensors are required to either respond to the extremely broad amplitude range, from less than 10^{-3} to greater than 10^4 nanoteslas, or to confine observations to the characteristic frequencies of a particular class of geomagnetic phenomena.

The limit has not yet been reached in the creation of more sensitive instruments to measure the geomagnetic field. Each new step into finer geomagnetic field resolution brings with it new, rewarding details of our geophysical environment. Fortunately, there are a variety of phenomena that respond to magnetic fields in such differing ways that a large variety of sensors have been produced, each with its own distinct advantages. In unfolding this chapter on instrumentation I will present both the simply understood older (but still in use worldwide) systems as well as some more complicated modern ones. Some information about observatories and satellites will be included.

4.2 Bar Magnet Compass

The lodestone magnetic compass was the earliest field-sensing device (Figure 1.1). To understand what makes the compass work, we must first look at the crystal properties of *ferromagnetic materials* (i.e., a magnetic matter that contains mainly iron). All atoms and molecules contain electrons that are responsible for the properties of chemical bonding. In the simplified model of atomic structure, an electron possesses an intrinsic angular momentum, or *spin*, whose associated magnetic field is dipolar in form (Figure 4.1a). Nearly all these electrons are paired off, and opposing spins leave no residual fields (Figure 4.1b). However, in special cases (e.g., ferromagnetic materials) this "cancellation" of paired spins doesn't occur. The angular momentum of the residual, "noncanceled," spinning electrons provides a magnetic moment whose direction is as one would expect for a rotating charged sphere. In addition to the effect of the electron spin, there can be some field effect from the orbital motion

Figure 4.1. Classical picture of (a) field direction for the spinning electron and (b) paired electron spins and fields.

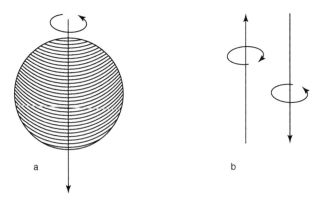

a   b

of the atom itself. Crystals are relatively rigid, ordered arrangements of atoms and molecules. In the crystal structure, atomic orbital motion is restricted and contributes little to magnetism.

There is an *exchange tendency* when a parallel axis alignment exists for spins of adjacent ferromagnetic atoms, so that *domains* (regions of only one direction of magnetization) of field-aligned atoms are established. An atom is about 1 to 2×10^{-10} m in size. Exceedingly small (about 10^{-7} m) crystal grains can have a *single domain* (SD), but larger grains (up to 10^{-3} m) usually have many domains (multidomain), separated by domain walls with some special wall physical properties that help control domain magnetic alignment. Natural minimum energy constraints for the crystal structure usually align adjacent domains in opposing directions under normal conditions.

For most ferromagnetic crystals there is a preferred (easy alignment) axis for magnetization (*magnetic anisotropy*), and nearly all the crystal domains are aligned in that direction. Impurities and imperfections divide these crystals within grains of rock so that our picture of an unmagnetized fragment of a rock sample may be visualized as in Figure 4.2a. In the figure the arrows indicate crystal domains of aligned spins, the light straight lines indicate the domain boundaries (walls) within the crystals, and the dark curved lines indicate grain boundaries. The study of the physics of magnetic energy, domains, and boundaries with respect to rock grain characteristics is called the *domain theory* of paleomagnetism (see Section 1.9). In recent years, with the explosion of research on

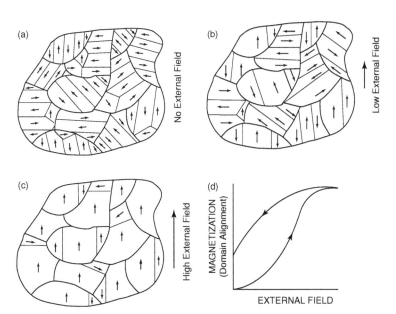

Figure 4.2. Illustration for visualizing (a) an unmagnetized rock fragment containing crystals (represented within heavy curved lines) with regions dividing domains of similar field alignment (within light straight lines), (b) realignment of domains when a small external field (arrow to right of fragment) is applied, and (c) high external field (arrow to right of fragment) saturation of rock fragment magnetization. Graph (d) shows the hysteresis curve for the fragment magnetization in a changing (arrow direction) external field.

data-recording devices, the science of "mesoscopic magnetization" has refined and revised some of the older domain-theory assumptions. The collection of articles in the April 1995 issue of *Physics Today* provides a good review of this subject.

At first, as a gradually increasing external field is applied to the rock fragment, those crystal domains with an easy-alignment axis that is close to (or 180° opposite) the direction of the external field will shift directions parallel to that field in a jerklike motion. Because the more easily turned domains move first, this part of the alignment is essentially reversible. However, as the external field increases even more, those domain walls, which have a more difficult time in alignment, are turned (Figure 4.2b) also in jerks. With each domain alignment change, domain boundaries change, the crystal itself shifts dimension slightly (*magnetostriction*), and tiny sound waves (*Barkhausen noise*) are created by the changing domain-wall energy.

With even greater external field, the further alignment of domains proceeds more and more slowly until all but the most rigid domains have been turned into the external field direction and a saturated magnetized state is reached (Figure 4.2c). The up-arrow trace of Figure 4.2d shows how the magnetization changes with increasing external field. As the external field is gradually reduced (down arrow in Figure 4.2d), the domain alignment decreases; but because of the energy already lost by the movements, the process is not quite reversible. At zero field, the domain wall movement, magnetostriction, Barkhausen effect, etc., do not allow the sample to return to its original state; considerable magnetization remains. We then say that the sample has *hysteresis* and is magnetized. Figure 4.2d represents the magnetization of the samples in an external field; it is called the *hysteresis curve*. The domain alignment magnetization is destroyed at a high temperature, characteristic of the crystals, called the *Curie point* (Section 1.9).

In the Earth's crust, most ferromagnetic materials are a molecular compound of iron, titanium, and oxygen. Magnetite, with the chemical formula Fe_3O_4, is the principal carrier of magnetism. Man-made substances exhibit magnetization; for instance, steel is ferromagnetic. To manufacture permanently magnetized materials, special techniques are used such that the new compound's domain boundaries are exceedingly difficult to move, and the crystals have very long grains aligned in the preferred magnetization directions. Cooled to a temperature below the Curie temperature in a strong magnetic field, a high magnetization is fixed into the compound. One such material contains 51% iron, 24% cobalt, 14% nickel, 8% aluminum, and 3% copper. Another manufactured "ferrite" substance has about 5 times more magnetic energy density (kilojoules per cubic meter) than steel. A recently designed permanent magnet called "neodynium-iron-boron" is said to contain about 5 times more magnetic energy density than ferrite. Such materials are used in industrial manufacturing when restricted conditions demand extremely high magnetization.

If a magnetized matter is elongated along the magnetic moment axis, regions are found near the ends where the magnetic effect is most intense. Such regions, called *magnetic poles*, occur only in pairs. The strength of a permanent magnet is defined such that a *unit pole* strength, p, will exert a force, F, upon an equal pole at a specified distance, d, in a vacuum. By standardizing the size of the force and the distance, a definition of a unit pole is obtained from

$$F = \frac{p^2}{\mu_0 d^2},$$
(4.1)

where μ_0 is a constant $4\pi \times 10^{-7}$ of the medium called the *permeability of free space*. If s is the distance between the poles, then the product of the pole strength and distance is called the *magnetic moment of the magnet*, M (see Equation (1.82)).

The region about a magnet where its influence may be detected is called a *magnetic field, H*. The intensity of this field is defined by the force that would be exerted upon a unit pole placed at the point of measurement.

$$B = \mu_0 H = \frac{\mu_0 F}{p}.$$
(4.2)

Combining Equations (4.1) and (4.2), the definition of field becomes

$$B = \frac{p}{d^2}.$$
(4.3)

This equation describes how the field decreases with the square of the distance from the unit pole. By convention, the magnetic *north pole* of a freely suspended magnet is that pole that points to the north geographic region of the Earth. The direction of a magnetic field corresponds to the direction of the force acting upon a north pole of a magnet. If the magnet is suspended to allow freedom of motion in the vertical direction, we have a magnetic dip meter for measuring inclination (see Equation (1.29)).

If a horizontal dipole magnet, freely suspended at its midpoint between the poles, is moved slightly from its favored position, it will oscillate in simple harmonic motion about the north–south direction before slowly coming to rest because of friction. The period, T, of this free oscillation is determined by the strength of the field and a constant of the magnet. This constant is the ratio of the magnet's mechanical moment of inertia, I, and the magnet's magnetic dipole moment, M. The moment of inertia is found from the distribution of mass of the magnet about its pivot point. The field, B, is found from

$$B = \frac{\mu_0 I}{M}\left(\frac{2\pi}{T}\right)^2.$$ (4.4)

To avoid a troublesome separate determination of I, the constant $(\mu_0 I/M)$, is found by timing the oscillatory period of the magnet in a field of known size. For many years before the advent of proton magnetometers, the oscillation of a calibrated magnet was used to find the Earth's field magnitude.

As late as the nineteenth century, researchers used long compass needles and careful observation of the variation in the needle's pointing angle (after oscillations are damped out) to verify the European simultaneity of geomagnetic storms. In 1840, Gauss improved the measurement technique by affixing a mirror to the compass magnet suspension. Tracking a light source bounced off the mirror, he thus allowed the angular motion of the compass to be detected at a distance with great accuracy. With the development of photographic paper in the early twentieth century the variometer was developed.

4.3 Classical Variometer

Figure 4.3. Design of variometer for magnetic observatories. Motion of mirror attached to magnet deflects a trace on to photosensitive paper that rotates in 24 hours. The fixed mirror is adjusted to show a baseline position about which the variations are traced. Calibrations of the baseline and scale are required periodically.

The classical *variometer* is similar to a recording compass and dip meter in many ways. The magnet (suspended in such a way as to resolve H, D, or Z field components) is adjusted to an appropriate zero position in the field to be measured by a number of techniques: by twisting the suspending fiber, by a secondary field of an adjustable auxiliary magnet, or by balance weights in the gravitational field (for vertical plane measurements). Usually, a light reflected from a mirror attached to the magnet registers the variation on photographic paper or film moving at a prescribed speed (Figure 4.3). In the middle of the twentieth century

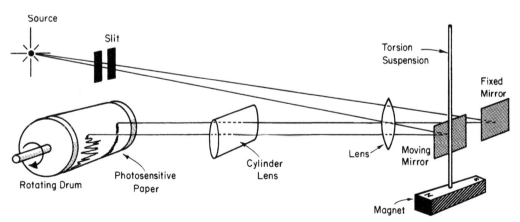

at some observatories, analog charts had been replaced with electronic output. This is done by using a wedge mask in front of a line of light reflected from the mirror; as the mirror moves, more of the light line is sensed by a photometer with an electric output proportional to the amount of light received.

Numerous fine corrections can be made to the operational magnetometer to compensate for the natural resonant oscillation of the suspension as well as the temperature and humidity effects upon the mechanical elements of the system. Special mounting is required to avoid seismic vibration of the elements. Accurate measurements to ±0.1 gamma (and 0.1 minute in declination) are claimed for sensitive instruments, but most are in the 1 to 5 gamma range. The problems associated with variometers are those mentioned above plus the instrument's sensitivity to physical vibration and the elastic fatigue of the suspension fiber. Proven reliability, ruggedness, low cost, and simplicity of operation had made the variometers an instrument of choice for early standard observatories. The instrument is still in use throughout the world.

4.4 Astatic Magnetometer

Paleomagnetic determinations require the measurement of extremely weak fields from remanent magnetic materials (Section 1.9) in rock samples. An ingenious adaptation of the variometer is used to determine these fields in the presence of the strong Earth's field. Two oppositely directed magnets, of identical moment and rigidly connected, are suspended from a torsion fiber like the variometer suspension (Figure 4.4).

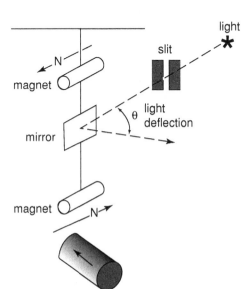

Figure 4.4. Astatic magnetometer for measurement of rock magnetization. Upper and lower magnets move together with Earth's field changes but are deflected differently by the rock sample near the lower magnet.

For the short distance between the magnets, there is little difference in the Earth's field vector. That is, the Earth-field gradient, or change of field with distance, is negligible over the separation between the two magnets.

A rock sample, placed below the lower magnet, will cause a mirror to turn and measure the rock's magnetization vector because the lower magnet is substantially closer to the rock than to the upper magnet. The field gradient of the rock is significant over the scale of distance between the magnets. Changes in the Earth's field during the measurement will act equally upon both suspended, oppositely directed magnets and will not turn the mirror. Sensitivities to approximately 1 nT/cm^3 in rock samples are obtained. One problem is that the system is quite sensitive to physical vibrations. Also, there is usually a long recovery time for the variometer alignment following a sample measurement.

4.5 Earth-Current Probe

Every change in magnetic field, dH, has associated with it a change in the electric field, dE. Although this law is certainly well remembered at radiowave frequencies, students occasionally overlook the fact that the relationship between dH/dt and dE/dt (where dt is the change in time; see Section A.7) also exists throughout the longer wavelengths of geomagnetic interest. In the conducting surface of the Earth, the electric field counterpart of geomagnetic variation may be readily measured using the Earth itself as an antenna. Table 4.1 presents a list of some conductivities of Earth surface materials.

Relative to the values in Table 4.1, the air we breathe is not at all a conducting substance. Therefore the Earth's surface represents a discontinuity (a change from nonconducting to conducting) of the medium in which electromagnetic fields exist. It can be shown, as a direct consequence of Maxwell's laws, that the transition of the tangential components of the electric field intensity vector, **E**, and the magnetic field intensity vector, **H**, (for **H** = **B**/μ) are continuous through the Earth–air boundary. As a result, we can appropriately measure the natural electric fields on the ground beneath us.

To obtain an idea of the relative magnitudes of the electric fields that are involved when one makes Earth-surface measurements at the low frequencies of interest in geomagnetism, let us consider the simple example of a plane electromagnetic wave traveling in the vertical (z) direction. Then Maxwell's equations lead to field relationships, orthogonal to the direction of propagation, which have the form

$$\frac{E_x}{H_y} = \sqrt{\frac{\mu 2\pi f}{\sigma}}, \tag{4.5}$$

Table 4.1. *Typical conductivity values*

	σ (Siemens/meter)
Ocean water	1 to 10
Wet Earth surface	10^{-2} to 10^{-3}
Fresh water streams and lakes	10^{-2} to 10^{-3}
Distilled water	10^{-4}
Dry Earth surface	10^{-4} to 10^{-5}

where E_x is the electric field intensity parallel to the Earth's surface in the positive x direction, H_y is the magnetic field intensity parallel to the surface in the positive y direction, f is the frequency in cycles per second, and σ is the conductivity in Siemens/meter. The permeability, μ, is a constant ($4\pi \times 10^{-7}$ Henrys/meter) representing the magnetic properties of free space (essentially that of air). Taking some typical values of $B_y = \mu H_y$ to be about 3 gamma (3×10^{-9} Tesla), for the case of a dry Earth surface ($\sigma = 4 \times 10^{-5}$) and for long waves with periods of about 30 seconds ($f = 1/30$), we obtain approximately $E_x = 194$ mV/km. The measurement would be about 1.9 millivolts over 10 meters, an easily detected size. In practical application the local geological features, Earth conductivity–depth profile, and topography modify the direction and size of E_x. See Section 5.8.3 for further information on the magneto-telluric method of profiling the Earth's crustal conducting materials.

In the usual method of Earth-current (telluric current) measurement, two contact probes are set into the Earth that define the azimuth of field detection; their separation is adjusted to the recorder sensitivity. In most observation conditions there is some noise generated by differing contact potentials between the probes and the soil. Some researchers use probes with an intermediate electrolytic salt solution such as copper sulfate to assure proper contact with the ground. However, these "wet" electrodes can be sensitive to temperature and diffusion rate.

Many observers of rapid field variations are unconcerned with the very slow contact-potential drifts in the baseline voltage level. They use simple lead-strip electrodes buried about a meter below the Earth's surface. The contact potential, which varies with the nature of the soil and surface area of the probe, is then biased out of the readings by regular adjustment of a bucking voltage through the probe circuit (Figure 4.5). Either current or potentiometric amplifiers are used. The latter have the advantage that because no current is drawn, the measurements are less dependent on ground circuit resistance. Such a system often has the disadvantage of poor response to higher frequencies.

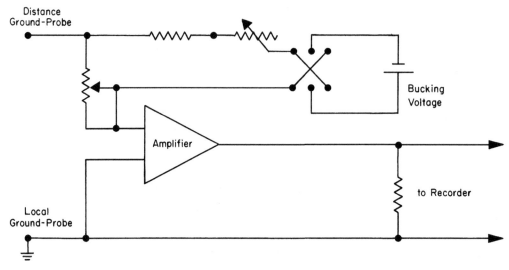

Figure 4.5. Earth (telluric) current probe system measures voltage variations between two grounded locations. Adjustable bucking voltage is provided to compensate for a steady contact potential difference at the probes.

Earth-current probes are simple detectors of geomagnetic disturbances. At typical sensitivities, the system is about equivalent to a magnetometer capable of detecting 10^{-4} to 10^{-5} gammas at 1 cycle/second. The two major problems with the electric-probe system are (a) the baseline drift due to contact and polarization potentials and (b) the sensitivity of the system to Earth surface conductivity changes.

4.6 Induction-Loop Magnetometer

In N loops of wire the voltage, V, induced by a uniformly varying field, B, is obtained from the rate of change of magnetic flux, Φ, through the loops:

$$V = -N\frac{d\Phi}{dt}, \tag{4.6}$$

where Φ is the normal (perpendicular to area of loop) component of B summed over the area, A (in square meters), of the loop. If we consider the field to be oscillating at frequency, f, and amplitude B_0, then

$$|V| = 2\pi f\, NAB_0, \tag{4.7}$$

where $|V|$ is the magnitude of the output voltage (in volts) from the N-turn loop of area A (Figure 4.6).

Natural variations of the geomagnetic field faster than 0.003 c/s (sometimes written as 3 mHz) generally decrease in amplitude with increasing frequency (see Figures 3.44 and 3.45). However, Equation (4.7) tells us that the response of the loop antenna increases with increasing frequency. The two features compensate, so the induction loop is a favorite detector for pulsation studies. Source signal amplitude restrictions

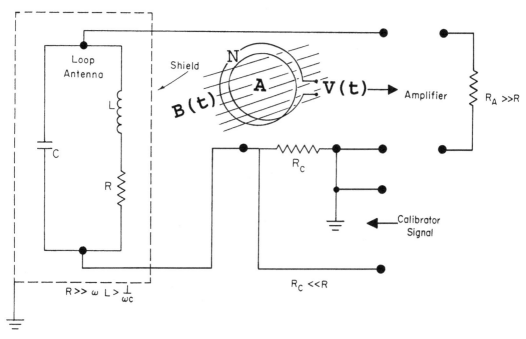

Figure 4.6. Oscillating field $B(t)$ causes induction of voltage $V(t)$ in loop antenna with N turns about area A. Equivalent circuit of antenna is shown within the dashed rectangle. A calibration circuit is also shown.

usually limit application of this magnetometer to field fluctuations with periods shorter than a few minutes.

Some induction loop designs involve a ferromagnetic core over which a long solenoidal winding is placed. The relative dimensions of the core shape as well as the encompassing windings together with the core composition determine an effective permeability. The attendant increase in voltage response of the antenna is proportional to the effective permeability of the core material. Such antennas are simply wound and relatively portable. However, the ferromagnetic material may introduce some problems, principally associated with the nature of the hysteresis curve. The desired fields cannot be separated from higher frequency natural field changes and man-made electromagnetic radiations before they are mixed in the high-permeability core. These extraneous fields may move the operating range to the more nonlinear part of the hysteresis curve, resulting in addition of undesirable distortion components to the geomagnetic signal.

Induction antennas are usually made with the natural resonance response well above the highest desirable operating frequency. Such antennas have a linear amplitude and phase response in the detection range. Intrinsic resistive noises limit the low-amplitude response of the antenna. A circular shape provides a maximum coil area for minimum conductor winding. A typical circular, open-center antenna of 2-m diameter and 16,000 turns of 0.127-mm diameter copper wire has a natural resonance frequency of about 140 c/s and produces 0.3 microvolt for a 1-mγ signal at 1 c/s. The usual core-type antennas have high permeability rod centers with diameters of about 2.5 cm, lengths of about 2 m, and 20,000 or more winding turns. With such systems the response to 1 mγ at 1 c/s would be about 0.1 mV.

Induction magnetometers are limited in application to higher frequency geomagnetic field changes. The principal drawbacks in their use are their large size and heavy weight.

4.7 Spinner Magnetometer

From Equation (4.7) we see that the voltage from a loop antenna increases with the frequency of the field variation. To measure the field of a paleomagnetic rock sample of very weak magnetic properties, we can rotate the sample within a loop antenna (Figure 4.7) at an extremely high frequency. The amplitude of the output depends upon the intensity of magnetization component in the sample directed along the axis of spin perpendicular to the induction loops. The phase of this field variation is measured relative to a reference point on the supporting sample rod. The signal level is integrated over a long measurement time to overcome any noise background.

Spin rates are often 5–100 c/s, or 150–500 c/s with air turbines. Sensitivity of typical instruments is expressed as dipole moment measuring ability, about 10^{-6} to 10^{-10} Am^2. The inherent problems with the system include mechanical rotation noises and electrostatic charge pickup.

Figure 4.7. Spinner magnetometer for rock sample measurement. Induced field in loops is a function of spin rate, sample magnetization, and pickup-loop characteristics.

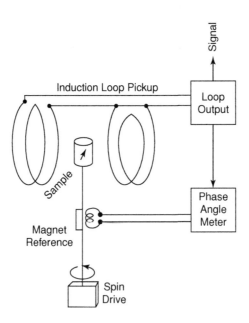

4.8 Fluxgate (Saturable-Core) Magnetometer

In instruments of this type, the nonlinearity of magnetization properties for the high permeability of easily saturated ferromagnetic alloys provides an indicator of the local field strength. The word "fluxgate" typically applied to these magnetometers refers to the method of periodic switching (gate) of magnetic flux in the detector.

The term *saturable-core* means that a highly permeable material is used to amplify the magnetic field signal picked up in a tiny (few centimeters or less) loop antenna that is much like the cored loop antenna system described above. The difference between the two systems is not only the loop size but also that the hysteresis (Section 4.2) property (saturation) is utilized by the imposition of a strong oscillating field. This field is offset, from what would be the symmetrical + and − directions, by the natural local field. The geomagnetic field strength is obtained from the generation of distortion harmonics in the output field, measured by secondary loops about the core.

In the most-used version of the saturable-core method the quantity measured is the second harmonic component of the excitation frequency, generated in the nonlinear saturation range of the sensor. In Figure 4.8

Figure 4.8. Response of a saturable core to a excitation field of amplitude A and period T in the presence of an external field H_0 (lower left in figure). Because of hysteresis (upper left), the core material of the sensor exhibits an asymmetric nonlinear response field, B (at upper right), to the combined fields such that the harmonics of the output voltage (lower right) become a measure of the external field H_0.

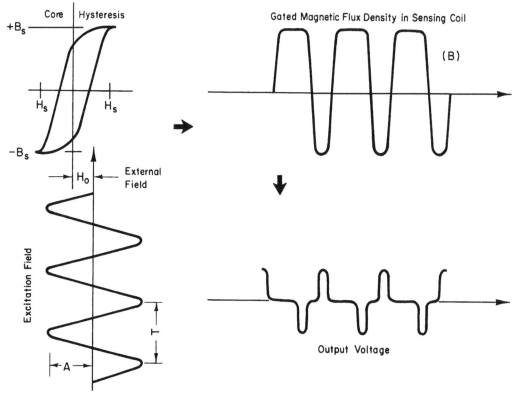

we see an external, ambient geomagnetic field H_0 superposed upon a sinusoidal excitation field of amplitude A. The relative sizes are such that $A \gg H_0$, in which A is adjusted to be larger than needed for saturation of the high-permeability core. The asymmetrical field offset caused by the addition of the ambient field, H_0, results in a distortion of the form of the magnetic flux density variation, B, linking a secondary sensing coil wound about the core. This distortion of the B variation with respect to time, t, which is asymmetric with respect to the time axis, has a high content of harmonics in its Fourier components. We can approximate the function of the input field as

$$B(t) = a(H_0 + H_e) + b(H_0 + H_e)^2 + c(H_0 + H_e)^3, \qquad (4.8)$$

where H_e is the excitation field and a, b, and c are constants.

The third and higher harmonics of $B(t)$ are considerably smaller than the second. With narrow-band filtering, this second harmonic of the output is selected to become a measure of the ambient field, H_0, after comparing it to an "artificial" undistorted second harmonic generated from frequency doubling of the original excitation oscillation. Various second harmonic voltage and phase-sensing techniques of registration are used to give H_0.

With an excitation oscillation of about 700 c/s, a 1-γ variation of the ambient field yields an output level of about 25 μV. For measurement of weak signals, special steps are taken to enhance the second harmonic detection. In one method, two parallel cores are wound in opposing directions in the excitation field circuit, and a single secondary coil encompasses both. The two primary contributions into the pickup windings are thereby canceled. The example shown in Figure 4.9 illustrates

Figure 4.9. Fluxgate magnetometer for directional (vector) geomagnetic field measurements (see text for description).

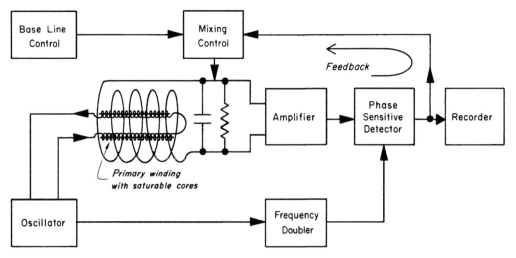

such a system, which utilizes second harmonic phase detection. With specifically selected cores and careful construction, sensitivities of a few milligamma have been obtained. In recent years, the cores have been fashioned in the shape of a ring, just about a centimeter in size, and the excitation and pickup windings have been made toroidally about the ring.

Fluxgate instruments are quite rugged. When electronic registration of the field is needed, the fluxgate systems are usually chosen; they are found on many satellites and at most modern observatories. Some typical problems with fluxgates are the slight temperature sensitivity of the response, the requirement for a skilled electronic technician when repairs are demanded, and the need for absolute calibration. This last problem has been solved at observatories by periodic calibration against a proton magnetometer (Section 4.9).

4.9 Proton-Precession Magnetometer

The physics of a proton magnetometer operation is usually described in simple classical terms rather than the more detailed quantum-mechanical terms. A *proton* is a hydrogen atom stripped of its orbital electron. In a hydrogen-rich liquid, the protons are not tied in place by a crystal lattice. We can consider the proton (nucleus of the hydrogen atom) to be a spinning charged sphere having an inherent *magnetic moment, m_p,* and an intrinsic *spin angular momentum, I_p* (Figure 4.10). The ratio of these two vector quantities is called a scalar *gyromagnetic ratio, γ_p,*

$$\gamma_p = \frac{m_p}{I_p} (\gamma \ \text{sec})^{-1}. \tag{4.9}$$

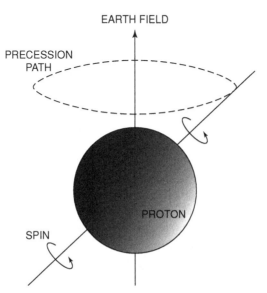

EARTH FIELD

PRECESSION PATH

PROTON

SPIN

Figure 4.10. The proton spin axis offset with respect to the Earth's field causes precessional motion that can be detected with an induction coil placed around a liquid proton cell.

An external magnetic field ($\mathbf{H} = \mathbf{B}/\mu_0$) will exert a torque on the spinning proton to align its magnetic moment and cause a precession of the spin axis. The whole process is much like the familiar precessional behavior of a toy top as it slows. The top precesses because of the interaction of gravitational and gyroscopic forces; the top's precessional rate changes with its spin velocity because the gravitational force is constant. For the top, friction at the point of spin surface contact retards the spin, thereby slowing the precession frequency as the top slows down. For the proton, the spin angular velocity is an atomic constant; it is the force of the magnetic field that changes. There is no friction for the proton spinning, just a scrambling of the proton orientation in time because of collisions between the molecules carrying the protons.

The angular frequency of the proton precession, ω_p, called the *Larmor frequency*, is equal to the product of the gyromagnetic ratio and the magnitude of the total field,

$$\omega_p = 2\pi f_p = \gamma_p H. \tag{4.10}$$

Thus, knowing the gyromagnetic ratio, the local field strength is measured by the frequency, f_p, of the precession. In the more exact quantum-mechanical representation of the atomic structure, ω_p is the quantized rotation representing the energy state splitting into sublevels in the presence of a magnetic field (see Zeeman magnetometer below).

For a normal sample of hydrogen-rich material the protons are randomly oriented so that the precession of nuclei in an external field cannot be detected by an induction coil in the vicinity of the sample. However, if the sample is first subjected to a strong magnetic field at an angle to the external field, $\mathbf{H}_0$, which is to be measured, the ensemble of proton magnetic moments exhibits a degree of alignment. Then if the polarizing field is abruptly reduced, in a time much less than a Larmor period, to a value considerably less than $\mathbf{H}_0$, the protons will precess about $\mathbf{H}_0$, in unison, inducing a signal in a pickup coil. Often the polarizing coil is also used as a pickup coil and connected to an appropriate band-pass filter and amplifier.

For a simple proton sensor, such as that illustrated in Figure 4.11, the signal amplitude is dependent upon the magnetometer orientation and is proportional to $\sin^2 \theta$, where θ is the angle between the polarizing coil axis and the field direction. Recently, toroidal windings about a donut-shaped proton cell have been used. For this arrangement the maximum signal output is obtained at axial alignment of the field, whereas the minimum signal is only half this value when the field is aligned perpendicular to the cell axis.

Hydrogen-rich liquids such as water, alcohols, oils, and kerosene have been used in the sample cell. In a cylindrical water cell of 300 to

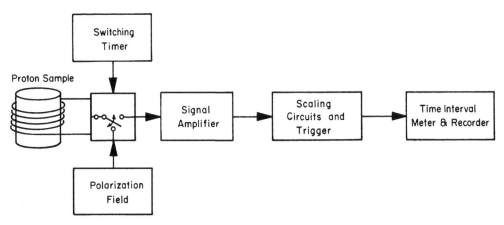

Figure 4.11. Proton magnetometer for total field (scalar) geomagnetic field measurements. The coil around the sample is used to align protons and to detect the proton precessional period.

500 cm^3, a polarizing field of 1 to $4 \times 10^7 \gamma$ applied for 2 to 5 sec will give a detectable microvolt signal in a fine wire coil of 500 to 1,000 turns. The generally adopted proton gyromagnetic ratio is 0.26751525 $(\gamma \text{ sec})^{-1}$. At the Earth's surface, where the largest fields range from 3 to $7 \times 10^4 \gamma$, the observed proton frequencies are from 1.2 to 2.0×10^3 c/s.

A variant of the proton magnetometer uses a special property called the *Overhauser effect*. Selected chemical salts added to the proton cell cause some of the orbital electron spin energy to be transferred to the protons in high-frequency fields. This allows the proton precessional frequency to be continuously measured.

The sensitivity of a typical proton magnetometer is limited by the accuracy of the gyromagnetic ratio determination, the integration time of the frequency counter, the duration of the detectable signal, and the field gradient across the sample. Accuracies of 1 γ are easily obtained; 0.05γ requires rather special systems.

The inability to isolate the field direction and the delay time between excitation and reading limit proton magnetometer applications. Usual instruments are of two types: (1) a simple field-survey, hand-held unit, from which the operator may obtain an immediate readout of the local field; and (2) an observatory-integrated unit that is operated periodically in conjunction with the station fluxgate system.

4.10 Optically Pumped Magnetometer

Optically pumped magnetometers are similar to the proton variety in that the atoms are given a measurable net magnetic moment. However, in the optical system this moment is obtained by using certain unique features of the energy levels and quantum selection rules for the critical electrons of selected gases upon absorption and reemission of light energy. An impressed oscillating field at the appropriate frequency gives a coherence to the atom's magnetic moments whose Larmor frequency, a measure of the magnetic field, is determined by selective absorption of light.

To tune our thinking to the concept of energy levels that are forbidden, permitted, ground, or excited, and to appreciate the physics of light energy absorption and emission, let us consider the following analogy. Assume that there is a hurricane approaching the Miami Beach, Florida, tourist area. It is a balmy evening and many summer strollers are walking past the luxury hotels, particularly in front of the Miami Holton Hotel. Receiving a notice from the City Disaster Alert Center, the Holton's assistant manager, who is an officer in the Florida Emergency Preparedness Unit, suddenly orders strollers to leave the streets and go into the hotel ground-level lobby.

The hotel's front door is then closed and locked. "Guests' " passage through the lobby to the hotel back door, which is steps away from a waiting city bus, is not allowed. The hotel bellhop, seeing a way to obtain extra tips, absorbs money from the "new guests" by taking small groups of them from the ground level to the excited-level penthouse where a "hurricane party" has been quickly arranged.

As the newly arrived party guests sample the food and drinks, they gradually tire of the noise (in a time called the "half-life at the excited level"); they wander back downstairs by two routes: (1) Most go by the down-elevator back to the ground-level lobby. There the persistent bellhop absorbs more tips by taking some of them as part of his next up-elevator group to the excited-level penthouse again. (2) A few go by the back stairs (a supposedly forbidden route) that leads to a service door and to the city bus that has parked at a distance from the hotel because of storm-downed trees.

Only a small portion of those leaving the excited-level penthouse select the stairs–back-door route. Because the bellhop brings fewer and fewer arriving guests in the lobby to the penthouse, in time, the party stops for lack of guests and everyone arrives at the bus parked outside, a location originally forbidden to the "guests."

Looking through a window, the Holton's assistant manager notices the guests at the bus. He sends out a warning, by the outside loudspeaker, for the "guests" to return. Because of the distance to the bus and the increasing storm noise, the warning is not loud enough, so the volume is gradually turned up until the passengers hear that an ocean tidal wave is imminent and that they must return to the safety of the hotel lobby. The necessary sound-volume level of the loudspeaker is a measure of the storm intensity. The people reluctantly "unload" and return to the hotel. The lobby bellhop now can absorb money again, leading groups of guests to the high level penthouse, etc., and the process restarts.

In optical pumping, light of the proper wavelength is absorbed by an alkali vapor cell. With absorption, higher energy levels load and reemit energy to lower levels; part of the reemission follows a "forbidden" route to a level that is known to be sensitively shifted by the local magnetic field strength. When absorption stops because of complete loading of that one particular forbidden ground state (the cell becomes transparent), then radio-frequency waves, at a frequency determined by that forbidden level, can unload that particular level (depump the cell) so that it can become light-absorbing (opaque) again. The rubidium cell flickers between transparent and opaque, responding to a light source oscillating at the proper Larmor precession frequency that corresponds to the selected energy level. It is the physical property of that forbidden energy level to be able to shift position with the local geomagnetic field.

More specifically, a rubidium magnetometer illustrates the procedure. In this description there are some special terms, unique to quantum mechanics, that I will leave undefined in order to avoid a major diversion. Even unskilled readers should be able to get some idea of the process without such explanation. The description here uses an energy-level diagram in which position upward indicates greater energy and the spacing between horizontal levels represents an energy transition (or wavelength of light emission).

Consider a rubidium gas cell with quantum energy levels of ^{85}Rb, as displayed in Figure 4.12. Along the direction of the field, there is a preferential absorption of a 7947.6 Å source light that causes transitions from the $^2S_{1/2}$ ground state to the $^2P_{1/2}$ first excited state. F is the total angular quantum number (which is obtained from the quantized vector sum of the total electron angular momentum and the nuclear spin quantum numbers). In the presence of a magnetic field the F levels are split into fine emission divisions of magnetic quantum number, m, according to the Zeeman effect (see Section 4.11). The quantum mechanical selection rule requires, during excitation to the excited state, either $\Delta m = +1$ or $\Delta m = -1$, depending upon the sense of circular polarization and direction of the field component along the optical axis. For a fixed geometry, with one or the opposite direction of polarization, this Δm selection rule can prevent transitions out of either the ground state $\Delta m = +3$ or $m = -3$, respectively. Consider a polarization related to $m = +3$. Only atoms that are not in the $m = +3$ state of the $^2S_{1/2}$ level can absorb enough light energy to make transitions into the various excited levels of $^2P_{1/2}$. Then collisions between atoms cause disorientation within these high-energy levels and there occurs a subsequent decay back to all ground levels, including the forbidden $m = +3$. In this way an appreciable loading of one particular ground state, the

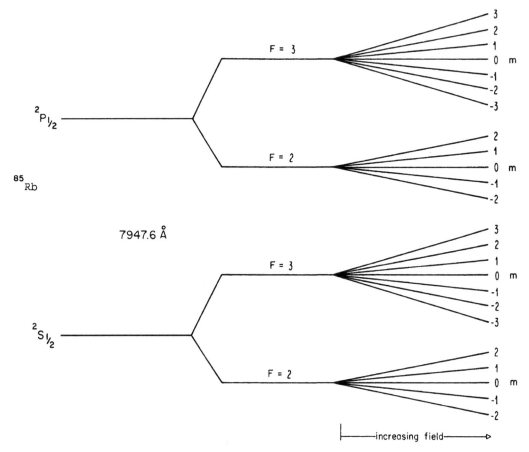

Figure 4.12. Quantum energy levels, and their alphanumeric names, for the ^{85}Rb used in the cell of an optically pumped magnetometer. See text for explanation of symbols.

$m = +3$ level, is obtained and, after a while, no further absorption of light is possible because there are no populated lower energy levels. Because of the Zeeman effect, the energy level of this loaded ground state is dependent on the local magnetic field strength.

The rubidium gas absorption cell is "optically pumped" when all the atoms are trapped in the forbidden level. It is unloaded by the field energy of the variable radio frequency (RF) reaching a frequency appropriate to $m = +3$. The cell is opaque while the level is being filled; it is transparent when full. In Figure 4.13 the filtered excitation lamp provides the energy for the loading. The circular polarizer determines the selected m level. The absorption cell fluctuates from transparent to opaque at each RF cycle at the frequency appropriate to the energy level of the ambient geomagnetic field because the pumping process is much faster than the RF unloading. A photo cell detects the frequency at which the unloading occurs and locks-on the RF oscillator. The counter translates frequency to geomagnetic field strength.

Table 4.2. *Optical pumping gases*

	^{85}Rb	^{87}Rb	^{133}Cs	^{39}K	^{4}He
Wavelength of absorption (Å)	7948	7948	7944	7699	10830
Response (c/s per γ)	4.7	7.0	3.5	7.0	28.0

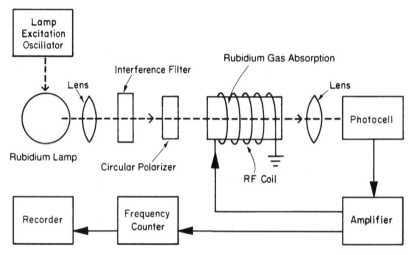

Figure 4.13. Optically pumped rubidium vapor magnetometer. See text for detailed description.

The process is complicated by the fact that there is a nonlinear term in the relationship of field strength to the separation of the hyperfine levels. In addition, the modulation concerns the precession of many sublevel states from which an average value of 4.66737 c/s per gamma of ambient field results. In recent years more accurate "single-line" rubidium magnetometers have been developed that isolate a single sublevel state. Table 4.2 presents a comparison of some atoms that have been employed in optically pumped magnetometers.

The usual rubidium magnetometers have a limiting sensitivity of about 0.2 γ. The magnetometers measure the total field and variations along the total field direction. Bias fields produced by large mutually perpendicular coils about the cell are required to resolve vector components of the main field. The advantages of the instruments, that they operate well in low magnetic fields and have a rapid response time, have given them prominence in satellite usage. The principal problems attending the general observatory use of these instruments are the limited lifetime of the absorption cells and the complicated electronic circuitry of the instrument.

Helium is also used in optically pumped magnetometers. In a weak radio frequency electric field these atoms can be excited to a metastable state that has a relatively long life and is split into sublevels by an ambient field. From this pseudo ground state the pumping and geomagnetic field detection is similar to that described above for the rubidium cells. Sensitivities of about 0.01 γ are claimed for commercial systems.

4.11 Zeeman-Effect Magnetometer

Zeeman discovered that in a strong magnetic field a single atomic spectrum line is split into three lines whose separation distance is a measure of the field-strength magnitude (Figure 4.14). Viewed at right angles to the field, the middle (original position) line is polarized in a direction parallel to the field, whereas the two symmetric outer lines are polarized at right angles to the field. Viewed parallel to the field, only the two outer components are seen; these two lines are circularly polarized, in opposite directions, with their sense (clockwise or counter-clockwise) determined by the field direction.

To describe the process, a magnetic quantum number, m, is used in which $mh/2$ is the component of total angular momentum in the field direction (h is a number called *Planck's constant*). The Zeeman spectral lines of transition between energy levels of the atom are equally spaced at the Larmor frequency. These lines are restricted to the quantum selection rule $\Delta m = 0$ or ± 1 in the direction perpendicular to the field and $\Delta m = \pm 1$ parallel to the field. At low field strengths more than three lines occur; this phenomenon is explained in terms of the interaction of electron orbital and spin momenta. The size of the Zeeman wavelength shift varies for different atomic elements and different parts of the spectra.

Optical detection of the Zeeman effect for magnetic field measurements is restricted to very strong fields. It is effective for regions where spectrally emitting atoms exhibit a relatively large wavelength shift for

Figure 4.14. Zeeman spectral-line splitting and polarization in strong magnetic fields. The horizontal position represents light frequency, which increases from left to right. **B** is the directed local field at the emission site.

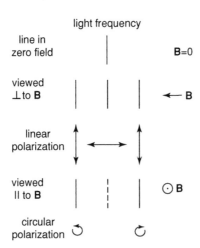

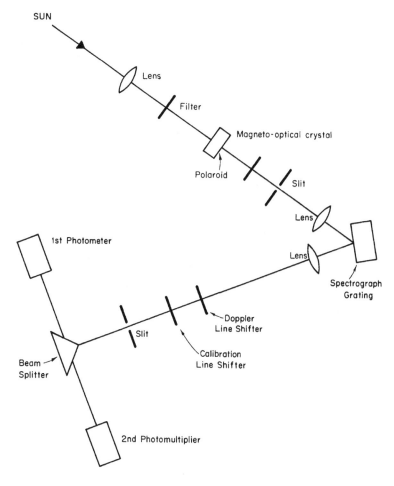

Figure 4.15. Zeeman line-splitting magnetometer used for solar field measurements. This system is designed to detect both the polarization and wavelength shift of iron lines in a spectral display of the solar surface.

the field size being studied. The Zeeman effect for the bright FeI iron lines near 5250.216 Å from the solar surface emissions can be detected on Earth. The magnetometer used for these observations primarily depends upon the ability of a high-dispersion spectrograph grating to separate adequately the fine-line splitting. In one model, Figure 4.15, the right- and left-circularly polarized emissions are alternately selected using special magneto-optical crystals that have the remarkable property of alternating their selective transmissions with the sign of the applied voltage. A beam splitter causes the line images to oscillate between two photomultipliers whose entry slits are wedges appropriately constructed to allow more light signal with increased line displacement. The amplitude difference from the photometers becomes a measure of the original field at the FeI emission location. Zeeman-effect magnetometers have made possible routine surveys of the Sun's magnetic field profile to a sensitivity of about 10^4 γ (see magnetograms in Figure 3.7).

4.12 Cryogenic Superconductor Magnetometer

At particular critical temperatures, usually below 20 degrees K, the resistance of many conductors such as lead (7.2 K), niobium (9.2 K),
tantalum (4.4 K), vanadium (5.4 K), and certain alloys approaches a
zero value. In this superconducting state a number of unique, submicro-
scopic, quantum-mechanical *cryogenic properties* have been discovered
to have manifestations at the macroscopic level. The low temperatures
are obtained in a bath of liquid helium (4.2 K), an easily acquired, inert
substance that is in a gas phase at room temperatures. At superconducting temperatures the chilled conductor completely expels the external
magnetic field up to some maximum value of field.

Wave properties of electrons require that in a ring of superconducting
material there should exist an integral number of wavelengths (de Broglie
wavelengths or quantized angular momentum states) so that a standing
wave pattern is established. Any angular momentum state not satisfying
this requirement destructively interferes with itself. The flux through
a superconducting ring is quantized into units of $\phi_e = h/2e$ with h as
the atomic constant (Planck) and e as the charge on the electron. See
Feynman et al. (1965), Volume III, Chapter 21, Sections 5 to 17 for a
good introduction to the special physics of superconductivity.

In a SQUID (super conducting quantum interference device) magnetometer (Figure 4.16a) the ring is weakened by a point contact (Josephson
junction); then it is possible to force non-integral multiples of ϕ_0 through
the ring with an externally applied field. The extra energy is localized
at the point of contact and is explained by the physics of quantum mechanical "tunneling effects" using the wave properties of electrons. The
value of the applied field corresponding to the flux quantum ϕ_0 is just
$B_0 = \phi_0 A_{\text{eff}}$, where A_{eff} is the effective aperture of the SQUID. Practical designs of the SQUID put several loops in parallel about a single
Josephson junction (Figure 4.16b). The loops are then twisted so that the
apertures of the loops are aligned in a single direction for field detection
(Figure 4.16c).

A radio frequency (RF) oscillator (near 27 MHz) applied to a loop
encircling the SQUID causes a critical current to flow in the Josephson
junction, forcing the junction to oscillate from a normal resistive to a
superconducting state. Such oscillation allows the geomagnetic field flux
through the area of the SQUID loop to be captured periodically. Then
the superconductivity response of the SQUID to the external field, B,
alters the mutual inductance in a resonant circuit. A voltage output from
this circuit is what is finally measured. A "phase-lock" servo system
feeds back a sufficient direct current to hold the SQUID response to a
fixed characteristic cryogenic oscillation wavelength.

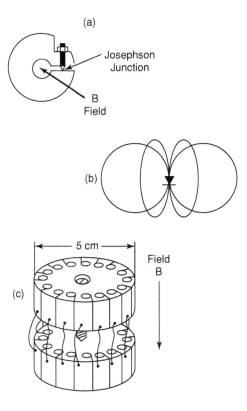

Figure 4.16. (a) Single-loop SQUID magnetometer. (b) Connection for 4-loop SQUID. (c) Practical design for 16-hole (loop) SQUID magnetometer sensor.

Typical systems have the astounding sensitivity of 10^{-5} gammas for measurements of either the geomagnetic field or paleomagnetic rock samples. The magnetometer response is linear up to 10 kHz. High-frequency cutoff is obtained through mechanically shielding the sensor. Figure 4.17 illustrates the typical system. The principal limitations are the need to recharge the liquid helium flask every 3 days to one week (the light helium atoms escape the walls of a container) and the need for complicated electronic circuitry.

Recently, higher superconducting operating temperatures have been obtainable at Lawrence Berkeley Laboratory for specially designed multilayer film deposits of $YBa_2Cu_3O_{7-x}$-$SrTiO_3$-$Yba_2Cu_3O_{7-x}$ (called "YBCO") SQUIDS, less than 0.5 mm square, with grain-boundary Josephson junctions. Input coils, patterned in the YBCO film, couple to large area field-sensing loops to form the magnetometer.

4.13 Gradient Magnetometer

In many situations the measurement of importance is the change of field with distance from the source. This *gradient-of-the-field* has a rather

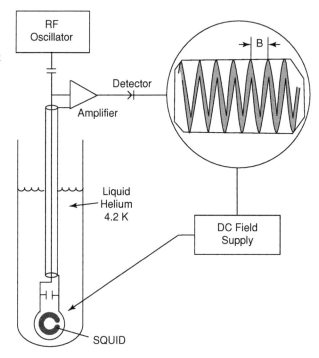

simple mathematical form for fields from a simple current configuration,
such as a sheet, straight line, or a loop. We have seen that the loop of
current results in a simple dipole field, similar in form to that of a bar
magnet. Calling B_r the value of the field in a radial direction from the
dipole center, at a distance that is large with respect to the apparent
separation between the poles,

$$B_r = \frac{\mu_0 M \, \cos(\theta)}{2\pi r^3}, \tag{4.11}$$

where μ_0 is the free-space permeability constant ($4\pi \times 10^{-7}$ Henrys/m),
M is the dipole moment, and θ is the colatitude of the measurement with
respect to the dipole axis (as in Section 1.3). The *gradient* of the field in
the radial direction is the change in B with respect to a change in r (or
the "derivative of B with respect to r"; see Section A.7):

$$\frac{dB_r}{dr} = \frac{-3\mu_0 M \, \cos(\theta)}{2\pi r^4}. \tag{4.12}$$

This equation tells us that the gradient in the vertical direction varies as
the fourth power of the distance.

To see what this means for the Earth's field, let us rearrange this
equation thus:

$$dB_r = -6 \left[\frac{\mu_0 M}{4\pi r^3} \right] \cos(\theta) \left(\frac{dr}{r} \right). \tag{4.13}$$

Because we know (text following Equation (1.27)) that the value of $[(\mu_0 M)/(4\pi r^3)] = 3.1 \times 10^4$, we can evaluate Equation (4.13) as

$$dB_r = -18.6 \times 10^4 \cos(\theta) \left(\frac{dr}{r}\right). \qquad (4.14)$$

As an example, consider the Earth $r = 6.37 \times 10^6$ meters. At $30°$ latitude, $\theta = 60°$, so $\cos(\theta) = 0.5$ over a distance of $dr = 1$ m in the Z direction; we obtain $dB_r = 0.0146$ gamma decrease. This is admittedly a small change with respect to the usual magnetometer sensitivity.

One application is to determine the gradient of field resulting from an electric current flowing in a linear pipeline at the Earth's surface, in the presence of both the Earth's main field and fields from ionospheric or magnetospheric currents. The method is to measure the Z component of the field at two defined locations along a line directed perpendicular to the pipeline. If the sensors are set in opposing Z directions, the sum of the sensor outputs becomes our gradient (the change of field over the sensor separation). Because the gradient of the Earth's natural dipole field is so small, and because the distant ionospheric and magnetospheric contributions will be identical at the two measuring sites, we can safely neglect any contribution by the main field. This gradient magnetometer arrangement is also used reliably to detect the fields of embedded magnetized materials or channeled electric currents within our Earth's environment. The very sensitive SQUID magnetometer can be connected in a gradient arrangement by effectively wiring half of the field loops (Figure 4.16c) in an opposing direction to the other half.

4.14 Comparison of Magnetometers

A comparative evaluation of magnetometers should be carried out within the constraints of their intended use. Not one of the above systems has proved to be an all-purpose detector. Table 4.3 lists the magnetometers with their sensitivities, some limitations, and typical applications.

4.15 Observatories

Studies of geomagnetic phenomena of magnetospheric origin have special station location requirements. Auroral zone electrojet and field-aligned currents show sufficient variability to require observations at dawn, dusk, midday, and midnight meridians and at latitudes from an L-value of 4 to at least 8. Low-latitude observatories away from the equatorial electrojet seem best for registration of global disturbance levels. Polar-cap measurements are important at regions above about $L = 12$ where dayside field lines are pushed by the solar wind to the tail of

Table 4.3. *Comparison of magnetometers*

Magnetometer	Resolution (γ)	Comments
Zeeman optical	10^4	Used for strong Sun fields.
Classical variometer	10^{-1}	Used at many observatories. Typically less sensitive. Poor response to pulsations. Simple design is reliable.
Fluxgate	10^{-1}	Found at most modern observatories. Satellite use for weak fields. Sensitive to stress and temperature. Needs proton calibration. Low noise version for pulsations.
Proton	10^{-1}	Ideal total field measurement. Poor response to pulsations. Reference standard for observatory. Extensive use in geophysical surveys.
Optically pumped	10^{-2}	Total field measurement. Often used on satellites. Complex electronic design. Rb cells age easily. Responds rapidly. Broad frequency range.
Induction loop	10^{-4} at 1 c/s	Measures field rate of change. Poor response for $T > 1$ min. Used on some spinning satellites. Ideal pulsation measurements.
Earth-current probe	10^{-5} at 1 c/s	Good in geological surveys. Good pulsation measurements. Simple construction. Calibration difficulties.
SQUID	10^{-5}	Extensive use for rock samples. Complex electronics. Liquid helium evaporation problem.

the magnetosphere. Dip equator and eccentric geomagnetic pole locations have important geophysical applications. Conjugately located stations at latitudes corresponding to the magnetospheric cusp regions and to the ionospheric feet of magnetospheric field-aligned currents have special research value.

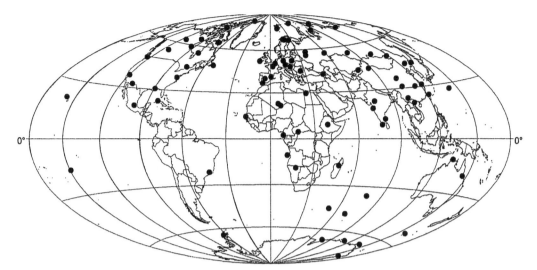

Figure 4.18. Locations of observatories. Solid circles are observatory locations known to exist in 2001 or locations of operational observatories from which WDC received data for 1990 or later. Figure provided by S. McLean, NGDC/NOAA.

Ideally, some special sites are needed to track ionospheric current effects. Where the equatorial electrojet is important, regions representing the coincidence and maximum separation of the dip and geographic equators are of special interest. A grid of stations equally distributed over a range of longitudes in the American, Euro-African, and East Asia–Australian zones would help the understanding of ionospheric, tidal, and wind processes that generate the Sq, flare and eclipse effects, and lunar tidal currents.

For measurements of the main field, a global grid of observatories would be the goal. Figure 4.18 shows that presently the high latitudes, Southern Hemisphere, Asia, and the ocean areas are lacking adequate coverage. Island observatories often present special problems because of unusual induction effects. For field modeling and charting, additional geomagnetic measurement sites (*repeat stations*) are often occupied for several days on a three- to five-year repetition schedule. Special efforts are made at these locations to obtain the best quiet-condition absolute field levels. To explore the observatory locations go to the IAGA website www.ngdc.noaa.gov/IAGA. Then under "Divisions" select "V Geomagnetic Observatories, Surveys, and Analysis". Follow with selection of "working Group V1 Geomagnetic Observatories, Instruments, and Standards" and then scroll down to the alphabetical listing of the world's observatories.

Significant geological, political, educational, and economic factors determine site locations for magnetic observatories. International efforts are now under way to improve the present global geomagnetic observatory distribution both by the addition of new sites and upgrading existing sites to modern digital recording standards.

Table 4.4. *Magnetic disturbance of 1 nT produced by common objects at given distances*

Safety pin	0.1 meter	Belt buckle....	1 meter
Watch...........	1 meter	Metalic pen....	1 meter
Knife	2 meters	Screwdriver....	1 meter
Hammer	4 meters	Bicycle........	7 meters
Motorcycle......	20 meters	Car............	40 meters

Source: Newitt et al., 1996.

Sites selected for magnetic observatories should be as free as possible from local geologically introduced irregularities in the magnetic field. Although this necessitates a complete local magnetic gradient survey, one rapid check can be performed by interrelating compass bearings at two points more than 100 meters apart at opposite directions from the proposed site. A site is generally considered unsatisfactory if horizontal intensity gradients are discovered to be greater than a gradient of about 2 nT/m or 10 sec/m in direction. Electric railways as far as 15 km from the site have interfered with magnetic measurements in some areas. Water and sanitation facility service pipes of iron or steel should be kept at more than 200 meters distance. Power line interference depends upon the service load and the direction of the site from the supply line. Ideally, the station should be supplied with power through its own isolation transformer and be located at the end of a spur line at least 1 km from a main supply line. Local field strengths as large as 50 nT from nearby 50 c/s or 60 c/s lines are usually acceptable at the measurement site.

Table 4.4 shows how common objects can cause artificial magnetic field responses of one nanoTesla at various distances from station magnetometers. Nonmagnetic materials are used in magnetic observatory structures. Construction may even require special selection of the cement and gravel for the concrete foundation. In general, even the use of conducting materials should be minimized because seismic and wind-generated vibrations of the metal will induce eddy currents that can be detected by the magnetic sensors. A stable platform (pier) of large mass is necessary for most magnetometers. A method for regularly determining the pier tilt is available at many modern observatories. The International Association of Geomagnetism and Aeronomy has published two observatory operation manuals: "Guide for Magnetic Observatory Practice" (Jankowski and Sucksdorff, 1997) and "Guide for Magnetic Repeat Station Surveys" (Newitt et al., 1996).

4.16 Location and Direction

A recording of the geomagnetic field requires an accurate knowledge of the geographic directions from which H and D (or X and Y) field component directions are determined. Years ago the technique for finding geograhic north involved the sighting of stars and use of astronomical tables. The geographic directions, location, and altitude of an observatory has been simplified in recent years with the advent of the Global Positioning System (GPS). Initiated in 1990 for military purposes, the GPS is now available for public use in surveying and navigation. The system depends upon determination of propagation-time delays in the reception of spread-spectrum radio-wave signals near 1.575 and 1.227×10^3 MHz transmitted by twenty-four satellites (Hoffman-Wellenhof et al., 1994). These satellites are in six separate, near-circular orbital planes at 55° inclination and at an altitude of approximately 20,200 km. The arrangement of the satellites is such that transmission from at least four can be received at any given point near the Earth's surface. The signals, detected simultaneously by the user's GPS receiver, are automatically processed to determine the global altitude, latitude, and longitude of the receiver. Absolute position accuracy at a receiving site is expected to be within eight meters and the relative position accuracy (between two nearby locations) within three meters. With the price of receivers now less than 300 $US, the absolute position of magnetic observatories and repeat stations can be simply defined. Location of two nearby survey positions, using the GPS, can establish the the true geographic north azimuth necessary for the magnetometer setup and determination of magnetic declination. Note that errors in the readout can occur during magnetic storms (see Section 5.7).

Some magneticians prefer the use of a local map for location and a north-seeking gyroscope (called a gyro-theodolite) for accurate determination of azimuth (cf. Barreto, 1996). In that instrument a 2200 rpm gyro-rotor is pulled from its preferred spin plane by the Earth's rotation. The north direction is found from the midpoint of the resulting precession oscillation. Procedures for determination of the local magnetic declination with the gyro-theodolite (as well as the GPS) are described in Newitt et al. (1996).

4.17 Field Sampling and Data Collection

In recent years a consortium of national geomagnetic observatory leaders has arranged a cooperative, near "real time" data recovery system, from about 80 observatories, called INTERMAGNET for "INTErnational Real-time MAGNETic observatory network". Figure 4.19 shows the

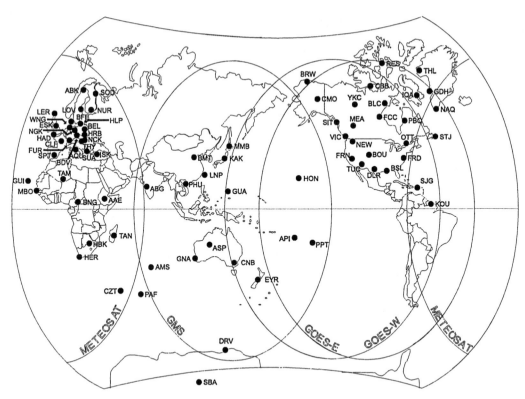

Figure 4.19. INTERMAGNET observatories operating in 2001. Figure provided by D. Herzog, USGS. For the station coordinates go to website www.intermagnet. org/ to select "geomagnetic observatories" and then "list of IMOs".

2001 distribution of contributing stations and the data-relay satellites. One-minute digital data from the INTERMAGNET observatories are broadcast, at 12-minute or one-hour intervals, to the geostationary satellites within the observatory transmission signal reception window. At selected locations, called GINs (for *geomagnetic information nodes*), these data are collected using electronic mail and file-transfer protocol and then distributed to the participating governmental agencies. The data are classified into three Categories: (1) "Reported Data" that have not been cleaned, (2) "Adjusted Data" that have had preliminary cleaning, and (3) "Definitive Data", containing the final error cleaning, available (with about a year delay) on a CD-ROM from INTERMAGNET offices. Usually the GINs hold the Reported Data. Immediate appraisal of selected INTERMAGNET recordings provides an important part of evaluation (by space environment forecasting centers – Section B.6) of the present and expected "space weather" affecting satellites, communication, and electric power transmission (Chapter 5).

INTERMAGNET representative D. Herzog of USGS has provided the following information on the observatory standards:

An INTERMAGNET Magnetic Observatory (IMO) has full absolute field measurement control, that provides one-minute magnetic field values measured by a vector magnetometer, and an optional scalar

Table 4.5. *IMO Requirements*

Vector Magnetometer	
Resolution:	0.1 nT
Dynamic range:	6000 nT Auroral & Equatorial
	2000 nT mid latitude
Band pass:	DC to 0.1 Hz
Sampling rate:	0.2 Hz (5 sec)
Thermal stability:	0.25 nT/°C
Long-term stability:	5 nT/year
Accuracy:	±10 nT for 95% of reported data
	±5 nT for definitive data
Scalar Magnetometer	
Resolution:	0.1 nT
Sampling rate:	0.033 Hz (30 sec)
Accuracy:	1 nT

magnetometer, all with a resolution of 0.1 nT. Vector measurements performed by a magnetometer must include the best available baseline reference measurement. An IMO must try to meet the minimum requirements shown in Table 4.5.

World Data Centers (see Section B.2), started for the International Geophysical Year in 1958, are still the prime repositories for most national geophysical records in the form of charts, microfilm copies of charts, tables of values read from the charts, or computer files of digital recordings. Geomagnetic records are distributed, free of charge, from these centers to contributing sources and, for a small fee, to other individuals or organizations. Besides archiving and distributing geomagnetic records, individual centers perform a great number of other services of importance to national and research requirements such as producing indices, converting the geomagnetic data to application formats for easy digestion, and collating related solar–terrestrial phenomena. Geomagnetic data from most observatories are carefully edited, cleaned of extraneous noises, and quality checked before being archived at the centers. To sample geomagnetic data deposited at World Data Centers, you can go to website http://www.ngdc.noaa.gov and select "Geomagnetism" on the list to the left. Then select "Geomagnetic Data" under the digital data list. Finally look at the "1-min values" and proceed from there.

4.18 Tropospheric and Ionospheric Observations

Magnetic survey instruments towed by aircraft sample the field below 15-km altitudes. Difference measurements between the aircraft and fixed-observatory magnetometers yield the magnetic profile of the Earth's crust along a predetermined flight path. The flight stability of the airborne magnetometer usually restricts measurement accuracy to about 0.1 to 1.0 gamma on quiet magnetic days under stable atmospheric conditions (see Section 5.8.2).

Balloon-borne instruments can reach ceiling altitudes of about 20 to 30 km and drift with the local wind system. Fluxgate and proton magnetometers have been flown aboard balloon packages. High-latitude data telemetered to the ground have been used to monitor phenomena simultaneous with changes in the auroral electrojet such as aurora and bremsstrahlung X rays from precipitating electrons (see Figure 3.40). The radio horizon for the balloons is about 650 km. Only about ten percent of the balloons are recovered. The high cost of lost instrumentation and the marginal requirement for such in situ magnetic determinations have limited extensive field measurements of this type.

Almost 100 successful rocket magnetometers have been launched through the high-conductivity layers of the atmosphere for the purpose of measuring ionospheric currents (see Figure 2.17). Most of these flights were designed to study the equatorial electrojet; the others were about equally divided between the auroral electrojet, Sq current, and main field measurements. Difficulties in obtaining platform stability, accurate orientation determinations, and exact trajectory tracking information have largely restricted the instrumentation to simple proton magnetometers. On occasion, fluxgate or optically pumped magnetometers have been used.

Penetration of current-carrying layers by the rockets is indicated by the difference between the measured field and that expected from the inverse-cube-law decrease of the Earth's IGRF model (Section C.12) field with increasing height (Equation (4.11)). Thus far rocket observations have been limited to magnetometer sensitivity levels of a few nanoteslas. Successful flights reach well above the 100- to 120-km current layer to about 250 km. The E-region location of ionospheric dynamo currents as well as the auroral and equatorial electrojet currents were verified by rocket-borne measurements. Such information has led to better use of surface observatories as ionospheric disturbance monitors.

4.19 Magnetospheric Measurements

Equatorial and polar satellites are generally restricted to altitudes far above 300 km because the atmospheric drag will pull the satellite out of orbit at lower altitudes; 600- to 2,000-km altitudes and approximate 1.5- to 3.5-hour orbits seem to be typical arrangements. Geostationary satellites are limited to about 6.6 R_e distance in the equatorial plane. On occasions of a major coronal mass ejection, the dayside magnetospheric boundary may be pressed in past this geostationary position, providing shock-boundary crossing measurements for the in situ magnetometer (Figure 3.20).

Polar orbiting satellites make significant contributions to global magnetic-field mapping because of the completeness of global coverage (Figure 4.20). The problem in space is to detect a small ambient field in the presence of spacecraft noise and then to separate spatial from temporal field changes. The inclination of these satellites (the angle between the Earth's spin axis and the direction perpendicular to the plane of the orbit) should be close to 90° to provide full coverage of the Earth.

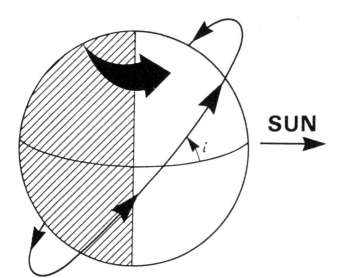

Figure 4.20. (Top) Path of a spacecraft at inclination, i, in an orbit around the rotating Earth. (Bottom) Typical ground track of MAGSAT satellite for one day. Figure from Langel and Baldwin (1992).

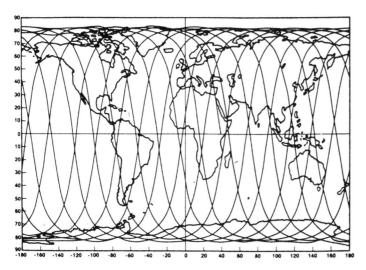

Table 4.6. *Early magnetic field satellites*

Satellite	Instr. (nT)	Type	Inclin.	Alt. (km)	Operation
COSMOS-49	Proton(22)	scalar	50°	261–488	10/64–11/64
OGO-2	Rubidium(6)	scalar	87°	413–1510	10/65–09/67
OGO-4	Rubidium(6)	scalar	86°	412–908	07/67–01/69
OGO-6	Rubidium(6)	scalar	82°	397–1098	06/69–07/71
MAGSAT	Fluxgate(22)	vector	97°	325–550	11/79–05/80
DE-2	Fluxgate(100)	vector	90°	309–1012	08/81–02/83
UARS	Fluxgate(20)	vector	57°	574–579	09/91–present
POGS	Fluxgate(30)	scalar	90°	apx. 700	01/91–10/93
FREJA	Fluxgate(30)	vector	63°	595–1759	10/92–1996

Source: Information provided by R. Langel and J. Heirtzler at Goddard Space Flight Center of NASA.

An inclination greater than 90° indicates a rotational path of the satellite opposite to the Earth's rotation direction. An error in the pointing direction of the field sensors of 5 arc seconds can introduce a 1.5 nT error in measurements. Trajectory errors are mostly due to limitations in tracking-station timekeeping and to changes in satellite acceleration with varying magnetospheric conditions (see Section 5.3). At a typical satellite velocity of about 7 km/sec, a one-tenth second time accuracy translates to 700 m location accuracy along the track and as much as 4.2 nT difference in total field.

Early satellite geomagnetic field measurements were of the total-field (scalar) type made with "absolute" field measuring rubidium (optically pumped) or proton magnetometers. In recent years, satellites have included three-axis (vector) fluxgate magnetometers, calibrated with colocated absolute sensors to about 2 to 5 gamma. The sensors are placed on a boom, about 2 to 8 meters from the body of the spacecraft, whose orientation in space is determined with a star sensor and/or GPS (global positioning system; see Section 4.16) receivers. A mechanism is then needed to monitor the magnetometer location with respect to the body of the spacecraft. Table 4.6 and Figure 4.21 list some of the features of the magnetic field satellites that have, to date, provided significant global field mapping data. Neubert et al. (2001) describe the remarkable successes that scientists have obtained using Oersted satellite to record high precision geomagnetic field data for resolving Earth surface anomalies.

Satellite observations have the great advantage of global coverage that is unmatched by surface observatories. However, there are several drawbacks that limit the prospect for total reliance on satellite

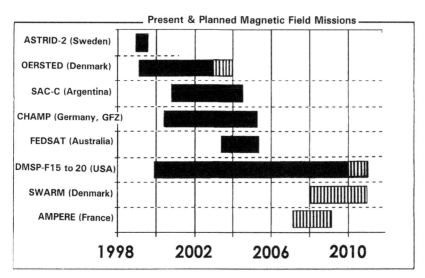

Present & Planned Magnetic Field Missions

ASTRID-2 (Sweden)
OERSTED (Denmark)
SAC-C (Argentina)
CHAMP (Germany, GFZ)
FEDSAT (Australia)
DMSP-F15 to 20 (USA)
SWARM (Denmark)
AMPERE (France)

1998   2002   2006   2010

Figure 4.21. Magnetic field missions for recent and proposed satellites. See website http://denali.gsfc. nasa.gov/research/mag_field/ purucker/mag_missions.html

measurements for charting the Earth's field. Interference generated by on-board electronic systems can disrupt the sensitive geomagnetic sensors. Observed field values from magnetometers, that are placed on a long beam to be away from the main satellite package, can experience undesirable loss of direction accuracy due to physical vibrations with respect to the fast-moving space platform. Individual satellite field measurements are considerably less accurate than those measurements obtained from the best surface observatory system. Because of the satellite position above the current-carrying ionosphere, the external and internal parts of Earth's field, computed from spherical harmonic analysis, place the ionospheric currents interior to the satellite and thereby degrade main-field determination. Secular change is better determined from long duration records of the more stable Earth observatories. As a result, the best world-field models are obtained from a combination of satellite and surface station records. Satellite magnetometers are also used for solar–terrestrial disturbance monitoring and forecasting (see Section 5.16 for some specific details about these systems).

4.20 Instrument Summary

Geomagnetic measurement techniques have included the magnetic and hysteresis properties of materials, the relationship of orthogonal electric and magnetic field variations, the induction of field variations within conducting loops, the atomic optical field response of gases, and the quantum-mechanical behavior of conductors at cryogenic temperatures. The principal instruments for geomagnetic field measurements throughout the world are still the older bar-magnet compass and a self-recording version, the classical variometer. The astatic, spinner, and cryogenic magnetometers are used by geologists for sensitive rock-sample measurements. Induction-loop antennas are ideal as sensors for geomagnetic micropulsations because the response increases linearly with the source frequency. The inexpensive Earth-current probes make use of the relationship of changing electric to changing magnetic fields over a conducting Earth. The fluxgate magnetometer is presently the favorite instrument of modern observatories. However, such fluxgates rely on companion absolute field measurements by proton magnetometers. Optical pumping, particularly of rubidium gas, utilizes the selective properties of atomic optical emission states to evaluate magnetic field strength; such systems have not found popular use. Zeeman shifting of the frequency of optical emissions in strong magnetic fields is employed for solar surface magnetic observations. SQUID magnetometers, which utilize special superconductivity properties of some metals and alloys at extremely low temperatures, are the most sensitive instruments yet produced; their application in science is rapidly expanding.

The large number of Earth-surface geomagnetic observatories and satellite programs cooperate in supplying records to international World Data Centers (Section B.2). The expanding INTERMAGNET program has provided magnetic field measurements that serve the modern interest of both the solid Earth and space physics programs. Archived data at these centers provide the basis for global maps and for continuing research into geomagnetic processes.

4.21 Exercises

1. Go to NASA Goddard Space Flight Center website http://image.gsfc.nasa.gov/poetry/workbook/page9.html and construct your variation of the "soda bottle" magnetometer described therein. Keep records of the magnetic field changes.

2. Suppose that you are measuring the electric field with probes placed into the surface of the following conductors : (a) ocean water and (b) dry earth. Assume that a magnetic field (B_y) has a magnitude of 10 nT and a variation period of 30 seconds. Determine the magnitude of the electric fields (E_x) associated with the magnetic field for the two cases. Express the electric fields in millivolts detected for a 10-meter probe separation.

3. What is the voltage output from a loop antenna magnetometer of 10,000 turns of copper wire wound about an area of 2 meters when the natural field amplitude of 10 nT (perpendicular to the loop area) is varying with a period of 30 seconds?

4. Determine the Larmor period of the protons (in a proton magnetometer cell) in the Earth's field measured for the main field strength at your home city. Use program **GEOMAG** in Appendix C to find the reported field magnitude for this computation.

5. You will need to use the information from the website listed with Figure 4.19 to locate the INTERMAGNET data. At the website, select "English" language at the left corner of the first screen. Next, select "geomagnetic observatories" in the first paragraph. Then select "Map of IMOs" to find the location of an observatory nearest to your home location. (I selected BOU for Boulder, Colorado, USA.) Then go back to the previous screen and select "list of IMOs" to highlight your selected stations for details. Describe the information you have found there.

6. At the geomagnetic field colatitude of your home city (program **GMCORD** in Appendix C) use Equation (4.14) to determine the change in field, dB, over a one meter radial distance starting out from the Earth's surface.

7. Pretend that you are given a bar magnet that is known to have a field of 1×10^8 nT adjacent to its pole. Suppose that you find that when you have your compass near that magnet's pole, it causes your compass needle to oscillate with a period of 0.1 seconds when jarred away from its normal pointing direction toward the bar magnet. What would be the period of oscillation of that compass needle when placed in the open environment (away from magnetized material) and responding only to the Earth's main field at your home city?

8. Go the website http://www.dmi.dk/projects/oersted to select "Oersted Satellite Mission". Then select "Instruments" so that you can discover the two types of magnetometers carried aboard the Oersted satellite. Return to the main page of this website and select "The Oersted News Letter". Then go down to the Oersted Newsletter #2, 2002 (February 21) and read about "Oersted 3 Years in Space" to see the satellite contributions that have been made to the IGRF 2000 field modeling.

9. A new magnetic observatory is to be established near your capital city. You are asked by your government to provide advice regarding the location, structure, and type of instrumentation. What should be your presentation?

Chapter 5
Applications

5.1 Introduction

In this chapter we will look at some of the ways in which geomagnetism finds utility in today's world. The main subjects are the impact of the geomagnetic field on modern technological systems and the application of geomagnetism to the discovery of the physical nature of our world. I also include interesting observations for which geomagnetic connections imply future application directions.

Each period range of natural geomagnetic field fluctuations can be identified with special utilization topics. For example, consider the following:

(a) For the period range from 0.25 seconds to 1 minute the primary subjects of interest are Earth crust exploration, detection of hidden conductivity anomalies, electric power transformer failures, studies of hydromagnetic wave propagation, and discovery of magnetospheric processes.

(b) For the range from 1 minute to 24 hours, studies include the structure of magnetospheric deformation and currents, thermospheric heating and winds, ionospheric currents and tides, and conductivity characteristics of the Earth's lower crust, mantle, and continental coastlines. Geomagnetic storms in this time scale affect a multitude of man-made systems such as satellites, communication systems, electric-power grids, and long pipelines (see Heirtzler et al., 2002).

(c) From the range 1 day to 1 year we obtain information about the fluid motions within the Earth's core and at the core–mantle boundary, solar activity and solar sector changes, tropospheric weather changes, and magnetospheric deformation. Our main field magnetic navigation charts are obtained from data in this period range.

(d) For the range from 1 year to 100 years, geomagnetism reveals changes in the dipole-field moment generated in the Earth's outer core, solar-cycle variability, and climatic variation in solar-weather relationships.

(e) From the range 100 years to 3,000 years, archeomagnetic and lava-flow magnetic samples provide evidence that tells of the Earth's polar wandering, nondipole outer-core drift patterns, and historic climatic changes.

(f) From the range 3,000 years to 200 million years, paleomagnetic studies reveal information about main-field reversals and dipole-field disappearances, paleomagnetospheres, and continental drift.

I have chosen to introduce the topic of applications with space environment and satellite damage subjects. The reader may wish to review those parts of Chapter 3 describing the magnetospheric structure and the nature of field disturbances. Next, I will focus on the damaging effects of geomagnetic storms on pipelines, electric-power grids, communication systems, and geographic position determinations. Geomagnetism is widely used to survey the Earth's composition, both for mineral discovery in the crust and for revelation of the Earth's structure. Geomagnetic measurements led to our understanding of the formation of continents. Magnetic charting for navigation is not as sensational as the other topics, yet this use quietly forms a preponderance of geomagnetism's applications to human lives. I will also present two subjects that may represent future interesting directions for geomagnetism: connections to global weather and with living organisms. The chapter will conclude with some information on geomagnetic activity predictions.

5.2 Physics of the Earth's Space Environment

We have learned that the Earth has a dipolelike magnetic field called the main field that originates from electric currents within the Earth's liquid outer core (Section 1.8). These currents appear to be driven by a gravitational growth of the Earth's inner core and are organized by the spin of the Earth. The dipolelike field defines geomagnetic coordinates at the Earth's surface whose poles are tilted by about 11° with respect to the geographic coordinate system (Section 1.5). All this information comes from years of carefully archived records of our Earth's field and its changes.

We saw in Chapter 3 that the geomagnetic field of the Earth extends its control over charged-particle motions far into space. The magnetosphere is clearly dipolar in form out to several Earth radii, but it assumes a more elongated teardrop appearance as its outer boundary is approached. The outer shape of the magnetosphere is fashioned by a constant solar wind plasma of ionized particles and their associated magnetic fields from the Sun. The earliest information on this space behavior came from geomagnetic observatory records interpreted early in the last century before the advent of satellites (Chapman and Bartels, 1940). Our present knowledge of the space environment about the Earth (Figure 3.19) depends upon both satellite discoveries and magnetospheric models calculated from the characteristic surface magnetic field changes.

5.3 Satellite Damage and Tracking

On-board computers in space systems orbiting the Earth have lifetimes that are mostly determined by the accumulated radiation damage to their circuitry from energetic particles in the magnetosphere. Solar-cell arrays powering the satellites also lose a few percent of their efficiency during each year of exposure to the solar–terrestrial environment. Storm-time high latitude field-aligned currents heat the thermosphere, causing it to expand upward and move toward the equator (see Section 3.10). These thermospheric winds and density changes modify the drag on satellites at their typical equatorial and polar orbiting altitudes near 500 km. The attending decrease in orbital velocity can cause transitory tracking loss and eventually shorten satellite lifetimes.

During some storm-time solar particle events, the magnetospheric compression by the solar wind forces the magnetospheric boundary inward past the geostationary satellite position at 6.6 R_e; such transitions correlate with numerous synchronous-orbit satellite operation anomalies. The offset of the eccentric dipole's midpoint from the Earth's geographic center, determined from accurate global field charting (Section 1.6), causes the South Atlantic/South American anomaly in the total field (Figure 1.23d) where a magnetospheric charged particle concentration is abnormally high. Satellites traveling through this region often experience computer memory upsets (Figure 5.1). The Hubble space telescope has been switched off during passage through this region. Placement of spacecraft (to minimize the energetic particle exposure) becomes a compromise location, with geomagnetic considerations determined from global field modeling that relies upon surface geomagnetic observatories.

Allen and Wilkinson (1992) summarized the space effects of the geomagnetic storms of 19–21 October 1989:

> During the particle events and magnetic storms of this period in October both GOES-5 and -6 experienced SEUs [single particle upsets], polar orbiters experienced multiple switching off of microwave transmitter units – operators quit 'counting' anomalies and just left the system turned off until conditions improved. TDRS-1 had 50 RAM [random access memory] hits on the 19/20th in the radiation susceptible memory chips. Even the hardened

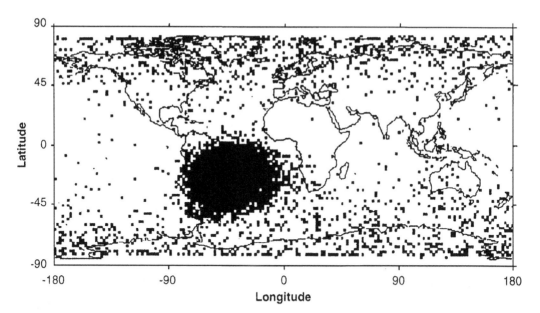

Figure 5.1. The Surrey Space Centre, UK, tracked the position of satellite memory upsets occurring on the UoSat-2 satellite from September 1988 to May 1992. These upsets were concentrated in the South Atlantic/South American field anomaly region (Figure 1.23d). Some enhancements in the polar region are a result of the arrival of cosmic ray particles guided there by the Earth's field when they have not been swept away by the solar wind. Figure supplied by Underwood, SSC.

TDRS-2 and -3 experienced SEUs. Polar orbiting UoSat-2 had many SEUs, particularly in the expanded South Atlantic Anomaly [low value of geomagnetic main-field shown in Figure 1.23d] region. Significant power panel degradation occurred on GOES-5, -6, and -7. Commercial geostationary satellites had power panel degradation, pitch glitches, and SEUs (137 events reported). Bright red aurora were seen at unusually low latitude over Japan during these events. On 20–21 October, $Ap^* = 162$ [Ap^* is the maximum 24-hr running mean of the ap index]... on board ATLANTIS [space shuttle] the astronauts reported irritating 'flashes' in their eyes as energetic protons penetrated the optic nerves. Although they are reported to have 'retreated' to the innermost shielded part of the Shuttle, the eye flashes did not subside until the proton event ended.

Satellites have been completely disabled during a period of special solar–terrestrial disturbance activity. As an example of the major problems for geostationary satellites, Wrenn (1994), of the UK Defence Research Agency, reported that "On 20 January, 1994, the ANIK E1 and E2 communications spacecraft suffered serious failures of their momentum wheel control systems. It is likely that the satellites were subjected to reduced lifetimes and to bulk revenue losses running into tens of millions of dollars." Baker et al. (1994) attributed these satellite failures to "deep dielectric charging effects" in the momentum wheel control circuitry due to a particularly intense, long-duration flux of high-energy (>1 MeV) electrons in the outer magnetosphere. Such relativistic electron enhancements of the magnetosphere seem to be controlled by the storm-time arrival of a high-speed solar wind.

Presently, it is thought that many failures of spacecraft circuitry are caused by disturbance-time internal dielectric charging of satellite components (Baker et al., 1987) and the resulting subsequent discharge. Walpole et al. (1995) showed the correpondence of such high-voltage spikes and large Kp indices. Satellite shielding (a major weight–cost consideration) is limited by useful-lifetime estimates that rely on the accurate interpretation of past solar–terrestrial disturbance behavior extrapolated from geomagnetic storm information provided by surface observatories.

Satellite drag changes with the atmospheric density modification attending thermospheric heating during a geomagnetic storm (Fuller-Rowell et al., 1994). Figure 5.2 shows the number of satellite tracks lost following a period of high geomagnetic activity. The flywheel effect of thermosphere heating (Section 3.10) and the accumulating effect of the drag on the satellite orbit places the orbit losses of approximately 9000 objects tracked in space to over 1,500 in number during the days immediately following a substorm in 2001. Although most of these trackings represent space debris, their location is important for the operational safety of all spacecraft. Working satellite lifetimes are shortened during storms as fuel is expended to recover orbital altitude.

Figure 5.2. Impact of geomagnetic storm on satellite tracking. Daily Ap index (connected line segments; scale to right) and number of satellite tracks lost (heavy vertical lines; scale to left) are displayed from 1 to 31 March 1989, overlapping the period of a geomagnetic storm. Figure from Gorney, Koons, and Waltersheid (1993).

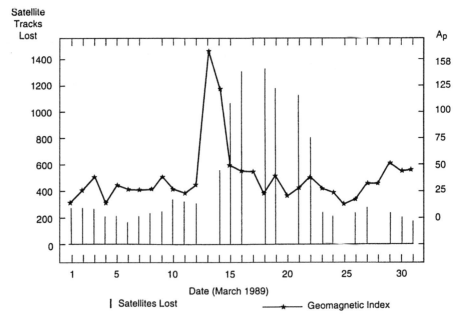

Adequate geomagnetic storm warning systems allow implementation of tracking alerts, protective procedures, or for some spacecraft, delay of command transmission to more favorable times. Space shuttle programs have plans to abort extravehicular activities (or even the flight itself) during major storms to protect astronauts from exposure to storm-time particles. In 2000 the National Academy Press in Washington DC published "Radiation and the International Space Station: Recommendations to Reduce Risk" (ISBN 0-309-06885-1), a 76-page report of convened representatives from seven USA national science organizations. The report indicates that the astronauts' health can be at risk from particle radiation during large geomagnetic storms, especially during space walks. The publication advises that special radiation protection and forecasting measures (Section 5.16) be taken. Supersonic airplanes, such as the Concorde, have arrangements to be brought to lower altitudes in extreme, storm-time, exposure-threatening conditions. "Real-time" geomagnetic information from INTERMAGNET observatories (Section 4.17) is an important part of the alarm system used at space environment forecasting centers.

5.4 Induction in Long Pipelines

From an electrical point of view, the Alaska oil pipeline is a 769-mile (1,280-km) long, surface-grounded conductor extending from about 69° geomagnetic latitude at the Arctic Ocean to about 62° geomagnetic latitude at the North Pacific Ocean. Those geomagnetic latitudes include the auroral zone of intense field-aligned and ionospheric electrojet currents. The resistance per unit length of the pipe is about 4.8×10^{-6} ohms/meter. Approximately half of the pipe length is buried. The central third of the pipeline parallels the geomagnetic latitude directions favored by the auroral electrojet current. Because of the high pipe conductivity with respect to the ground (to which it is electrically connected by zinc grounding cables), the strong fluctuating ionospheric currents (in the preferred geomagnetic east–west direction) at geomagnetic storm times induce currents to flow in the pipeline. When the pipeline was originally placed in the ground, small cuts in the electrically insulating plastic surface coating occurred. A fluctuating current, induced from the magnetic disturbances, travels between the pipe and ground; when directed appropriately, that current causes pipe corrosion. The amount of corrosion is dependent upon the frequency and amplitude of the storm-time source current, the exposed area of the pipe, the material in which the pipe is embedded, the frequency dependence of the corrosion process, and the frequency dependence of the local Earth induction (Campbell, 1986; Trichtchenko and Boteler, 2001).

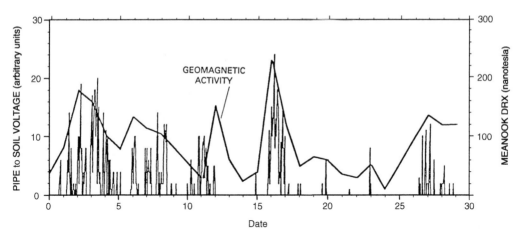

Figure 5.3. Comparison of pipe-to-soil voltage (vertical line segments in arbitrary units) on the Leismer Gas pipeline system (northeast of Edmonton, Canada) and the geomagnetic field changes (DRX – daily range of X component measured at Meanook, Canada, observatory). Figure adapted from Shapka (1993).

Figure 5.3 shows the correspondence of the pipe-to-soil voltages and geomagnetic activity. For high-latitude pipelines such as those in Alaska, Canada, and Siberia the effect of corrosion from the induced geomagnetic storm current maximizes with the 5- to 30-minute period field fluctuations. During large storms, induced currents in the pipelines can be as high as 1,000 Amperes. Estimates of the corrosion effects over the lifetime of the Alaska pipeline were obtained by comparison of presently measured pipeline currents with the USGS College and Barrow Observatory magnetic field records and an extrapolation of the field's historic behavior. It is now believed that high-latitude pipelines are corroding faster than originally anticipated because induced currents were not taken into consideration in the original design.

At all latitudes, pipeline corrosion is also caused by the steady currents generated either from the difference in contact potential between grounded points along the pipeline or from the induction of currents from nearby man-made systems. Most pipelines are protected from this harmful corrosion by applying an overriding current onto the pipe to make all exposed areas negative (a cathode) with respect to ground. Corrosion engineers regularly schedule surveys of their pipeline's potential to adjust their cathodic protection devices. Geomagnetic storms give rise to induced currents in all pipelines throughout the world. If the corrosion engineer's surveys are performed during geomagnetic storms, the measured cathodic protection voltage levels may be incorrect and cathodic protection settings, made improperly, may actually cause an increase in corrosion.

Near the dip equator locations, the strong daytime electrojet Sq fields and enhancement of low-latitude geomagnetic storm fields (Figure 2.18) induce strong currents to flow in pipelines that parallel the equatorial electrojet. The great intensity of these equatorial source currents exacerbates the corrosion in this region, compared to middle latitudes, and requires greater attention to the cathodic protection devices. The equatorial regions of intense induced currents are determined by geo-magnetic surveys.

Geomagnetic data from global observatories help make it possible for the space environment forecast centers (Section B.6) to warn pipeline management companies of the present and expected activity levels that are of importance to corrosion protection activities.

5.5 Induction in Electric Power Grids

During severe magnetic storms ($Kp \geq 7$), damaging currents can be induced in power grids, particularly at higher latitudes (Albertson and van Baalen, 1970; Kappenman and Albertson, 1990; Kappenman, 1993; Zanetti et al., 1994). The effects vary from simple tripping of circuit breakers, resulting in temporary electric power blackouts of cities, to destruction of expensive power station transformer banks and major economic loss from extensive private and industrial power disruption. The problem occurs when the storm-induced currents appear in three-phase transformers that are electrically connected by long transmission lines. Destructive localized heating occurs in the windings, capacitor banks become overloaded and trip-out, protective relays fail, and power transmission is degraded or lost completely.

As an example, on 4 August 1972, a large geomagnetic storm ($Kp = 9_0$) caused the failure of a 230-kV power transformer at the British Columbia Hydro and Power Authority of Canada. Also consider the great magnetic storm (Kp index of 9+) that occurred on 13 March 1989 (Allen et al., 1989), with major field fluctuations centered near southeast Canada and the northeast United States that caused a nine-hour blackout of the 21,000 MW Hydro-Quebec electric power system. A vivid description of that failure has been provided by G. Blais and P. Metsa (1993) of Hydro-Quebec:

> Telluric currents induced by the storm created harmonic voltages and currents of considerable intensity on the La Grande network. Voltage asymmetry on the 735-kV network reached 15%. Within less than a minute, the seven La Grande network static var compensators on line tripped one after the other … With the loss of the last static var compensator, voltage dropped so drastically on the La Grande network (0.2 p.u.) that all five lines to Montreal tripped through loss of synchronism (virtual fault), and the entire network separated. The loss of 9,450 MW of generation provoked a very rapid drop in frequency at load-center substations. Automatic underfrequency load-shedding controls functioned properly, but

they are not designed for recovery from a generation loss equivalent to about half system load. The rest of the grid collapsed piece by piece in 25 seconds.

Power pools serving the entire northeastern United States also came perilously close to a comparable calamity with similar cascading system failures during the same geomagnetic storm that affected Hydro-Quebec. In addition, the storm destroyed transformers at the Salem Nuclear Plant of the Public Service Electric and Gas Company (PSEGC), at a replacement cost of about $12,000,000. During this loss of power output, the PSEGC replacement energy cost was approximately $400,000.

The power-grid vulnerability seems dependent on its nearness to the region of maximum auroral electrojet currents, the interconnection pattern of the power-grid system, and the regional geology of high-resistance igneous rock. Figure 5.4 illustrates the regions of concern for North America. The rapidly fluctuating storm-time induced currents are thought to enter and exit power systems through the grounded

Figure 5.4. Areas of igneous rock (shaded areas) and auroral zone (cross-hatched area) locations. Interconnected power systems are vulnerable in the overlapping regions during geomagnetic storms. Figure adapted from Kappenman (1993).

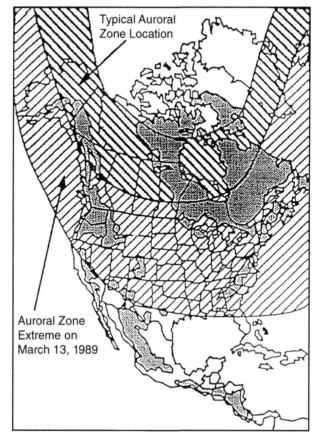

connection of transformers, causing a high level of half-cycle saturation, a destructive version of the process described for the fluxgate magnetometer (Section 4.8). What follows is a dramatic increase of reactive power consumption and localized destructive heating, then an intolerable system voltage depression, unusual transmission line flow, and relay malfunction.

Protection strategies that power companies presently use to deal with *geomagnetically induced currents* (GIC) involve system redesign, based upon the history of geomagnetic field levels at the critical locations over a solar cycle and upon adequate warning of an impending storm onset and recovery time. Improvements in geomagnetic storm forecasting mostly depend on an expansion of the global satellite and geomagnetic observatory real-time network as well as on the careful reanalysis of past records.

5.6 Communication Systems

Early realization of major communication problems began with unmanageable induced voltages on telegraph lines during the severe geomagnetic storm period of early September 1859 (Stewart, 1861, and Figure 3.11). Modern communication systems rely primarily on satellite transponders, radiowave links, oceanic and land-based cables, telephone-line connections, microwave transponders, and fiber-optic cables. Although not all these systems are sensitive to geomagnetic storms, the universal interconnection of transmission facilities can cause trouble throughout the network.

Eighty percent of all long-distance telephones in Minneapolis were silenced by the great ($Kp = 9_0$) magnetic storm of 24 March 1940. The major ($Kp = 9_0$) geomagnetic storm of 10 February 1958, caused 2.7 kV induced voltage on the Bell System transatlantic cable from Newfoundland to Scotland and caused voice communications to fluctuate from "squawks to whispers" (Lanzerotti and Medford, 1989). The 4 August 1972 magnetic storm, described above for its damage to the British Columbia power net, shut down a "long-haul" coaxial communications cable between the US states of Illinois and Iowa.

At high latitudes, satellite transionospheric radio-wave signals during storms suffer from refraction and rotation of the signal plane of polarization that are related to severe changes in total electron content along the propagation path. Storm-time signal phase and amplitude radio-wave scintillations disrupt both satellite (at 10^9 Hz) and surface high frequency (HF) communications. The scintillations arise from scattering by ionization irregularities in the altitudes above 200 km when the ionosphere is unstable during geomagnetic storms.

Geomagnetic storm conditions upset the expected pattern of received signals at those radio-wave transmission frequencies that depend upon

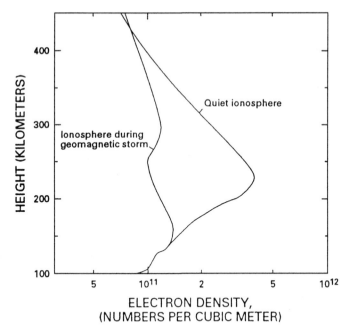

Figure 5.5. Example of electron density profiles during daytime quiet and disturbed geomagnetic conditions for summertime at middle latitudes. Figure adapted from Norton (1969).

the ionosphere as a reflecting medium. This problem is particularly severe at auroral and polar latitudes (see Section 3.11) where the ionospheric conductivity is greatly enhanced during storm conditions. At lower latitudes, the storms are responsible for phase changes in VLF (very low frequency) navigation systems, GPS reception, fadeouts of shortwave communication links, and major modification of usable radio frequencies.

Global ionospheric models used to predict propagation conditions all employ geomagnetic indices (Section 3.15) for critical adjustments. However, the disturbed ionospheric F region (Figure 5.5) and total electron content (TEC) are not easily predicted because of the multitude of conflicting geophysical processes in the ionosphere during a geomagnetic storm. For fixed-frequency station transmissions, Figure 5.6 illustrates the degradation of high-frequency radio propagation quality over six representative transmission paths during a severe magnetic storm period.

Some transmitters have flexibility in selection of broadcast frequencies. For these, the usable frequency is predicted from Ap index values that have been computed from geomagnetic records and disseminated by forecasting centers. In addition to government and industrial broadcasters, more than one million amateur radio operators worldwide make use of the geomagnetic "nowcasts" and forecasts (see Section 5.16).

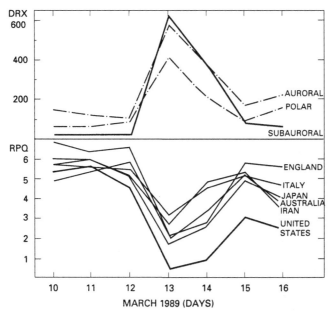

Figure 5.6. Correspondence of (top) the daily range of the *X* field component (DRX) in subauroral, auroral, and polar cap zones to (bottom) the radiowave propagation quality index (RPQ) for paths from western Germany to the United States, Iran, Australia, Japan, Italy, and England during a storm period in March 1989. Figure from Hruska et al. (1990).

5.7 Disruption of GPS

During a geomagnetic storm, a locally disturbed ionosphere introduces, into the propagation of encoded signals, irregular delays that are roughly proportional to the total ionospheric electron content along the propagation path and inversely proportional to the square of the carrier signal frequency. Although the use of dual-frequency transmissions greatly lessens the Global Position System (Section 4.16) sensitivity to electron density variations, "ionosphere free" positioning is realizable only between two sites that are closely located. Propagation delays due to geomagnetic storms can represent position errors of several tens of meters at a single receiver (Kleusberg, 1993). Storm notifications by warning centers using real-time data from geomagnetic observatories alert subscribers to positioning error periods during which GPS locationing is unreliable and corrective measures are needed. Ionospheric modelers use geomagnetic activity indices to predict the ionospheric response changes for computation of propagation path delays. Complete loss of GPS signals can occur during some *F*-region disturbance scintillation conditions. Figure 5.7 shows, for one month in 1987, the occurrence of GPS problems in Canada during times of high geomagnetic field levels.

5.8 Structure of the Earth's Crust and Mantle

Geomagnetic fields measured near the Earth's surface are a mixture of the main field of the Earth, fields from current systems in the ionosphere

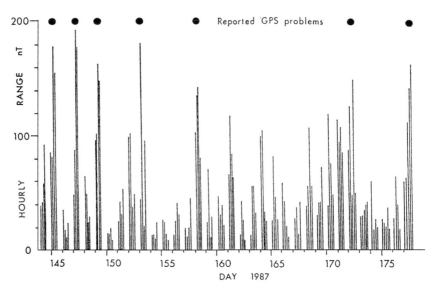

Figure 5.7. Occurrence of Global Positioning System (GPS) survey failures (solid circles) and large hourly range (nanoteslas) of geomagnetic field variation (vertical lines) reported at Yellowknife, Canada, for a month (day number at bottom) of activity in 1987. Figure provided by R.L. Coles, Geological Survey of Canada.

and magnetosphere, fields from geologic materials that have been magnetized, and fields from currents induced to flow in the electrically conducting portions of the Earth. Polar explorers "discovering" the Earth's "magnetic pole" location from the vertical dip in magnetic field are, in fact, reporting only the position of a mixture of all these sources, not the true dipole field location.

Two interrelated geomagnetic methods are used to reveal the Earth's hidden structure. In one method, magnetic irregularities caused by the various crustal deposits are determined by an area survey at quiet geomagnetic times. Typically, the International Geomagnetic Reference Field (Table 1.2) representation of the main field in the study region and an accurate value of field variations measured at a calibrated surface station site nearby are subtracted from the data set. Surface traverses, aeromagnetic surveys, and ocean-bottom surveys are of this type.

Hydrothermal alteration, metamorphism, and cataclysmic events can modify the Earth's crustal magnetization. Magnetic anomaly maps showing these changes are used in the detection of energy and mineral resources. Aeromagnetically detected anomalies over oil-bearing layers have been interpreted as an effect of magnetite that formed from chemical processes attending the microseepage of petroleum. The oil and gas production fields of the Alaskan Navarin Basin and the European North Sea Basin were discovered using aeromagnetic surveys. An off-shore aeromagnetic petroleum survey in the region of the northern Yucatan peninsula in 1991 revealed a 120-mile impact crater of an ancient asteroid that is considered to be the source of the global catastrophe that caused the dinosaur extinction 65 million years ago at the end of the Cretaceous period in the Earth's evolution.

In the other study method, the natural geomagnetic field fluctuations due to source and induced currents are interpreted to give a multidimensional crustal conductivity structure or a one-dimensional profile of the upper mantle. A disturbed field is required to provide the frequencies for probing the Earth's crustal layers. The long-wavelength quiet-field variations are used to probe deeper into the Earth's mantle.

5.8.1 Surface Area Traverses for Magnetization Fields

Magnetic irregularities occur because the Earth varies in the percentage of magnetite (Fe_3O_4) or its related minerals (which can be included in the "magnetite" name; see Section 4.2) and in the type of natural paleomagnetization processes (see Section 1.9). Magnetite is present in most rocks, although often in just very small amounts. With sensitive magnetometers one can detect the magnetization fields from the crustal embedded materials. Surveys have to be carried out in geomagnetically quiet times. Compensation for the quiet time Sq (Chapter 2) field variations and small disturbance fluctuations are obtained by subtracting the variations determined from a stationary, local field magnetometer used as a baseline. The profiling method depends upon changes in magnetic response resulting from variations in amount of buried magnetite. Occasionally, horizontal gradient measurements are appropriate. Proper interpretation of the surveys depends on the accurate collation of magnetic information with surface geology, gravity records, and seismic profiles.

Some small-area traverses are simply interpreted from the change in field strength and direction as a sensor is moved along a given line. Figure 5.8 illustrates the different responses of a magnetometer for a simple dipole source alignment. In addition to surveys for mineral discovery and geological structure, buried archeological formations have been revealed by magnetic surveys of this type. Detection is facilitated by the facts that man-made objects are typically of simple geometric design, consistently constructed of one material that differs from that of the immediate general area, and located at shallow depth. The contrasting magnetic properties of archeological interest are, as a rule, easily detected in the uniform overburden of soil, water, or rocks.

5.8.2 Aeromagnetic Surveys

A magnetometer attached to a boom extending from an aircraft is subject to contaminating magnetic fields from five sources within the plane (assuming that the flight personnel take care with their apparel; see

Figure 5.8. Surface vertical and horizontal magnetic field response from buried magnetic material exhibiting dipole magnetization (with field lines only partially shown). Arrows at top section indicate the observed field direction. For a surface traverse the expected vertical and horizontal field variations that would be measured are indicated at the bottom.

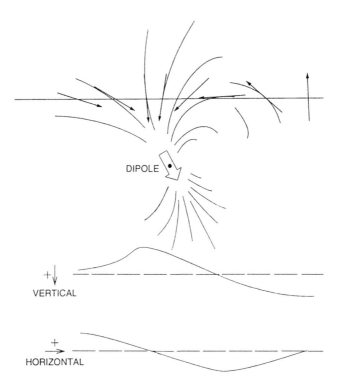

Table 4.4): (1) permanently magnetized aircraft parts, (2) parts whose magnetization is induced from the presence of the Earth's main field, (3) currents generated in electrically conducting materials (eddy currents) as the plane moves through the Earth's main field, (4) fields from the changing static charge arising from the plane in flight, and (5) currents from the aircraft electrical systems. Adequate compensation for each of these contaminants is a serious concern for all aeromagnetic programs. Compensation methods involve computer program adjustments to the output registration or the placement of magnetic strips and current-carrying coils near the geomagnetic field sensor head.

To determine the compensation values for aircraft, prescribed air swings are made over a major magnetic observatory. The air-borne magnetometer responses are evaluated with respect to the speed, yaw, pitch, and roll of the aircraft in the four cardinal directions. Calibration flights, as well as surveys, are made on geomagnetically quiet days, which are determined by standard observatories and solar–terrestrial disturbance forecasting centers (Section B.6).

The first and fifth types of magnetic field contamination source described above should require permanent adjustment for an unchanged plane configuration, but aircraft magnetization must be regularly reevaluated because physical shock to the aircraft (such as a hard landing) can shock magnetization to a new local-field arrangement (see Figure 4.2). Periodic recalibration air swings over the observatory are important. The second type of contamination (induced magnetization) can be evaluated from calibration flight paths, but the induction effect changes for each survey region in which the Earth's main field is significantly different. The third type of contamination should be evaluated in a manner similar to that for the second with the addition of calibrations for the usual aeromagnetic cruising speeds. The fourth type of contamination, fields from in-flight static charge buildup and dissipation, is difficult to compensate for because the charge variations follow weather conditions. However, the range of contamination can be discovered through a comparison of magnetometer readings from overflights of the fixed observatory location in a variety of weather situations.

The most extensive regional aeromagnetic surveys have been made by the US Navy to support the production of global magnetic charts of importance for navigation. Figure 5.9 illustrates the coverage of this global aeromagnetic data set from the "Project Magnet" program from 1950 through 1990. The data is now available from the magnetic field archives at World Data Center, Boulder (see Section B.2). Surveys by the US Navy were made at "high levels," usually between 15,000 and 25,000 feet (4.6 and 7.6 km) elevation. Over the years, navigational accuracies have improved from ±5 nautical miles (±9.3 km) to ±100 meters. Three-component fluxgate magnetometers calibrated with an optically pumped metastable helium magnetometer gave Project Magnet field-determination accuracies of approximately ±15 gamma. Changes in aeromagnetic techniques over the years have increased the mapping contour intervals from a "standard sensitivity" of 10 gamma to

Figure 5.9. Tracks of US Navy Project Magnet aeromagnetic flights for which data were deposited at NOAA National Geophysical Data Center. Figure provided by R. Buhmann, NGDC/NOAA.

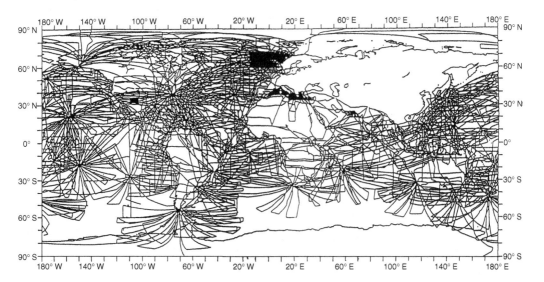

Figure 5.10. Schematic representation of the detection of seamounts using aeromagnetic techniques. Figure from US Naval Ocean Research and Development Activity report, 1993.

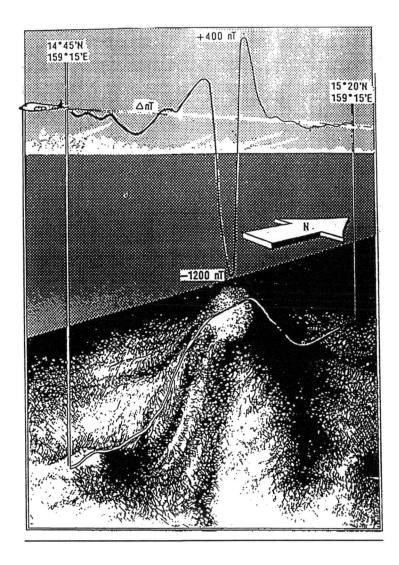

a "medium sensitivity" of near 1 gamma, to a "high sensitivity" of about 0.1 gamma. Figure 5.10 illustrates the aeromagnetic evidence of a major seamount.

High-resolution aeromagnetic surveys typically are flown at altitudes between 150 and 300 meters (about 500 to 1,000 feet). For example, the Australian national coverage was completed at a flight elevation of 150 meters and line spacings of 1.5 to 3.2 kilometers, but special areas are being surveyed at about 80 to 100 meters elevation with line spacings of 400 to 500 meters. Figure 5.11 illustrates the magnetic contouring used in a determination of the regional geology of the United States. Canada's aeromagnetic survey program used a flight elevation of

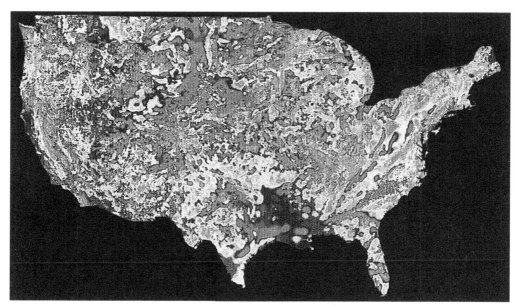

Figure 5.11. Grey scale version of colored crustal magnetic anomaly map of USA. Original color reproduction has spectral scale representing magnetic field magnitudes after removal of main Earth field model. Data from NGDC/NOAA "Geophysics of North America" CD rom that was plotted by D. Hastings.

300 meters at a line spacing of 800 meters. Finland, having been surveyed at an elevation of 150 meters with 400-meter spacing, has begun more detailed 40-meter terrain clearance flights using a 200-meter line spacing.

Ideal flight-line separations are generally about twice the distance from a position B at the geological "magnetic basement" (for crustal studies, the depth at which the Earth may be assumed to have an infinite conductivity) to the aircraft position A. The sample interval is, by rule, less than one quarter of A–B. Flights are usually restricted to the quieter geomagnetic activity days by using disturbance information from the nearest geomagnetic observatories and activity predictions from space environment forecasting centers (Section B.6).

Three methods are used for adjustment of flight-time measurements to variations in geomagnetic activity: (1) Mapped fields are taken to be the difference between the observation and the field measured at a base-station magnetometer. (2) With a "checker-board" grid flight pattern, adjustments are applied to bring the cross-track values into accord; these adjustments define linear alterations (called "leveling") of the remaining data samples. (3) Best estimates of the "transient variations" (small disturbance fields and quiet-day, Sq, variations) defined for each data sample location and time are removed from the data. The transient fields can be obtained from local magnetic observatories or (in the absence of such base stations) from a quiet-field synthesis program such as **SQ1MODEL** (Section C.3). Errors increase with the distance between the flight path and base station.

Because of the difficulty in identifying seasonal and secular variations of the main-field measurements, problems arise when any attempt is made to match boundaries of adjacent regions surveyed at different periods. As a result, long-wavelength distortions can be expected in composite magnetic anomaly maps of sizes much greater than the original survey areas (typically several hundred kilometers). Continental maps showing major field trends should be viewed with caution.

Survey problems arise because the upper atmospheric sources induce currents in the conducting Earth materials; their field effect, detected in the aeromagnetic flight, represents a noise background or a distorted quiet daily variation. The induced fields depend upon the conducting-body geometry, the source current configuration, and the field fluctuation frequency. Short period induction sources (Section 3.13) can be avoided with the selection of quiet geomagnetic conditions for the day of flight. For the longer-period daily variations of Sq (Chapter 2), it is usually assumed that the induction region varies little over the survey area so that a nearby base-station field daily variation (external plus internal Sq) can be subtracted from all the flight records. Occasionally special problems arise when the Sq currents are channeled into unique conductivity alignments within (or near) the survey area or the base station.

5.8.3 Conductivity Sounding of the Earth's Crust

A precise electromagnetic technique for determining the location of anomalous structures in the Earth is called *geomagnetic depth sounding* (GDS). The GDS method uses measurements of all three orthogonal magnetic field components over a large-area array of temporary observation sites. Appropriate equations relate the vertical Z-component field to the horizontal H and D (or X and Y) components for each source frequency. Conducting structures are interpreted from special plots of vector direction, response contours, etc., within the region (see Gough and Ingham, 1983, for more details).

In another important survey technique, the ratio of the orthogonal surface electric and magnetic fields is determined to obtain the electric conductivity (in Siemens/meter or its reciprocal, the resistivity in Ohm-meters) as a profile of the Earth beneath the measurements. To provide some understanding of the method, let us consider the following. The currents that are induced to flow in the conducting Earth decrease with the oscillation period and with the depth into the conductor. The parameter, δ, called the "skin depth," defines the distance within a uniformly conducting Earth at which the diminishing field of oscillating period, T seconds, reaches $1/e$ (about 37%) of its surface amplitude (for the constant e see Section A.3). The attenuation of source signal with depth into the Earth is similar to the radio-wave fading experienced by an auto driver listening to his radio after entering a tunnel. In addition to a period–amplitude relationship, there is a phase rotation of the electromagnetic wave as it penetrates the conducting medium.

Measurements at the Earth's surface give the sum of the induced and source fields. The ratio of the phase and amplitude of the surface electric and magnetic fields provides the specific information on the interior conducting Earth structure. In operation, this method of determining electric conductivity has been named "MT" or "magneto-telluric" sounding for the combination of magnetic and telluric fields that are used.

To a first rough approximation, the apparent conductivity, σ Siemens/meter, in the region of the measurement is

$$\sigma \approx \frac{5}{T} \left| \frac{B_y}{E_x} \right|^2 , \tag{5.1}$$

where T is the field fluctuation period in seconds, B_y is the magnetic field in gammas, and E_x is the electric field in millivolts per kilometer (the subscripts x and y refer to horizontal orthogonal directions). The relationship between the two electric (E_x, E_y) and two geomagnetic (H_x, H_y) components is called the *plane wave transfer function*. This function is computed using special complex impedance elements $(Z_{xx}, Z_{xy}, Z_{yx}, Z_{yy})$. The "plane wave" name arises because the method for deriving this relationship assumes a source represented by a two-dimensional field (plane) with oscillations of period T extending to infinity in the x and y directions. This is a valid model when the region to be explored is small with respect to the actual source extent. Problems would arise beneath an electrojet current.

In a given conductivity half-space, longer-period field fluctuations penetrate to greater depths. The apparent conductivity determined in this way is assumed to be valid to a penetration depth of Z kilometers approximately given by

$$Z \approx \frac{1}{2\pi} \sqrt{\frac{5T}{\sigma}}. \tag{5.2}$$

Present analysis techniques are considerably more complex than these equations would indicate. The relationship of both the amplitude and phase of the electric and magnetic fields enters more exact computations. With observations at a multitude of frequencies, a depth profile of the apparent conductivity for the region is obtained. This profile is taken to be a smoothed representation of the true conductivity. A model structure of the crust is then adopted with an appropriate depth scale

Figure 5.12.
One-dimensional inversion of magneto-telluric data to show conductivity (or resistivity) as function of depth. Dashed lines are smoothed from computations (dots). Solid lines represent conductivity model structure derived from the analysis. Figure adapted from Schmucker (1970).

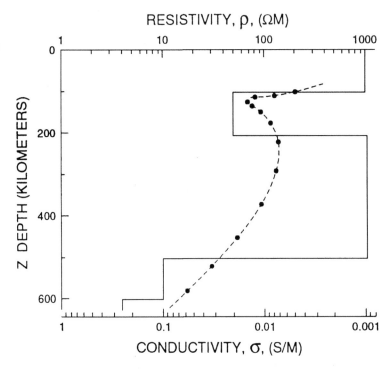

to fit the MT profile (Figure 5.12). Processing the data in this way is called "MT inversion" (see Oldenburg, 1990, for more details). Before proceeding with an inversion of the MT data, the scientist must consider the numerous electromagnetic distortions such as the galvanic (static shift, W-effect) and other special induction effects. Small near-surface three dimensional conducting bodies are accommodated by unique decomposition methods (see Groom and Bailey, 1989).

The interrelationship of the magnetic and electric (telluric) field (MT) vectors becomes substantially more intricate for the realistic anisotropic and laterally inhomogeneous Earth in which the conductivity must be represented as a function of directions other than depth. The problem resolves into modeling minimum-structure arrangements of conducting masses of defined boundaries in two or three dimensions (2D or 3D models). Figure 5.13 illustrates the 2D model of the Juan de Fuca plate subduction zone of oceanic crust near southwestern Canada.

There are always measurement errors, so the data cannot be fitted exactly. Often more than one model will satisfy the MT responses. Special techniques have been developed to evaluate the misfit between the data and the model. With the ever-increasing level of undesirable electromagnetic sources resulting from increasing population, special noise suppression methods using remote reference stations have become desirable (Goubau et al., 1989) to avoid false conductivity distribution

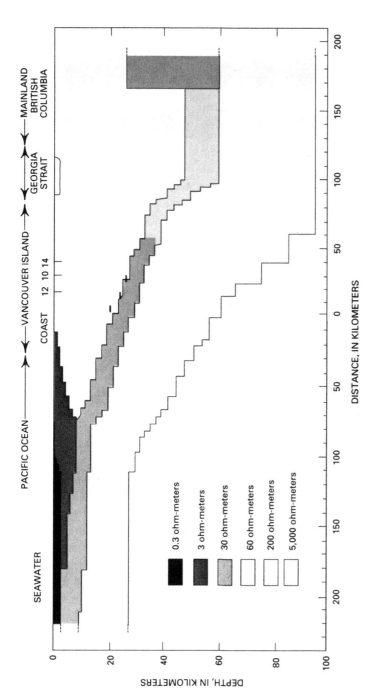

Figure 5.13. The 2D electrical resistivity model of the Juan de Fuca plate subduction region near Vancouver Island, Canada. Black horizontal bars beneath sites 10, 12, and 14 on Vancouver Island indicate the corresponding seismic reflector location. Figure adapted from Kurtz, DeLaurier, and Gupta (1990).

results. Usually, the true distribution of conducting materials is not obtainable, nor is the true error of the computations. Therefore there is always considerable reliance on a comparison of the conductivity with the seismic evidence of the Earth's substructure; major changes at specific depths must be co-located. However, with the seismic evidence alone we cannot compute the conductivity that is necessary for composition determination. Therefore, modeling accuracy also depends upon the detailed laboratory measurement of geologic material conductivities.

Induction workers often refer to forward and inverse problems or computations. By the term "forward" they mean one in which the condition of a conducting mass pattern in the Earth is established and there is an attempt to discover what would be the behavior of induced fields that could be measured. By the "inverse" they mean just the opposite; the source and induced fields have been measured and the Earth's conductivity profile is to be computed.

Special MT continuous monitoring is occasionally employed in active regions where crustal conductivity properties can change. Such measurements have proved useful in active volcanic and seismic fault zones and have the potential for providing warning precursors for major surface disruption (Hayakawa and Fujinawa, 1994). MT techniques have also been used to determine, from the Earth's surface, the burn-front location of fire-damaged coal mines because the hot dryer region has lower conductivity.

Natural micropulsation field variations (Section 3.13), which have a wide range of frequencies and great horizontal uniformity (at least at latitudes away from the auroral zones), are a convenient source field in the application of magneto-telluric prospecting techniques. Although crustal modeling usually focuses upon the first 10 km below the Earth's surface, reliable conductivity determinations have been extended to about 200 km. Earth conductivity profiles in regions from the middle crust into the upper mantle have been closely related to the regional surface heat flow (Adam, 1978). The MT method is used principally for mineral exploration, geothermal prospecting, and the determination of geological structure and tectonics (that can be indicated by graphitized rocks and fluid in geological shear zones or thrust sheets. See Adam, 2001). The technique is particularly valuable in cases where a sedimentary overburden might conceal an important substructure. Magnetic induction methods are used to detect buried pipeline ruptures. Shifts in conducting magma, as precursors of a volcanic explosion, have been detected with magnetic means. The New Madrid (Missouri, USA) seismic zone and San Andreas (California, USA) fault zones have been delineated by MT methods. For solving environmental problems and for near-surface exploration, the higher frequency modes of magneto-tellurics are now applied using sources of electromagnetic wave frequencies in the ELF, VLF, and even radio station broadcast ranges. (Revisions to Section 5.8.3. have been provided by A. Adam.)

5.8.4 Conductivity of the Earth's Upper Mantle

For application to deep-Earth conductivity modeling, the field that is measured must come from source currents completely external to the Earth and from source-induced currents arising within the Earth. The size of the geomagnetic current source must be large with respect to the skin depth in the region probed. The source must also be "well-behaved," meaning that it must change only slowly and smoothly, in linearly prescribed ways. The quiet daily variation, Sq, currents obtained from field records of standard magnetic observatories (Chapter 2) meet these requirements for determining the Earth's upper-mantle conductivity.

Let us explore this method in greater detail because we have already gone through the expressions for separating external and internal spherical harmonic analysis (SHA) field representation coefficients in the first two chapters. The special mathematical properties established for SHA allow each harmonic to be analyzed independently for an induction response. The method (Campbell et al., 1998) involves computation of the complex ratio of the internal to the external components, S_n^m, expressed in terms of phase and amplitude, and the establishment of a complex number transfer function, $C_n^m = z - ip$, for producing the conductivity profile from the spherical harmonic coefficients (see complex numbers in Section A.6). There are restrictions on the arguments (a word describing a feature of complex numbers) of C and S. Then, for each SHA coefficient (in nT) of degree n and order m, the depth (d in kilometers) to a layer of conductivity (σ in Siemens/meter) that could be responsible for the induced field measurements is obtained from

$$d_n^m = z - p \tag{5.3}$$

and

$$\sigma_n^m = 5.4 \times 10^4 / m \, (\pi p)^2, \tag{5.4}$$

where

$$z = \frac{R}{n(n+1)} \times$$

$$\left\{ \frac{A_n^m \left[n \, (\text{aex})_n^m - (n+1) \, (\text{ain})_n^m \right] + B_n^m \left[n(\text{bex})_n^m - (n+1) \, (\text{bin})_n^m \right]}{\left(A_n^m \right)^2 + \left(B_n^m \right)^2} \right\}$$

$$\tag{5.5}$$

and

$$p = \frac{R}{n(n+1)} \times$$

$$\left\{ \frac{A_n^m \left[n(\text{bex})_n^m - (n+1)(\text{bin})_n^m \right] - B_n^m \left[n(\text{aex})_n^m - (n+1)(\text{ain})_n^m \right]}{\left(A_n^m\right)^2 + \left(B_n^m\right)^2} \right\}$$

(5.6)

with

$$A_n^m = \left[(\text{aex})_n^m + (\text{ain})_n^m \right]; \quad B_n^m = \left[(\text{bex})_n^m + (\text{bin})_n^m \right],$$

(5.7)

where R is the Earth radius in kilometers, the (aex) and (bex) are the external cosine and sine coefficients, and the (ain) and (bin) are the internal cosine and sine coefficients from the SHA of the Sq field.

Because the profiles are simply the best fit of conductivities that can produce the same fields, the values are called "substitute" conductivities at "equivalent" depths. The reality of the computed values depends on independent corroborating seismic and laboratory evidence. In Figure 5.14 the upper-mantle density, obtained from seismic data for the same depth range, is plotted together with the substitute conductivity. The similarity in pattern changes supports the assumption of a true representation of the upper-mantle conductivity profile.

The measurement of the electrical conductivity by geomagnetic procedures, established independent of seismic evidence, helps define

Figure 5.14. Upper mantle conductivity profile (large dots) obtained from analysis of quiet-day geomagnetic records at North American observatories. Solid line segments indicate density values obtained from seismic data.

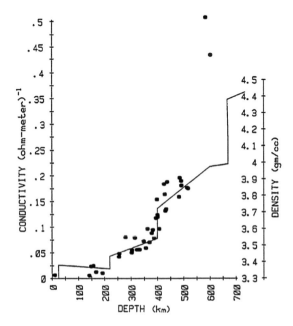

the candidates for the Earth's mantle composition in high-temperature/pressure laboratory modeling. Also, values from Equations (5.3) and (5.4) provide magnetic basement conductivity levels for the more shallow crustal conductivity determinations.

5.9 Ocean Bottom Studies

Unequivocal evidence of continental drift, tectonic motion, and seafloor spreading was obtained from marine magnetic survey records. In opposite horizontal directions, perpendicular to midoceanic ridges, ship-towed magnetometers have detected symmetric patterns of magnetization that were acquired during the cooling of hot magma that arose from the spreading region mid-oceanic ridges, just tens of meters in width. Since the discovery of the patterns (Vine and Mathews, 1963), some geomagnetic programs for ship-towed magnetic surveys (Figure 5.15) have had as their objective the global establishment of chronologies of seafloor geomagnetic polarity changes so that the structure and evolution of the seafloor can be revealed.

National economic interests have driven extensive continental-shelf exploration with ships for coordinated marine seismic and magnetic measurements. Such surveys have led the way to major aeromagnetic discoveries of oil and gas reserves. National military concerns center on the use of magnetic sensors for airborne, ship-towed, and subsurface detection of vessels, and the determination of natural continental-shelf magnetic anomalies where submarines may be hidden from

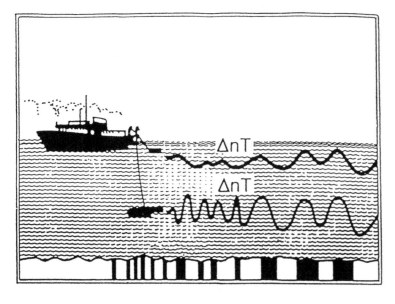

Figure 5.15. Schematic of ship-towed magnetometers responding to ocean-bottom magnetic striations caused by seafloor spreading of crust in which field alignments were frozen-in during the original magma cooling. Figure from US Naval Ocean Research and Development Activity report (1993).

aeromagnetic discovery. Most national naval facilities have arrangements for demagnetization (degaussing) of their vessels.

Ship-towed magnetometers are usually about 500 to 1,000 ft (152 to 305 m) from the vessel. Ocean surface measurements are made at about 50 ft (15 m) below the surface for stabilization. Deep-ocean magnetometers have been towed at a depth of about 3,000 ft (914 m). On occasion, gradient determinations are made with separated sensors towed in-line. Compensation for the permanent and induced magnetization of the ship and for eddy currents is similar to that for aircraft (although greatly different in magnitude) and relies on standard geomagnetic observatory calibration.

Both proton (total-field) magnetometers and the combination of fluxgate (vector) and proton (for calibrating the fluxgate) magnetometers are used in ship-towed systems. Ship-towed recordings have been valuable in augmenting the database for global magnetic field modeling. As knowledge of the geology of the continental margins has become increasingly important for oil exploration, ship-borne magnetic surveys of the coastal regions have gained further significance.

Paleomagnetic (see Section 1.9) ocean-bottom mapping of the mid-ocean ridges that represent the spreading zones of continental drift shows recurring field-direction patterns that extend perpendicular to the spreading axis. Molten mantle material, at temperatures above the Curie temperature (Section 1.8) at which magnetization is lost, slowly extrudes into the ridge and hardens, locking in (by TRM, see Section 1.9) a record of the main field at the time of cooling. The striping is a result of remanent magnetization that resulted from the Earth's field change in polarity. Assuming an average horizontal flow rate of about 1 to 3 cm/yr, this upwelling zone records the spreading of the oceanic and continental plates. The frozen magnetization is a "magnetic tape recording" of ancient magnetic field directions that clearly shows reversal patterns (Figure 5.16). For regions where the present spreading rate is established, a consistency assumption provides a preliminary dating of reversals that occurred many millions of years ago (see Opdyke and Channell, 1996).

The paleomagnetic reversals of field over time form an irregular pattern, much like tree growth rings, that can be matched to continental samples of unknown date to determine their time of formation or matched to similarly dated samples to fix their tectonic movement. Established polar-wandering curves (Piper, 1987) have been used to find the likely age of a group of rocks taken from the same tectonically stable block. Relationships have been established, dating backward about 200 million years (My), between the paleomagnetic field intensity changes, the Earth's orbital eccentricity, and climatic changes (Wollin et al., 1977).

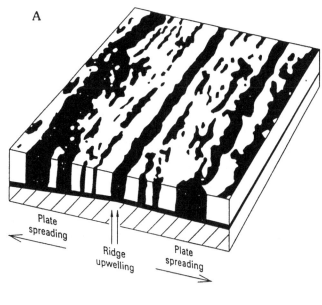

A

Plate spreading

Ridge upwelling

Plate spreading

Figure 5.16. Field reversal patterns for the Reykjanes Ridge south of Iceland (after Heirtzler, LePichon, and Brown, 1966): (a) schematic of pattern, (b) field strength from single traverse, (c) polarity time sequence assuming 1 cm/yr spreading rate. Figure adapted from Tarling (1971).

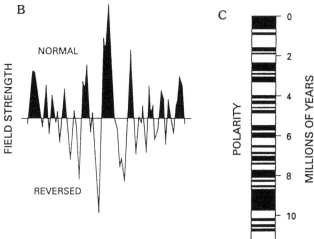

B

NORMAL

REVERSED

FIELD STRENGTH

C

POLARITY

MILLIONS OF YEARS

0

2

4

6

8

10

5.10 Continental Drift

We learned in Section 1.9 that a virtual pole was obtained from a rock sample using Equations (1.87) to (1.90). With many independently dated samples from a given region the apparent polar wander (APW) path could be determined. For paleomagnetic periods, the APW from a single continental plate could arise either from a change in location of the Earth's spin axis with respect to the continents (called *polar wander*) or from a drift of the continental plate itself. When, for two continental regions, a part of their APW paths are alike, it can be assumed that

Figure 5.17. (a) A single continent "Pangaea" existed about 225 million years ago during the geological Permian Period. (b) In the Triassic Period there was clear evidence of a supercontinent breakup that continued into the (c) Jurassic Period. By 65 million years ago, in the Cretaceous Period, the present continental plate locations had become recognizable. Figure from USGS website http://pubs.usgs.gov/publications/text/historical.html.

the continents were joined at that time. However, with different pole locations defined from a number of plates, an explanation in terms of *continental drift* is favored.

The validity of continental motions finds support in the reconstruction of geologic features from separated continents, paleoclimate evidence, and matching of animal and plant species. In recent years, unequivocal evidence for continental drift theory came from studies of seafloor spreading (Section 5.9) and the astoundingly accurate global positioning system (Section 4.16) determinations of continental motions. The discovery (Wegener, 1929) and verification of continental drift have been the monumental achievements of paleomagneticians. Figure 5.17 shows their reconstruction of the continental plates during four time stages reaching back 225 million years.

The paleomagnetic, ocean bottom, and seismic studies have shown how the Earth's crust is presently divided into a great number of plates. These plates, in motion with respect to each other, drift about one to

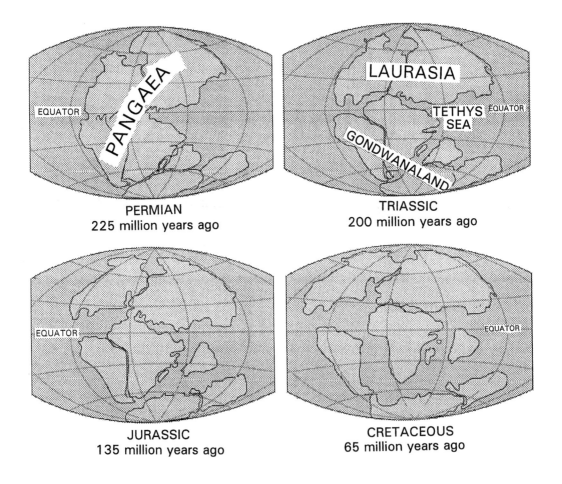

PERMIAN
225 million years ago

TRIASSIC
200 million years ago

JURASSIC
135 million years ago

CRETACEOUS
65 million years ago

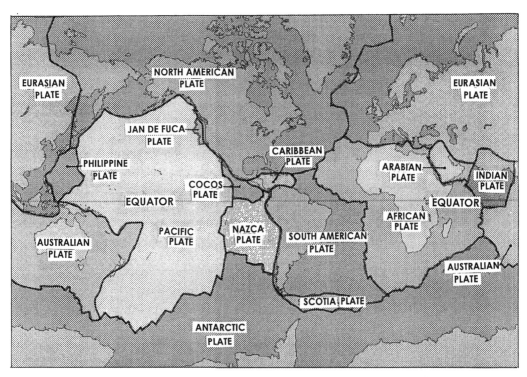

Figure 5.18. Location and names of the major tectonic plates of the Earth. Figure from USGS website http://geology.er.usgs.gov/eastern/plates.html.

three centimeters each year. Figure 5.18 shows the present distribution and naming of the major global plate system.

5.11 Archeomagnetism

Buried archeological formations, such as the Roman Wall in England, have been identified with ground magnetic surveys. An aeromagnetic survey of the ocean surface fields is used in the search for recoverable ancient shipwrecks. Sample archeological materials that acquired a magnetic remanence at the time that they were baked (e.g., bricks or clay pots) can be matched with the historical geomagnetic field vectors to establish the age of the sample. Dating capability, by this method, has been extended back 5,000 years. For the last century, about 1° in changed field direction represents about five years of age. Sample remanence is also used to reconstruct artifacts from pot shards and establish archeological structure positions from the parallel magnetic vector orientations.

5.12 Magnetic Charts

Charting the changing Earth's surface magnetic field has been a necessary and regular function of the major nations of the world since the early days of global exploration. Typically, contour-type lines of equal (isomagnetic) increments of field are plotted. Because mercator projections preserve azimuthal relationships, the magnetic field declination is best represented on such maps. For consistency, the other field components are similarly displayed (Figures 1.23a–1.23d). Dashed contours in Figure 1.25 illustrate the secular change in nanotesla/year over the United States for the 2000 epoch. Such values are used for a linear projection of the field levels up to five years following the mapping epoch.

Field-disturbance measurements by observatories play a role in the improvement of main-field models. Adjustments to the spherical harmonic coefficients representing a reference main field are made using empirical relationships to a global activity index, Dst (cf. Equations (1.83) to (1.86)). These adjustments are necessitated by the unavoidable inclusion of activity currents in the data samples used in modeling.

5.13 Navigation

Originally, the compass was a necessary part of maritime navigation. A ship's latitude could usually be determined at night from the elevation of stars in the region of the North or South Pole zenith. However, fixing the ship's longitude depended upon a daytime determination of the difference between the time of highest Sun elevation and the ship clock time, set when leaving home port. Between the night and daytime latitude–longitude determinations and in overcast conditions, the ship held its course with the compass. The prime years of global oceanic exploration fostered designs for improved ship clocks for better longitude fixes. The east–west distortions of charts from those times give evidence of early errors in longitude.

Since the eighteenth century, maritime global magnetic charting has been an important government-supported science in the world's major countries. Of course, aircraft and ships equipped with advanced technology now use GPS satellite positioning (Section 4.16). However, cruise headings are still given in magnetic directions and most of the world's aircraft and oceangoing small vessels still depend on the simple magnetic compass and charts for navigation. Air-flight maps contain prominent magnetic declination information so that aircraft can use their compasses to identify their bearings and make in-flight course alignments in the magnetic coordinates provided by the control towers. National airport runways, numbered with the magnetic east declination, must be periodically resurveyed. Airplanes and ships need calibrations of their onboard magnetic compasses to apply adjustments for the magnetic materials and fields from electronics that are part of their construction. Whether for navigation of the hikers and small boat/plane pilots who are all following a local chart with compass directions or for the major degaussing of naval vessels, the Earth's field measurements remain an important mission of modern nations.

Information about the local magnetic declination is in common use throughout the world. Almost all topographic maps of small areas contain, along with the date of map production, a graphical indication of the magnetic declination (and annual change) for the area. Seasoned backcountry travelers take along a map and compass for establishing their routes. Surveyors still make note of compass alignments; some older property lines, originally run with only compass headings, have had to be recreated from historic magnetic information. The assistance of magnetic field charts continues to be a necessary part of modern man's activities.

5.14 Geomagnetism and Weather

At selected places about the world the year-to-year changes in growing-season length, tree-ring separation, temperature, pressure, rainfall, thunderstorms, storm-tracks, winds, etc., have shown clear year-to-year fluctuations relative to sunspot (or double-sunspot) cycles (Figures 5.19 to 5.21) (Carapiperis, 1962; Roberts and Olson, 1973; Hines, 1974; Roberts and Olsen, 1975; King, 1975; Gnevyshev and Ol', 1977; Herman and Goldberg, 1978; McCormac, 1983). Northern Hemisphere annual mean temperatures are lowest near sunspot maximum and highest near sunspot minimum. Tropopause height over the western Pacific region varies, in phase, with the sunspot cycle. At equatorial latitudes, average annual rainfall is greater during years of solar maximum. Droughts in the western United States follow a 22-year alternate sunspot cycle. Courtillot et al. (1982) showed that there was a relationship between the secular variation of the geomagnetic field and the global temperature from 1860 to 1980.

There is a well-known covariation of sunspot (solar) activity and geomagnetic disturbances (see Figure 3.55). It may be possible that the thermospheric heating by electric currents associated with geomagnetic storms causes global modification of the atmospheric pressure, whose persistent patterns may alter seasonal weather conditions. Presently, an active research topic is the role of geomagnetic mechanisms in weather modification.

Some specific connections between geomagnetic storms and weather have been established. Figure 5.22 shows the recurrent high-latitude, Northern Hemisphere pressure changes, 3 days after geomagnetic storm commencements in winter months. The solar sector boundary crossings (registered as interplanetary magnetic B_y field changes at the

Figure 5.19. Smoothed
annual total rainfall at three
locations in South Africa
compared with double
sunspot cycle. (Line 1)
Rustenburg (26°S, 27°E), (Line
2) Bethal (27°S, 30°E), and
(Line 3) Dundee (28°S, 30°E).
Figure from King (1975)
(reported in Herman and
Goldberg, 1978).

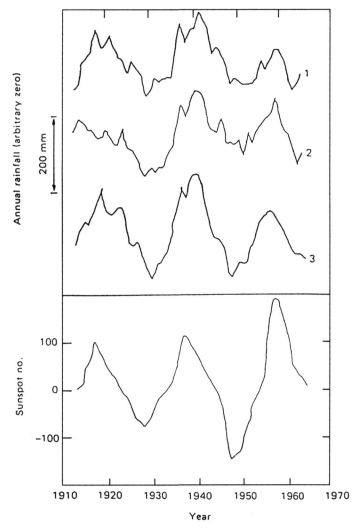

magnetospheric boundary) are associated with large increases in geo-magnetic activity.

Solar cycle effects on climate have been documented. Lawrence (1965) found that air surface temperatures seemed to be out-of-phase with the sunspot number (Figure 5.23). Friis-Christensen et al. (1991) found that there was a correspondence between the sunspot cycle duration and the Northern Hemisphere temperature change (Figure 5.24). Lassen and Friis-Christensen (1995) reported an association of terrestrial climate with the length of the solar cycle. Recall that Hoyt and Schatten (1997) discuss effect on the Earth's climate change caused by the solar ultra violet radiation reaching the upper atmospheric

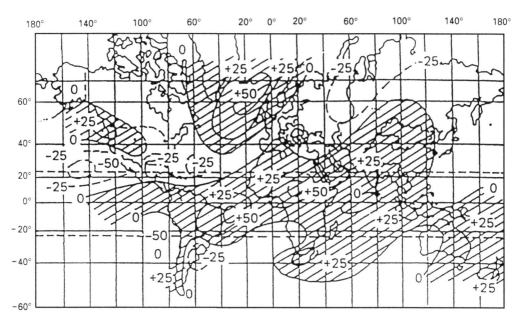

Figure 5.20. Global map of annual rainfall difference (in centimeters) between sunspot maximum and sunspot minimum, 1860 to 1917. Shaded area represents greater rainfall at maximum. Figure from Claton (1923) (reported in Herman and Goldberg, 1978).

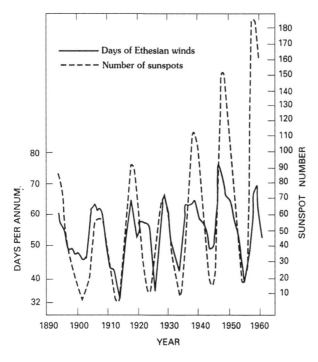

Figure 5.21. Occurrence of Ethesian winds of Athens, Greece, from 1891 to 1961 (scale to left) compared to the sunspot number (scale to right). "Ethesian" is the name for the recurring (summer and early autumn) dry northerly winds in the region. Figure from Carapiperis (1962) (reported in Herman and Goldberg, 1978).

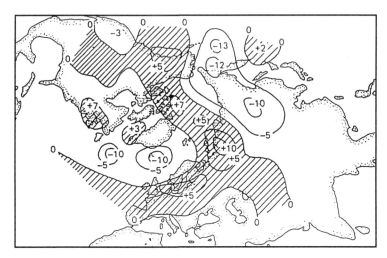

Figure 5.22. Polar view of averaged surface pressure changes three days following fourteen geomagnetic storm sudden commencements in the Northern Hemisphere for 1950 to 1970 winter months. Contour lines represent differences from the average level (in millibars), and regions of pressure increase are shaded. A reproducible, high-latitude, stationary pressure pattern is shown to be associated with the geomagnetic storms. Figure from Sidorenkov (1974) (reported in Herman and Goldberg, 1978).

ozone layer. Over a number of sunspot cycles there seems to be a gradual change in the balance between the radiation darkening effect caused by the sunspot region and the increase of radiation by the active region flux over the solar active years.

Figure 5.25 shows the systematic decrease in Northern Hemisphere high-latitude atmospheric pressure surfaces about 4 days after a sector boundary (Figure 3.41) crossing. A low-pressure vortex index was determined that represents approximately the high-latitude Northern Hemisphere area enclosed within the 300 millibar atmospheric pressure contour. Figure 5.26 shows that the vortex index regularly goes to a minimum value about one day after the sector boundary crossing (Wilcox et al., 1983).

5.15 Geomagnetism and Life Forms

The atomic precession physics used for the proton magnetometer (Section 4.9) has been applied in medical science. Because hydrogen atoms (protons) in random spin make up a large portion of the molecules in a human body, an externally applied intense field can cause about 0.0003% of these atoms to align their natural spin and precess (with Larmor frequency, Equation (2.7)) in synchronization (resonance) with a period determined by the applied field. Because each group of human

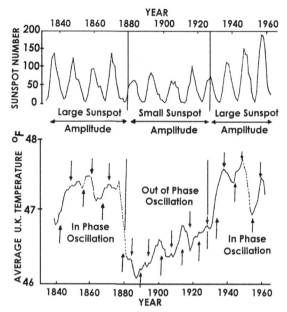

Figure 5.23. The sunspot number seems to be out-of-phase with the Earth's surface temperatures measured in UK from about 1880 to 1930 and in phase from about 1805 to 1964. Figure adapted from Lawrence (1965).

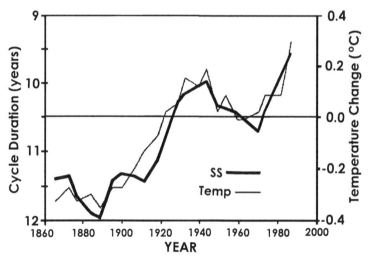

Figure 5.24. Correspondance of the sunspot cycle duration and Northern Hemisphere temperature change from about 1865 to 1987. Figure adapted from Friis-Christensen and Lassen (1991).

tissues has its own particular density of protons, there are characteristic responses of the differing precession signals. With special computer techniques (tomographic analysis) selected-section images of the patient's functioning biological system are recorded and displayed with a speed that can follow the ongoing bodily processes. This harmless medical nuclear magnetic scanning is named *Magnetic Resonance Imaging* (MRI) and is superior to X-rays in many ways.

In 1974, R.P. Blakemore of the University of New Hampshire discovered fresh water pond magnetotactic bacteria that aimed their movements

Figure 5.25. Variation of height difference of pressure surfaces between 40°N to 50°N and 60°N to 70°N latitudes before and after the arrival of an IMF sector boundary passage on day 0, based on a superposed epoch analysis of 54 crossings during the winters of 1964 through 1970. Approximate heights (kilometers at left) of pressure surfaces (millibars at right) are indicated. For high latitudes a sudden decrease in atmospheric pressure follows the time of a sector boundary change and is shown to appear consistently at eight atmospheric pressure altitudes. Figure from Svalgaard (1973) (reported in Herman and Goldberg, 1978).

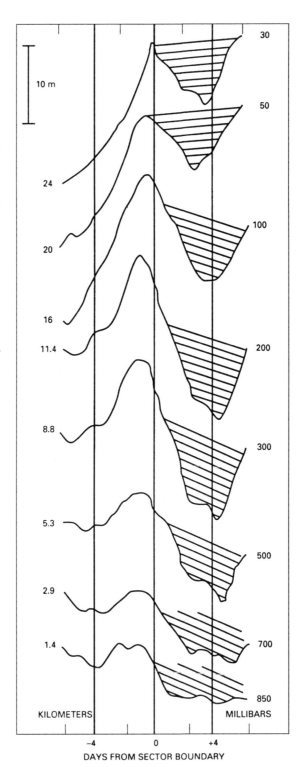

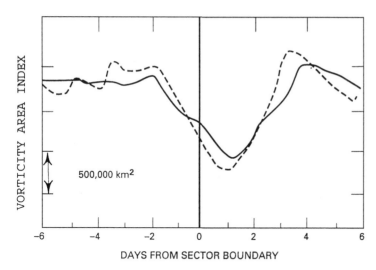

Figure 5.26. Average response of vorticity area index for magnetic sector boundary passage on day 0. Dashed curve represents 28 boundaries in 1967–1970. Solid curve represents 26 boundaries in the interval from 1964 to 1966. Ordinate axis is in units of 5×10^5 km^2. Figure adapted from Wilcox et al. (1973) (reported in Herman and Goldberg, 1978).

by sensing the local magnetic field direction (Blakemore and Frankel, 1981). During their growth, these bacteria biologically synthesized as many as twenty highly purified cuboidal magnetite crystals, each approximately fifty nanometers in size, arrayed in a dipolar string that aligned along the long axis of their bodies. Other magnetotactic bacteria as well as magnetotactic green algae were subsequently discovered. The importance of the algae was that although bacteria do not have a cell nucleus and other cell features, algae and higher organisms do have them.

The search for similar biological examples of crystalline magnetite subsequently led to discoveries of magnetite in the abdomens of honey bees and in the brains of homing pigeons, tuna, blue marlins, green turtles, dolphins, whales, etc. Honey bees dance to describe locations of their feeding sources; the dances are sensitive to the local magnetic field direction as well as to the Sun's direction (Gold, Kirschvink, and Deffeyes, 1978). There are also indications that the navigation of birds involves a sensing of geomagnetic field direction (Wiltschko and Wiltschko, 1988); homing-pigeon rally organizers, wary of disturbance fields, depend upon forecasts of no geomagnetic disturbances to schedule their programs. Quinn (1980), at the University of Washington, showed that a 90° directional shift in the horizontal component of the Earth's magnetic field caused approximately 90° changes in the mean direction of movement of migratory sockeye salmon fry at night.

Maugh (1982) reviewed the findings of magnetite in organisms. He reported that in mammals, "the investigators found that the magnetic particles appeared to be surrounded by nervous tissue, suggesting the possibility of interaction between the particle and the brain." The first extensive English-language review of the biological effects of the geo-magnetic field had been published by Dubrov (1978). By 1985 there was a sufficient accumulation of research on the topic that Kirschvink, Jones, and McFadden (1985) edited a collection of articles on magnetoreception in organisms (see also Kirschvink, 1994).

For many years the special superconducting properties of metals required study in liquid helium at temperatures near absolute zero. Re-cently, experimenters have developed materials that become supercon-ducting at the significantly higher temperature of liquid nitrogen (77 K or −196°C). High-temperature SQUID magnetometers (Section 4.12) are being used in medical research to map the fields associated with hu-man high-order mental functions using biomagnetic imaging. Specific response areas of the brain have been identified for cognitive functions, epileptic seizures, Alzheimer's anomalies, etc. Brain-wave frequencies span the range of geomagnetic micropulsations and geomagnetic storm oscillations; the geomagnetic fields are considerably more intense than brain waves (Figure 5.27; Williamson et al., 1977). The question of whether human brain processes respond to external field stimulation has not yet been definitively answered.

Baker et al. (1983) of the University of Manchester discovered mag-netite in the ethmoid cavities of humans. There are reports of geomag-netic disturbance effects upon man. A four-year study (Becker et al., 1961; Friedman et al., 1963) showed a positive correlation between the monthly sums of geomagnetic K indices from Fredericksburg and the monthly admissions to two mental hospitals in Syracuse, New York, USA; the probability of obtaining such a relationship by chance was one in a thousand. Nikolaev et al. (1976) described an extremely interesting study of the psychopathic behavior of inmates at a Moscow mental hos-pital. From April 1975 through January 1976 each hospital worker con-tacting patients rated the degree (0 to 5) to which an inmate evidenced his or her psychosis. All numbers were averaged daily for each patient and a hospital daily average "disturbed condition" index, S, was obtained. The daily values of S were compared to the daily geomagnetic A_p index, to the polar region geomagnetic field strength, and to polar sector structure indices. Figure 5.28 shows the increase in disturbed mental behavior of the patients during disturbed geomagnetic field conditions.

Novikova and Ryvkin (1977) of the Sverdlovsk Medical Institute reported a 1961 through 1966 study of the deaths due to myocardial infarction in Sverdlovsk, USSR. Table 5.1 shows the results of about 3,000 cases of infarction and about 1,000 deaths. On magnetically active days in all years, both morbidity and mortality were higher than on magnetically quiet days, with a probability of 0.005 that such a situation was random. For the same study group, Gnevyshev et al. (1977) found

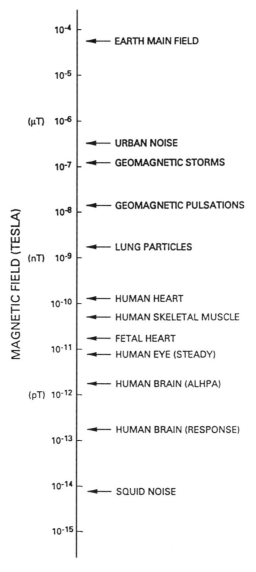

Figure 5.27. Magnetic fields of the human body compared to geomagnetic field levels and sensitivity of SQUID magnetometer. Figure adapted from US Naval Research Reviews (1977).

that the largest number of sudden deaths from cardiovascular disease occurred within the first twenty-four hours of a geomagnetic storm.

Rajaram and Mitra (1981) found a correlation between convulsive seizures and geomagnetic activity. Persinger (1988) and Randall and Randall (1991) reported an increase in human hallucinations with increasing geomagnetic activity. Bureau and Persinger (1992) reported "a significant correlation of Pearson $r = 0.60$ between mortality during the critical 4-day period that followed induction of limbic seizures in rats and the ambient geomagnetic activity during the 3 to 4 days that preceded

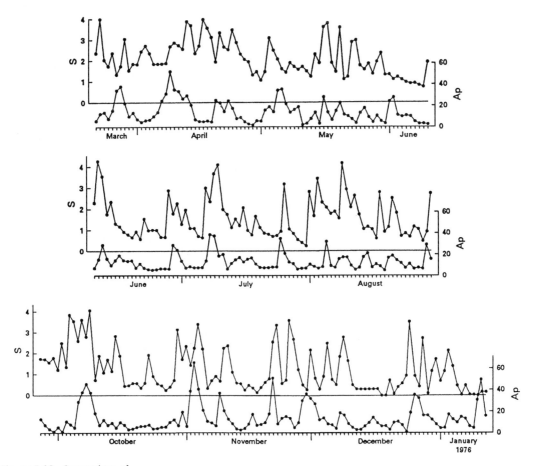

Figure 5.28. Comparison of group psychopathological syndrome expression index, S, at a Moscow mental hospital and geomagnetic index Ap for nine months of 1975 and 1976. Figure adapted from Nikolaev et al. (1976).

death; the risk increased when the 24 hr geomagnetic indices exceeded 20 nT for more than 1 to 2 days." Villoresi et al. (1994) studied the 1979 to 1981 Moscow ambulance calls for brain stroke and myocardial infarction and found significant increases on geomagnetic storm days.

As we look at the special effects that have been connected to geomagnetic storms we should keep some cautions in mind. Biological effects can respond to weather changes which may depend on geomagnetic disturbances. Although associations and correlations between phenomena are important initial steps in reaching a basic understanding of the physics of any relationship, correlation does not mean dependence. Many of the related phenomena may exist on separate branches of the development tree or may simply represent a similar secondary variation. For example, the annual number of Sydney Chapman's publications from 1910 through 1967 had a significant 0.49 correlation coefficient with sunspot number (Campbell, 1968). Most of us would believe that his viewing of the observatory recordings of spectacular solar–terrestrial storm events

Table 5.1. *Myocardial infarction in Sverdlovsk –*
percentage of days with cases of disease or death
separated for active and quiet geomagnetic conditions

Year	Morbidity		Mortality	
	Active	Quiet	Active	Quiet
1961	78.8	70.4	24.5	18.8
1962	73.5	65.3	35.0	25.7
1963	77.6	73.2	24.9	20.6
1964	62.0	57.0	36.4	21.7
1965	79.1	75.0	27.9	25.3
1966	90.8	75.4	59.5	40.9

at sunspot maximum would more likely stimulate Chapman's productivity than some human magnetic–biological connection. Correlations should not be taken as functional relationships, but rather as interesting possible subjects for further independent investigation. However, Roederer (1995) recommended "Given the results so far available, it would be irresponsible to dismiss biogeomagnetics as being too implausible."

5.16 Solar–Terrestrial Disturbance Predictions

As the applications of geomagnetism to human activity increase, there is a corresponding growth in the international appetite for reliable predictions. Sufficient archived geomagnetic information is stored at the World Data Centers to allow accurate long-term statistical studies (such as the number of daily disturbances of more than 100 gamma in a 40-year period in a specific region of the world). Solar-cycle predictions of the time and level of next year's average geomagnetic activity also can be provided with reasonable certainty. Forecasters do well in their appraisal of the average activity level for the next rotation of the Sun (about 27 days) and for the possible arrival of activity seen on the solar limb. All these predictions are important for protective designs of man-made systems sensitive to geomagnetic disturbances.

Problems arise with the prediction of the onset time, activity level, and disturbance duration of specific events. From the tracking of solar activity and coronal mass ejections, considerable success has been accorded to space environment forecasters in their predictions of quiet conditions for the next hour, day, week, and year. However, storm prediction is still difficult. Table 5.2 lists the results of daily forecasts of storm conditions (daily Ap index greater than 50) for the period 1989 through

Table 5.2. *Daily forecast and*
subsequently observed conditions

	Observed	
Forecast	no-storm	storm
no-storm	1052	23
storm	12	7

1991 (values courtesy of K.A. Doggett of the Space Environment Center, NOAA).

The 97.9% accuracy of the no-storm prediction is related to the high occurrence of quiet days and the fact that easily observed quiet-Sun conditions guarantee a low in geomagnetic variations. For that reason, the predictions of suitable days for observatory baseline measurements, aeromagnetic flights, pipeline cathodic protection measurements, etc., are highly reliable. However, of the thirty storm days, 77% were not predicted. This problem is related to the fact that the detection of solar disturbance is necessary in order to warn us of possible geomagnetic storms but other conditions, necessary for the magnetospheric interaction to trigger turbulence, are not yet monitored. Nevertheless, as long as temporary movement to a protective operational mode is not costly, many storm-sensitive systems are now benefiting from even the present storm-prediction accuracy. Forecasters are developing ways to improve their products.

Although disturbance forecasting is an evolving science, equally important to users is detailed "nowcasting," an accurate appraisal of present space conditions. Relying on these data, operators of space programs, satellite tracking, communication systems, power distribution facilities, global position systems, etc., take protective actions that save millions of dollars of public and private funds. A vital part of nowcasting capability is the input from global magnetic observatories, especially via the INTERMAGNET network (Figure 4.19).

Disturbance forecasting capability was greatly improved by the placement of special satellites downstream in the solar wind. There is a position, called the *Lagrangian point, L_1*, about 1.5×10^6 km from the Earth (at 235 R_e it is about $1/100$ of the distance to the Sun or four Earth–moon distances) where a satellite can circle the Sun in a year, remaining in-line between the Sun and the Earth. The spinning (5 rpm) satellites are given a halo orbit (about 1.5 to 3.0×10^5 km) about this L_1 position. Two space warning satellites are located at this location: ACE (for Advanced Composition Explorer) receives continuous particle and field information (Horbury et al., 2001) and SOHO (for Solar Helio-spheric Observatory) makes measurements of the solar surface proper-ties (Judge et al., 2001). The GENESIS spacecraft, launched in August 2001, is also at the L_1 location. It will collect solar wind samples for over two years and then return them to Earth for analysis in September 2004 (Wiens et al., 2002). Knowledge of the incoming wind composition, ve-locities, and field directions is necessary for determining the reaction of the magnetosphere to solar–terrestrial disturbances. Data from ACE and SOHO at the L_1 location allow forecasters to provide about 1/2 to 1 hour warning of an approaching geomagnetic storm (Gopalswamy et al., 2001). In 2002 a satellite Solar Mass Ejection Imager (SMEI) was launched. For solar–terrestrial disturbance forecasting the SMEI instrument will measure the coronal mass ejections (CME) and de-termine those that are likely to hit the Earth (Webb et al., 2002). Operating in conjunction with the ACE satellite, a significant improve-ment of the short-term warning prediction of space weather hazards is anticipated.

Special Space Weather Disturbance tables have been developed by the NOAA Space Environment Center to provide the public with a more efficient scaling of a geomagnetic storm's impact at the Earth. These tables (5.3a,b,c) represent three categories. One is for the geomagnetic field effects based upon five Kp index levels. A second table is for the particle radiation effects measured by five levels of solar–terrestrial ion flux. The third is for radiowave propagation effects measured by five levels of X-ray radiation. By announcing notices of the separated cat-egories of possible environmental damage the SEC warning system is better utilized throughout the world.

5.17 Magnetic Frauds

Over the ages, magnetism has inspired the active minds of both the sci-entist and the charlatan. Albert Einstein said "A wonder of such nature I experienced as a child of four or five years, when my father showed me a compass. That this needle behaved in such a determined way did not at all fit into the nature of events, which could find a place in the uncon-scious world of concepts (effect connected with direct "touch"). I can still remember – or at least I believe I can remember – that this experience made a deep and lasting impression upon me. Something deeply hidden had to be behind things." Einstein's curiosity led to scientific discover-ies. However, a dark side of magnetism studies, magnetic frauds, comes from charlatans in today's world. An interesting website is http://skepdic. com where the reader can find detailed debunking of pseudo-scientific ideas in the everyday public belief. A supposed magnetic

Table 5.3a. *NOAA Space Weather Scale*

Category		Effect	Physical measure	Average Frequency (1 cycle = 11 years)
Scale	Descriptor	Duration of event will influence severity of effects	Kp values* determined every 3 hours	Number of storm events when Kp level was met; (number of storm days)
Geomagnetic Storms				
G 5	Extreme	*Power systems:* wide-spread voltage control problems and protective system problems can occur, some grid systems may experience complete collapse or black-outs. Transformers may experience damage.	Kp = 9	4 per cycle (4 days per cycle)
		Spacecraft operations: may experience extensive surface charging, problems with orientation, uplink/downlink and tracking satellites.		
		Other systems: pipeline currents can reach hundreds of amps, HF (high frequency) radio propagation may be impossible in many areas for one to two days, satellite navigation may be degraded for days, low-frequency radio navigation can be out for hours, and aurora has been seen as low as Florida and southern Texas (typically 40° geomagnetic lat.)**.		
G 4	Severe	*Power systems:* possible wide-spread voltage control problems and some protective systems will mis-operate, tripping out key assets from the grid.	Kp = 8, including a 9-	100 per cycle (60 days per cycle)
		Spacecraft operations: may experience surface charging and tracking problems, corrections may be needed for orientation problems.		
		Other systems: induced pipeline currents affect preventive measures, HF radio propagation sporadic, satellite navigation degraded for hours, low-frequency radio navigation disrupted, and aurora has been seen as low as Alabama and northern California (typically 45° geomagnetic lat.)**.		

G 3	Strong	*Power systems*: voltage corrections may be required, false alarms triggered on some protection devices. *Spacecraft operations*: surface charging may occur on satellite components, drag may increase on low-Earth-orbit satellites, and corrections may be needed for orientation problems. *Other systems*: intermittent satellite navigation and low-frequency radio navigation problems may occur, HF radio may be intermittent, and aurora has been seen as low as Illinois and Oregon (typically 50° geomagnetic lat.)**.	Kp = 7	200 per cycle (130 days per cycle)
G 2	Moderate	*Power systems*: high-latitude power systems may experience voltage alarms, long-duration storms may cause transformer damage. *Spacecraft operations*: corrective actions to orientation may be required by ground control; possible changes in drag affect orbit predictions. *Other systems*: HF radio propagation can fade at higher latitude, and aurora has been seen as low as New York and Idaho (typically 55° geomagnetic lat.)**.	Kp = 6	600 per cycle (360 days per cycle)
G 1	Minor	*Power systems*: weak power grid fluctuations can occur. *Spacecraft operations*: minor impact on satellite operations possible. *Other systems*: migratory animals are affected at this and higher levels; aurora is commonly visible at high latitudes (northern Michigan and Maine)**.	Kp = 5	1700 per cycle (900 days per cycle)

*Based on this measure, but other physical measures are also considered.

**For specific locations around the globe, use geomagnetic latitude to determine likely sightings (see www.sec.noaa.gov/Aurora).

Table 5.3b. *NOAA Space Weather Scale*

Category		Effect	Physical measure	Average Frequency (1 cycle = 11 years)
Scale	Descriptor	Duration of event will influence severity of effects		
Solar Radiation Storms			Flux level of ≥ 10 MeV particles (ions)*	Number of events when flux level was met**
S 5	Extreme	*Biological*: unavoidable high radiation hazard to astronauts on EVA (extra-vehicular activity); high radiation exposure to passengers and crew in commercial jets at high latitudes (approximately 100 chest X-rays) is possible. *Satellite operations*: satellites may be rendered useless, memory impacts can cause loss of control, may cause serious noise in image data, star-trackers may be unable to locate sources; permanent damage to solar panels possible. *Other systems*: complete blackout of HF (high frequency) communications possible through the polar regions, and position errors make navigation operations extremely difficult.	10^5	Fewer than 1 per cycle
S 4	Severe	*Biological*: unavoidable radiation hazard to astronauts on EVA; elevated radiation exposure to passengers and crew in commercial jets at high latitudes (approximately 10 chest X-rays) is possible. *Satellite operations*: may experience memory device problems and noise on imaging systems; star-tracker problems may cause orientation problems, and solar panel efficiency can be degraded. *Other systems*: blackout of HF radio communications through the polar regions and increased navigation errors over several days are likely.	10^4	3 per cycle

		Flux*	Number of events when flux level was met**
S 3	Strong	10^3	10 per cycle
	Biological: radiation hazard avoidance recommended for astronauts on EVA; passengers and crew in commercial jets at high latitudes may receive low-level radiation exposure (approximately 1 chest X-ray).		
	Satellite operations: single-event upsets, noise in imaging systems, and slight reduction of efficiency in solar panels are likely.		
	Other systems: degraded HF radio propagation through the polar regions and navigation position errors likely.		
S 2	Moderate	10^2	25 per cycle
	Biological: none.		
	Satellite operations: infrequent single-event upsets possible.		
	Other systems: small effects on HF propagation through the polar regions and navigation at polar cap locations possibly affected.		
S 1	Minor	10	50 per cycle
	Biological: none.		
	Satellite operations: none.		
	Other systems: minor impacts on HF radio in the polar regions.		

*Flux levels are 5 minute averages. Flux in particles $\cdot$ s$^{-1} \cdot$ ster$^{-1} \cdot$ cm^{-2}. Based on this measure, but other physical measures are also considered.

**These events can last more than one day.

Table 5.3c. *NOAA Space Weather Scale*

Category		Effect	Physical measure	Average Frequency (1 cycle = 11 years)
Scale	Descriptor	Duration of event will influence severity of effects	GOES X-ray peak brightness by class (and by flux*)	Number of events when flux level was met; (number of storm days)
Radio Blackouts				
R 5	Extreme	*HF Radio*: Complete HF (high frequency**) radio blackout on the entire sunlit side of the Earth lasting for a number of hours. This results in no HF radio contact with mariners and en route aviators in this sector. *Navigation*: Low-frequency navigation signals used by maritime and general aviation systems experience outages on the sunlit side of the Earth for many hours, causing loss in positioning. Increased satellite navigation errors in positioning for several hours on the sunlit side of Earth, which may spread into the night side.	X20 (2×10^{-3})	Fewer than 1 per cycle
R 4	Severe	*HF Radio*: HF radio communication blackout on most of the sunlit side of Earth for one to two hours. HF radio contact lost during this time. *Navigation*: Outages of low-frequency navigation signals cause increased error in positioning for one to two hours. Minor disruptions of satellite navigation possible on the sunlit side of Earth.	X10 (10^{-3})	8 per cycle (8 days per cycle)
R 3	Strong	*HF Radio*: Wide area blackout of HF radio communication, loss of radio contact for about an hour on sunlit side of Earth. *Navigation*: Low-frequency navigation signals degraded for about an hour.	X1 (10^{-4})	175 per cycle (140 days per cycle)
R 2	Moderate	*HF Radio*: Limited blackout of HF radio communication on sunlit side, loss of radio contact for tens of minutes. *Navigation*: Degradation of low-frequency navigation signals for tens of minutes.	M5 (5×10^{-5})	350 per cycle (300 days per cycle)
R 1	Minor	*HF Radio*: Weak or minor degradation of HF radio communication on sunlit side, occasional loss of radio contact. *Navigation*: Low-frequency navigation signals degraded for brief intervals.	M1 (10^{-5})	2000 per cycle (950 days per cycle)

*Flux, measured in the 0.1–0.8 nm range, in $W \cdot m^{-2}$. Based on this measure, but other physical measures are also considered.
**Other frequencies may also be affected by these conditions.

effect causing shipwrecks in the "Bermuda Triangle" region and the problems of a magnetic validation of water witching (dowsing) are discussed therein. Let us look at two samples of magnetic frauds.

5.17.1 Body Magnets

Magnetic fields have a great variety of genuine applications. However, recent advertisements regarding the health values of body magnets aid only the hucksters seeking financial gain by promising false relief and preying on those afflicted with pain. The responsible American Medical Association requires careful, statistically significant, double-blind testing to validate and approve new health remedies. Not one of the magnetic devices on sale has passed such tests. The respected University of California, Berkeley, School of Public Health has called therapy magnets nonsense; readers of their *Wellness Letter* are advised (in March 2002 issue) "There is no good scientific evidence – or any logical reason to believe – that magnets can relieve pain".

Magnetic remedies originated with Viennese physician Franz Anton Mesmer (1733–1815). Western societies wisely created the word "mesmerize" from his name. Interest in magnetic treatments ("without a knife or pill") have reappeared cyclically over the last 200 years. Even Benjamin Franklin was on a French scientific panel that declared magnet treatments to be nonsense (see page 41 in November 2002 *Scientific American*). In the nineteenth century, in England, wealthy newlyweds could spend their nuptial night in rented bedrooms filled with tons of magnets to guarantee that conception would produce super children (see Livingston, 1996). The number of afflicted finding pain relief by magnets matches the expected placebo effect – not unlike the number using "Dr. Pinkerton's Snake Oil" with success.

5.17.2 Prediction of Earthquakes

In recent years the United Nations and one major world government have been financing an effort to predict earthquakes using records of geomagnetic field variations. The advocates have reasoned that the scientific basis for the effort lies in the fact the magnetic field induction concerns the same subsurface region of the Earth where earthquakes arise. The proponents have advertised their ability to select special geomagnetic signals that forecast the magnitude and location of earthquakes occurring one to twenty-two months later and up to distances of 408 km from the magnetic recording site (Campbell, 1998). So far the reports have not produced any physical model for earthquake prediction but rather have stated that "geomagnetic methods are still in the stage of prediction by experience". The success rate statistics are exceedingly poor (not unlike random coincidences) and the advocates have covered their false predictions with arguments such as: Mistakes in prediction occur because there is "obvious precursory response to weather disasters" such as "big floods, droughts, high temperatures, freezes, etc."

Main (1997) has pointed out that "... modern theories of earthquakes hold that they are critical, or self-organized critical, phenomena, implying a system maintained permanently on the edge of chaos, with an inherent random element and avalanche dynamics with a strong sensitivity to small stress perturbations". In November, 1996, an international meeting, "Assessment for Schemes for Earthquake Prediction," was convened by the Royal Astronomical Society and the Joint Association for Geophysics in London. Geller (1997) reported: "The overwhelming consensus of the meeting was that earthquake prediction, in the popular sense of deterministic short-term prediction, is not possible at present. Most of the participants also agreed that the chaotic, highly nonlinear nature of the earthquake source process makes prediction (time, place, and magnitude of individual quakes) an inherently unrealizable goal". The public funds that build false hope of protection from earthquake occurrences could be better directed to hazard mitigation.

5.18 Summary of Applications

Applications of geomagnetism expand as society grows in its technological capabilities. With the increased use of the space environment, so increases the need to monitor those magnetic field changes that have an impact upon satellite operation. Large geomagnetic storms also affect location fixes, communication systems, power transmission networks, and pipelines. Paleomagnetic studies have established the formation of our present-day continents. Solid-earth geophysics relies upon geomagnetism for revelation of our Earth's structure. Magnetic surveying is vital to the discovery of mineral resources and specification of their emplacement. Magnetic charts continue to be used in navigation worldwide. Geomagnetism is beginning to play a role in atmospheric weather and climate modeling. A developing science of magnetoreception in organisms may offer an explanation for reported biological responses to geomagnetic fields and have future applications to benefit society.

The physics of the disturbed field and particle conditions from the Sun to the Earth's surface is still not fully revealed. Each new year and each new solar cycle brings us closer to a complete understanding of the solar–terrestrial processes and to a greater application of geomagnetism to societal needs. The national requirements for geomagnetic disturbance forecasting and warning create a growing demand for timely and accurate geomagnetic information on a global scale. National geomagnetic observatory networks, Space Environment Forecasting Centers (see Section B.6), and the geomagnetic data archival and retrieval facilities of World Data Centers work together to satisfy many of the national science and technology requirements of present and future programs.

5.19 Exercises

1. A consortium of business interests is considering the investment in a new communications satellite for your country. You are asked to attend a meeting to provide information regarding the space hazards that could affect the satellite's operation and how such damage could be mitigated. What is your advice to be presented at this meeting?

2. For the magnetic field fluctuations of 30 seconds period described in Exercise 4.2, what is the penetration depth of the magnetic field fluctuation signals into the dry earth and into the ocean water?

3. Go to one of the journal references given in Section 5.4 and determine the conditions that are responsible for pipeline corrosion. What problem of this type might exist in your home area?

4. Go to one of the journal references given in Section 5.5 and determine the conditions that are responsible for the geomagnetically induced damaging induction in electric power grids. What problem of this type might exist in your home area?

5. Discover what aeromagnetic surveys have been made in your country/state. Find out information on the resolution in field strength and the grid spacing that was used during the flights.

6. Go to the Gough and Ingham (1983) and the Oldenberg (1990) references and prepare a short review on the magnetic survey methods discussed therein.

7. Go to the USGS website http://geology.er.usgs.gov/eastern/plates.html then download and read the text. Next, select the highlighted "Information on Plate Tectonics" near the bottom of the page. Summarize how geomagnetic studies led to the understanding of earthquakes.

8. Go to the website http://skepdic.com. Select index letter "B", then "Bermuda Triangle" and read the misrepresentation of the magnetic field effect in that region. Next, return to the main page and select the index letter "M" to find "Magnet Therapy". Summarize and report on the results of these two magnetic subjects.

9. Use the Table 5.3 NOAA Space Weather Scales to report on the expected disturbance effects when the Kp index is at a value of 7_0. How many times is this level expected during a typical solar cycle according to this table? Use Figure 3.54 to determine the occurrence rate given there. How do the two rates differ? What are the two "descriptor" characterizations of this Kp storm? What would be the expected midlatitude field level corresponding to this Kp index?

Appendix A
Mathematical topics

This appendix presents a number of mathematical topics that arise in the book. The review is not meant to be comprehensive; it is limited to only items that could be helpful for understanding the flow of ideas in the chapters of this book.

A.1 Variables and Functions

Variable is the name we give to a value of something that changes. When we call a variable *independent* we mean that it can be any size within a prescribed physical *domain* of realistic values. A *dependent* variable is the value that we call a function of the independent variable. On a daily magnetogram, the magnitudes of the scaled H (a "dependent" variable) are dependent upon the selection of the "independent" variable of daily hourly time that we can take to be any value (in the domain) from 0 to 24. Maxwell's equations (Chapter 1) allow a unique field value to be determined from a given source-current distribution; however, given the field values, a number of possible currents might be the source. The dependent-variable field is a function of the independent-variable current. The extreme highest and lowest values of the dependent variable that occur over the domain of the independent variable define the *range* of the dependent variable.

The term *function* has a very special meaning in mathematics. When we say, for example, "the variable y is a function of the variable x" it is written as $y = f(x)$. "Function" means that for every value of x, there is a particular value of y. By contrast, for each y there may be quite a few values of x; x is not a function of y. For example, let us consider the values of the quiet-day H component values of field at some observatory for hours of a particular day in which there is a gradual variation of H between -5 and $+7$ nT. We would call t the hour and H the function of t and write $H(t)$ to indicate the relationship. For each time t there is one, and only one, value of H. But a given value of H, (for example 2 gamma), might occur at two or more different times t; so for each H there is not necessarily one, and only one, value of t. If the function is one that lets us transform values of one relationship to those of another relationship, we call the function a *transfer function*. In Section 5.8.4 there were transfer function equations that let us transform values of an external and internal magnetic field into depth and conductivity.

It is customary in science to plot a functional relationship with the independent variable along the horizontal axis and the dependent variable along the vertical axis. In the above example, time is plotted horizontally and the field H is plotted vertically. The domain of t is 0 to 24; the range of H is -5 to $+7$. Time is broken into 3-hr segments to determine the 3-hr range of H field components that are used to compute the ap and Kp geomagnetic indices (Section 3.15).

In Equations (1.40) to (1.45) the spherical harmonic function $P_n^m(\theta)$ Legendre polynomial was introduced. This dependent variable is a function of the independent variable θ for the domain of $0°$ to $360°$. The values of P_n^m are shown in Figure 1.11 only for the θ values from $0°$ to $180°$ because the symmetry of the function about the equator lets us understand the remaining values.

Often a function may be described by a precise mathematical relationship. For example, a function y is called the "*polynomial function of degree n*" of the independent variable x if y can be written as $y(x) = a_0 + a_1 x + a_2 x^2 + a_3 x^3 + a_4 x^4 + \cdots + a_n x^n$, where the a are constants and the n are integers. The degree of n is the highest integer value in the function. If n is 1, then the polynomial is called a *linear function* and the graph of y is a straight line; if n is 2, then the polynomial is called a "quadratic function" and the graph of y is a parabola. In geomagnetism, many smoothly varying dependent-variable functions of time $F(t)$, in the domain t_0 to t_1, can be represented by a fitted polynomial and thereby be used in predicting values of $F(t)$ at $t_n > t_1$ when $t_n - t_1$ is small with respect to $t_1 - t_0$ (see **POLYFIT** program in Section C.7). This method lets us use a polynomial representation of Gauss coefficients to predict the geomagnetic pole locations and geomagnetic coordinates many years past the present time (see **GMCORD** program in Section C.1).

A.2 Summations, Products, and Factorials

The Greek letter sigma, Σ, is typically used as a shortened representation of the *summation* of a series of mathematical terms; a subscript indicates the starting value of the summation and a superscript indicates the ending value. For example,

$$\sum_{n=1}^{5} a_n x^n = a_1 x^1 + a_2 x^2 + a_3 x^3 + a_4 x^4 + a_5 x^5, \qquad (A.1)$$

where the a_n are constants and the x^n are variables. In a similar way the product of terms can be written in a shortened way for mathematical formulas; for example,

$$\prod_{n=1}^{4} \frac{1}{n} = \frac{1}{1} \times \frac{1}{2} \times \frac{1}{3} \times \frac{1}{4} = \frac{1}{1 \times 2 \times 3 \times 4} = \frac{1}{4!} \qquad (A.2)$$

where $\times$ means multiplication and the exclamation symbol, !, is called a *factorial* indicator when used with a whole number. For example, 4! means that the positive integer 4 is multiplied by the sequence of all the lower positive integers; $4 \times 3 \times 2 \times 1 = 24$. By definition, $0! = 1$.

A.3 Scientific Notations and Names for Numbers

It is customary in *scientific notation* to represent large numbers as powers of ten. The "powers of ten" means $10 = 10^1$; $100 = 10 \times 10$ or 10^2; $1,000 = 10 \times 10 \times 10$ or 10^3; etc. Then $15 = 1.5 \times 10^1$; $150 = 1.5 \times 10^2$; $1,500 = 1.5 \times 10^3$; etc. The exponent (1, 2, or 3) is called the "power" of the number. We use the special expression "raised to the power of " to indicate the exponent of ten. For example, "ten raised to the power of three" to mean 10^3. Note that any number raised to the 0 power is equal to 1; thus, $10^0 = 1$.

Small numbers are handled in a similar way. But we must add the convention that a minus power means the reciprocal of the number raised to that power. That is, $10^{-1} = (1/10) = 0.1$; $10^{-2} = (1/100) = 0.01$; etc. In that notation, $0.15 = 1.5 \times 10^{-1}$; $0.015 = 1.5 \times 10^{-2}$; $0.0015 = 1.5 \times 10^{-3}$; etc. Note that the minus sign applies only to the power of 10, not the number in front of 10. Thus $(1.5) \times 10^{-1} = (3/2) \times 10^{-1} = 3/(2 \times 10^1) = 3/(20) = 0.15$ are equivalent.

Special names are given, as prefixes, to indicate the powers of ten of numbers that increase or decrease by a thousand. Thus, 10^3 is "kilo," 10^6 is "mega," and 10^9 is "giga" with corresponding letter representations of K, M, and G. Often k is used instead of K. For example, 1.5 MAmperes is 1.5×10^6 Amperes and 15 MHz is 1.5×10^7 cycles per second (because Hertz means cycles/sec). In a similar way the negative powers of 10 are given the special names "milli," "micro," and "nano" as prefixes (and corresponding letters m, μ, and n) to represent 10^{-3}, 10^{-6}, and 10^{-9} values. For example, 1.5 nTesla is 1.5×10^{-9} Tesla.

Numbers that can be represented as a ratio of whole numbers are called *rational* numbers. For example, 0.25 is a rational number that can be represented as the value of (1/4), or the number 0.33333 ... (where the ... indicates that there are more numbers of the previous type; here that means recurring 3s) can be represented as a value of (1/3). Numbers that cannot be represented as a ratio are called *irrational* numbers (nothing to do with the psychological use of that word). For example, the square root of 2, which is computed to be $\sqrt{2} = 1.4142135 \ldots$ (here the ... means that there are more special numbers, but differing in value) is an irrational number that cannot be represented by an exact ratio.

Unique names are used for two irrational numbers that often recur in science. The Greek letter pi, π, stands for the number 3.1415926.... Pi arises as the number we must use to multiply the diameter of a circle in order to obtain the circumference of the same circle; it is seen in mathematical expressions related to the geometry of a circle and sphere, such as in the trigonometric functions described below. The letter π recognizes the Greek mathematician Pythagoras.

The other number is the small letter e, which stands for 2.7182818.... The number, e, arises many ways in mathematics; one relates to the determination of the area under a curve representing the function $f(x) = 1/x$ in the x domain from 1 to any greater positive value of x. The depth of penetration of magnetic fields into the conducting Earth is defined by the point at which the field decreases to $1/e$ of its incident size (Section 5.83). The name *natural number* is often given to e, a letter that was selected to recognize the Swiss mathematician Euler. The value of e may be obtained by letting $x = 1$ in the series:

$$e^x = 1 + x + \frac{x^2}{2!} + \frac{x^3}{3!} + \frac{x^4}{4!} + \frac{x^5}{5!} \cdots = \sum_{n=0}^{\infty} \frac{x^n}{n!} \tag{A.3}$$

Together, the rational and irrational numbers make up the *real number system*. Each number of the real number system can be assigned a unique (ordered) position along a line extending (left to right) from the smallest possible negative number that can be written to the largest positive number that can be written.

A.4 Logarithms

For a relationship $x = a^y$, where the number we want is x, we call the number represented by the letter a the "base" and the number represented by y the "exponent" or "power" to which a is raised. To express y in terms

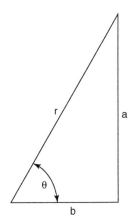

Figure A.1. Relationship of triangle sides for trigonometric functions.

of x, we use the form $y = \log_a x$ in which log is the abbreviation for "logarithm." If $a = 10$, then it is customary to omit the subscript on log and simply write $y = \log x$. The base can be any real number. Very often it is appropriate to use the base e (Equation (A.3)). When that is done, the logarithm notation for $\log_e x$ is shortened to $\ln x$. Tables and simple calculators are available to provide the values of the logarithms of numbers.

In Chapter 3 the geomagnetic index, Kp, was said to be "pseudo-logarithmic" in form. That means the magnetic field (range H) measured by the index changes somewhat like the expression (range H) $= (2.28)^K$ in the lower K values and somewhat like (range H) $= (1.95)^K$ in the upper K values.

A.5 Trigonometry

Relationships of the sides and angles of a "right" triangle (that has one $90°$ angle) are expressed as special *trigonometric functions*. In Figure A.1 for a specific angle, θ, we assign the following names for side a, b, and r ratios:

$$\sin(\theta) = a/r$$
$$\cos(\theta) = b/r$$
$$\tan(\theta) = a/b$$

where the names sin, cos, and tan are abbreviations for the words *sine, cosine,* and *tangent*. Note that $\tan(\theta) = [\sin(\theta)]/[\cos(\theta)]$. A principal relationship of the sides of a triangle that has one $90°$ angle is that $a^2 + b^2 = r^2$ (called the *Pythagorean Theorem*).

To depict the trigonometric functions another way we can write

$$\theta = \sin^{-1}(a/r)$$
$$\theta = \cos^{-1}(b/r)$$
$$\theta = \tan^{-1}(a/b).$$

These expressions are read as "theta is the angle whose sine is a/r," etc., and $\sin^{-1}(a/r)$ is called the "arcsine a/r" etc. Often the computer commands for the arctangent function are ARCTAN, ATAN, or ATN; and when the $+$ and $-$ signs of a and b are important for the direction of r, the command is ARCTAN2, ATAN2, or ATN2.

The angle values are given in units called "radians" (or "rad"). By definition, there are 2π radians in the $360°$ around a circle and π rad in $180°$. To convert radians to degrees, just multiply the rad by $(180/\pi)$. There are $\pi/2$ radians for $90°$, and $\pi/4$ radians for $45°$. This fact allows us to find the value of π from the simple relationship $\pi = 4 \times \text{ATN}(1)$. The reason is that when $(a/b) = 1$, the angle is $45°$ and $\text{ATN}(1) = \pi/4$.

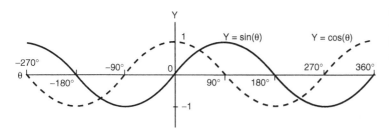

The values of the three trigonometric functions can be obtained from tables and simple calculators or can be computed from

$$\sin(x) = x - \frac{x^3}{3!} + \frac{x^5}{5!} - \frac{x^7}{7!} + \cdots \tag{A.4}$$

$$\cos(x) = 1 - \frac{x^2}{2!} + \frac{x^4}{4!} - \frac{x^6}{6!} + \cdots \tag{A.5}$$

Figure A.2 shows how the values of the $y = \sin(\theta)$ and $y = \cos(\theta)$ waves vary with changing θ (when $r = 1$) over the domain of $\theta = -270°$ to $360°$. Note that $\sin(90° + \theta) = \cos(\theta)$. If we introduce a whole number multiplier m before the angle θ, then we would have m oscillations of the wave in the original domain. It is the addition of the various m waves of sine and cosine functions, of selected amplitudes, that form the component parts of the Fourier representation of a recurring waveform. The Fourier components of a quiet daily field variation in H were illustrated in Figure 1.12 and may be determined from magnetogram values using the program **FOURSQ1** in Section C.8.

Some of the valuable relationships between trigonometric functions are the following:

$$[\sin(\theta)]^2 + [\cos(\theta)]^2 = 1 \tag{A.6}$$

$$\sin(-\theta) = -\sin(\theta); \quad \cos(-\theta) = \cos(\theta) \tag{A.7}$$

$$\sin(90° - \theta) = \cos(\theta); \quad \cos(90° - \theta) = \sin(\theta) \tag{A.8}$$

A.6 Complex Numbers

Special methods have been developed for mathematical treatment of problems that arise in the analysis of waves, vectors, etc. One of these methods is to form *complex numbers* that contain two parts, such as $a + ib$, where a and b are our usual real numbers. In the complex notation, the first part, a, is called the "real" part and the second part, b, is called the *imaginary* part when preceded by the letter, i (the naming has nothing to do with our mental perception). For mathematical reasons the imaginary symbol is given a value $i = \sqrt{-1}$.

Figure A.3. Graph of complex number $Z = a + ib$ in real and imaginary coordinates.

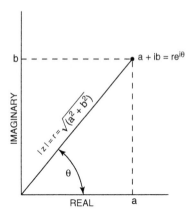

The complex number $Z = a + ib$ is plotted with the real part along the horizontal axis and the imaginary part along the vertical axis, as in Figure A.3 (note that the a and b are assigned differently than in Figure A.1). The magnitude of any variable is indicated by vertical bars on each side of the variable's letter. For example $|Z| = \sqrt{a^2 + b^2}$, which we can call r in comparison to Figure A.1. The value of θ is obtained from our trigonometric relationships of Section A.5 above. The complex number Z represents a quantity that has both a magnitude, r, and a direction θ; we call such a quantity a *vector*.

To express the complex number in another way, let us substitute $(i\theta)$ for x in Equation (A.3); then, because the value of $(i \times i) = -1$, we can write

$$e^{i\theta} = \left(1 - \frac{\theta^2}{2!} + \frac{\theta^4}{4!} - \frac{\theta^6}{6!} + \cdots\right) + i\left(\theta - \frac{\theta^3}{3!} + \frac{\theta^5}{5!} - \frac{\theta^7}{7!} + \cdots\right). \quad \text{(A.9)}$$

With Equations (A.4) and (A.5), the above equation can be written as

$$e^{i\theta} = \cos(\theta) + i\sin(\theta). \quad \text{(A.10)}$$

Now if we multiply both sides of the above by r and substitute b/r for $\sin(\theta)$ and a/r for $\cos(\theta)$ (our defined trigonometric relationships), we obtain

$$re^{i\theta} = r\cos(\theta) + ir\sin(\theta) = a + ib. \quad \text{(A.11)}$$

Thus the complex number Z can also be written as

$$Z = a + ib = re^{i\theta}. \quad \text{(A.12)}$$

where $r = \sqrt{a^2 + b^2}$, $\theta = \tan^{-1}(b/a)$, $a = r\cos(\theta)$, and $b = r\sin(\theta)$. In complex number nomenclature, r is called the "modulus of Z" (mod Z)

and θ is called the "argument of Z" (arg Z). In Section 5.8.5, we use a complex transfer function for determining depth and conductivity from a spherical harmonic analysis of the source field.

A.7 Limits, Differentials, and Integrals

Let us consider a dependent variable function $y = f(x)$ that varies smoothly over the domain (a to b) of the independent variable x (Figure A.4). Let c be some x value between a and b. If the trace of $f(x)$ approaches the same value $f(c)$ at $x = c$, moving in along the trace of the function toward $f(c^-)$ from $f(a)$ (i.e., from the left), as it does by moving in toward $f(c^+)$ from $f(b)$ (i.e., from the right); (using $-$ and $+$ to indicate a position just a little to the left or right of the point) and if that value, $f(c)$, is the same as the value of $f(x)$ that is obtained for the x position at c itself, then we say that the limit of $f(x)$, as x approaches c, is $f(c)$ and write the result symbolically as

$$\lim_{x \to c} f(x) = f(c) \tag{A.13}$$

Now let us call Δx a difference in two x values, $x_2 - x_1$, and $\Delta f(x)$ the difference in the corresponding $f(x)$ values $y_2 - y_1$. *Slope* is defined as the fraction obtained from "the rise over the run" (i.e., the change in dependent variable over the change in independent variable). Then the slope, m, of the curve determined Δx and Δy is just $m = \Delta y / \Delta x$. The slope, so defined, would change as different x values are selected along $f(x)$. Now we ask what the slope is as we approach the value of $x = c$ with our determinations and let the x_1 and x_2 values be separated by smaller and smaller amounts. We say, now, that we want the limit of the slope of $f(x)$ as x approaches c from the right and from the left,

$$\lim_{x \to c} \frac{\Delta f(x)}{\Delta x} = \frac{dy}{dx}, \tag{A.14}$$

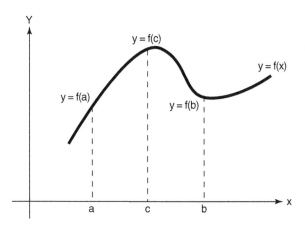

Figure A.4. Illustration for determining whether the limit of $f(x)$ exists at $f(c)$.

and call the value of dy/dx the "*derivative* of $f(x)$ with respect to x at $x = c$." Thus, for a smoothly varying function, the derivative at a particular point on the function (i.e., at a particular value of the independent variable) is simply the function's slope at that point. We call the dy/dx or $d[f(x)]/dx$ the "derivative of the function of x with respect to x." The d used here indicates that we have taken the Δ value to be exceedingly small.

If it happens that the function is dependent on more than one independent variable, for example, x and t, then the derivatives with respect to each independent variable (while the other independent variables are held constant) are determined separately. Slightly different symbols, using the Greek letter δ (e.g., $\delta y/\delta x$ and $\delta y/\delta t$), are used for these "partial derivatives."

To obtain some feeling for the mathematical *integration* operation, let us look at the area under the function $f(x)$ in the domain of the independent variable x from $x = a$ to $x = b$, as shown in Figure A.5. The area has been broken up into a number of strips, each a finite size, Δx, in width and approximately $f(x)$ in height, where $f(x)$ is the $y = f(x)$ value at the center of each strip. For these rectangles each has an area of $\Delta x \times f(x)$, where $f(x)$ is the value of the function for the particular strip location between a and b. The full area, A, between the curve and the x axis from a to b is approximately ($\approx$) thus:

$$A \approx \sum_{x=a}^{b} f(x) \times \Delta x \tag{A.15}$$

Now if we let Δx become exceedingly small (we say "in the limit, let Δx approach 0") we would have an exact expression for A.

$$A = \lim_{\Delta x \to 0} \sum_{x=a}^{b} f(x)\Delta x = \int_{a}^{b} f(x)dx, \tag{A.16}$$

Figure A.5. The sum of the areas within the rectangles represents the "integral-from-*a*-to-*b*" as an area A between $f(x)$ and x axis with x values between $x = a$ to $x = b$ in increments of Δx.

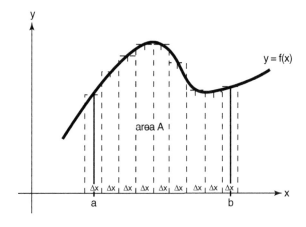

where the symbol $\int$, called the "integral," means that we have taken the limit. For computer computations, the summation version, Equation (A.15), is used and the Δx value is taken to be some number that is small with respect to a tolerable error size, but large enough not to expend computer time unnecessarily. In the spherical harmonic analysis of Section 1.4 we integrated over $d\theta$ by summing over $\Delta\theta$ increments of latitude.

A.8 Vector Notations

A function may be either a *scalar* (having a magnitude but not direction) or a *vector* (having both a magnitude and direction). For vectors, two coordinate systems that can be used are the *Cartesian system* and the *spherical system*. In the Cartesian system, the "unit" vectors (so called because they give direction but leave the magnitude unchanged) are labeled $\mathbf{i}$, $\mathbf{j}$, and $\mathbf{k}$ for the three orthogonal directions x, y, and z from the axis center. For example, calling the three field vector components B_x, B_y, and B_z the vector, $\mathbf{B}$, is written as $\mathbf{B} = \mathbf{i}B_x + \mathbf{j}B_y + \mathbf{k}B_z$ in Cartesian coordinates. This $\mathbf{i}$ is not to be confused with the imaginary number of Section A.6.

In the spherical system, the unit vectors can be labeled as $\mathbf{e}$ (here it is not the natural number e) with a subscript to indicate the direction r for radius out from the center, θ for the angle from a north pole along a great circle of longitude, and ϕ for the angle along a latitude circle measured from some reference longitude. The three components of $\mathbf{B}$ in the spherical system are similarly subscripted. In transforming the derivatives from Cartesian to spherical coordinates we use $\delta x = \delta r$, $\delta y = r\delta\theta$, and $\delta z = r\sin(\theta)\delta\phi$.

Important special mathematical functions for vector algebra occur in this textbook when the Maxwell's equations are discussed for performing spherical harmonic analysis. These functions are the gradient, divergence or dot product, curl or cross product, and Laplacian. A special symbol is used in these functions that is called "del," an upside-down Greek capital delta, defined as

$$\nabla = \mathbf{i}\frac{\delta}{\delta x} + \mathbf{j}\frac{\delta}{\delta y} + \mathbf{k}\frac{\delta}{\delta z}. \tag{A.17}$$

In Cartesian coordinates the *gradient* of a scalar function V, which is called "grad V" is defined as

$$\nabla V = \mathbf{i}\frac{\delta V}{\delta x} + \mathbf{j}\frac{\delta V}{\delta y} + \mathbf{k}\frac{dV}{\delta z}. \tag{A.18}$$

In the spherical coordinate directions the gradient of the scalar V becomes

$$\nabla V = \mathbf{e}_r \frac{\delta V}{\delta r} + \mathbf{e}_\theta \frac{\delta V}{r\,\delta\theta} = \mathbf{e}_\phi \frac{1}{r\,(\sin\theta)} \frac{\delta V}{\delta\phi}. \tag{A.19}$$

The *dot product* of del with the vector $\mathbf{B}$ is called the *divergence* of $\mathbf{B}$ (div $\mathbf{B}$); in Cartesian coordinates this is a scalar defined as

$$\nabla \cdot \mathbf{B} = \frac{\delta B_x}{\delta x} + \frac{\delta B_y}{\delta y} + \frac{\delta B_z}{\delta z}. \tag{A.20}$$

In the spherical coordinate system the divergence becomes

$$\nabla \cdot \mathbf{B} = \frac{\delta B_r}{\delta r} + \frac{1}{r} \frac{\delta B_\theta}{\delta\theta} + \frac{1}{r\,\sin(\theta)} \frac{\delta B_\phi}{\delta\phi}. \tag{A.21}$$

The *cross product* of del with the vector $\mathbf{B}$ is a vector called the *curl* of $\mathbf{B}$ (curl $\mathbf{B}$); in the Cartesian system this is defined as

$$\nabla \times \mathbf{B} = \mathbf{i} \left(\frac{\delta B_z}{\delta y} - \frac{\delta B_y}{\delta z} \right) + \mathbf{j} \left(\frac{\delta B_x}{\delta z} - \frac{\delta B_z}{\delta x} \right) + \mathbf{k} \left(\frac{\delta B_y}{\delta x} - \frac{\delta B_x}{\delta y} \right). \tag{A.22}$$

In the spherical system the curl of $\mathbf{B}$ becomes

$$\nabla \times \mathbf{B} = \mathbf{e}_r \left[\frac{1}{r} \frac{\delta B_\phi}{\delta\theta} - \frac{1}{r\,\sin\theta} \frac{\delta B_\theta}{\delta\phi} \right] + \mathbf{e}_\theta \left[\frac{1}{r\,\sin\theta} \frac{\delta B_r}{\delta\phi} - \frac{\delta B_\phi}{\delta r} \right]$$
$$+ \mathbf{e}_\phi \left[\frac{\delta B_\theta}{\delta r} - \frac{1}{r} \frac{\delta B_r}{\delta\theta} \right]. \tag{A.23}$$

It can be proved that the dot product of a vector cross product is equal to zero:

$$\nabla \cdot (\nabla \times \mathbf{B}) = 0. \tag{A.24}$$

The curl of a gradient is also equal to zero:

$$\nabla \times \nabla V = 0. \tag{A.25}$$

When a vector, such as the magnetic field $\mathbf{B}$, can be written as the gradient of a scalar, such as the potential function V (i.e., $\mathbf{B} = -\text{grad } V$), then Equation (A.25) applies and the expression becomes

$$\nabla \times \mathbf{B} = 0. \tag{A.26}$$

For the divergence of $\mathbf{B}$, using Equation (A.18), we can also write

$$\nabla \cdot (\nabla V) = \frac{\delta}{\delta x} \left(\frac{\delta V}{\delta x} \right) + \frac{\delta}{\delta y} \left(\frac{\delta V}{\delta y} \right) + \frac{\delta}{\delta z} \left(\frac{\delta V}{\delta z} \right). \tag{A.27}$$

A special symbol is assigned to the combination; it is given the name *Laplacian* and written as

$$\nabla^2 V = \nabla \cdot (\nabla V) = \frac{\delta^2 V}{\delta x^2} + \frac{\delta^2 V}{\delta y^2} + \frac{\delta^2 V}{\delta z^2}. \tag{A.28}$$

where $\delta^2 V/\delta x^2$ stands for $\delta(\delta V/\delta x)/\delta x$, etc. In spherical coordinates the Laplacian becomes

$$\nabla^2 V = \frac{\delta^2 V}{\delta r^2} + \frac{1}{r^2}\frac{\delta^2 V}{\delta \theta^2} + \frac{1}{r^2 \sin^2 \theta}\frac{\delta^2 V}{\delta \phi^2}. \tag{A.29}$$

The vector notations are efficient abbreviations of mathematical representations such as those for the electromagnetic field equations we encountered in Chapters 1 and 2.

A.9 Value Distributions

When a set of measurements with differing size are acquired during a sampling period, three values are typically important: the *median*, the *mean*, and the *deviation* of the values from the mean. The median for a set of measurements is that value for which the number of measurements with values less than the median value is equal to the number of measurement values greater than that median value. The median is found by setting all the n values in ascending or descending order and then just finding a value at the half-way ($n/2$) point of the count. The mean ($\bar{X}$) of a set (sometimes called the *average value*) is obtained from the sum of all the values divided by the number of measurements in the set.

$$\bar{X} = [\Sigma(X_1 + X_2 + X_3 + \cdots X_n)]/n \tag{A.30}$$

Often the mean value is considerably different from the median. As an example, our set of field measurements for one day may be recorded at mostly quiet times, but contain just a few minutes of very large amplitude occasions of a disturbance. Then the mean value, driven upwards by those few large values, would be considerably larger than the median. Comparison of the mean and median of a set of values indicates whether or not a few large out-lying values are included in the set.

The *standard deviation* (σ) of a data set is a measure of dispersion of the measured values about the mean value. Because the dispersion feature is to be independent of whether an individual value is larger or smaller than the mean, we square the difference between that value and the mean. These squared differences are then summed, divided by the number of values and then a square-root is taken to overcome our original squaring of the differences. The population standard deviation is shown as

$$\sigma = \sqrt{[(X_1 - \bar{X})^2 + (X_2 - \bar{X})^2 + \cdots + (X_n - \bar{X})^2]/n} \tag{A.31}$$

In Appendix C.9 and C.10 the computer programs **SORTVAL** and **ANALYZ** allow the student to determine the median, mean, and standard deviation of a data set.

We may plot the distribution of a large sample of data values by counting the number of samples that occur within equal increments of size (called a "bin" size). For example, Figure 3.58 displayed the number of hourly values of each index that reached a given field strength where each field strength was within a specified (bin) range. In this figure the bins were 1, 2, or 10 gamma in width depending on the index that was plotted.

Often, in geophysics, the plot of values in a data set forms a "normal distribution" which is symmetrical about the mean value and forms a bell-like shape. Figure A.6 shows an example, often displayed in science classes. The upper left column illustrates how marbles dropped through a funnel to a regular distribution of pins can be collected in bins. The count of marbles at each bin position describes a normal distribution.

Figure A.6. An illustration of normal and lognormal distribution functions resulting from a marble drop through uniform arrays of pins. See text for explanation.

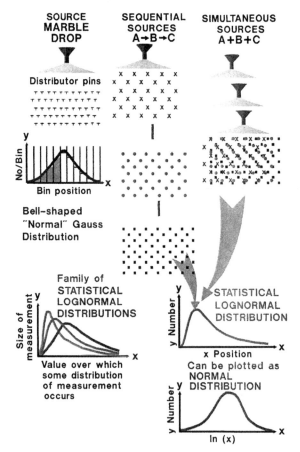

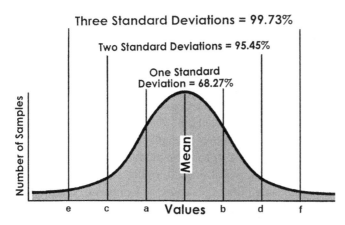

Figure A.7. A normal distribution curve with Y axis representing the number of values for each value along the X axis. The percent of values contained within the limits of one, two, and three standard deviations from the mean are indicated.

The funnel central position and pin arrangement (called the "source distribution mechanism") fix the distribution mean and standard deviation. These bell-shaped, normal distribution curves are often called "Gaussian distributions" because of Gauss's contributions to their mathematical study.

We can "normalize" the distribution curve by making the area under the Gaussian distribution curve equal to one. For example, in the case of the marble drop, if we used 250 marbles total to form the area under the curve, we would determine for each bin the fraction of the total number by dividing the number in each bin by 250. With that accomplished, the probability of obtaining marbles in any continuous range of bins from a to b is represented by the number of marbles (area) under the curve between a and b divided by the total of marbles.

The population standard deviation of a normally distributed data set is used to describe how well the set is centered on the mean. Figure A.7 shows a normal distribution curve obtained by plotting the number of samples found in each bin for series of measured values. The area under the curve between values a (= mean − σ) and b (= mean + σ) contains the values within one standard deviation and represent 68.27 percent of all values. Similarly, that between values c and d represents two standard deviations (mean ± 2σ) for 95.45 percent of the values. Also, between e and f the area represents three standard deviations (mean ± 3σ) for 99.73 percent of the values.

If a normal distribution results from plotting the logarithm of an independent variable measured for a statistical feature, then the probability distribution is said to have a "lognormal distribution" shape. Lognormal distributions are common in nature and arise when there are many simultaneous contributors (elementary processes) to a measurement or when the measurement is a result of a sequential series of elementary processes (Aitchison and Brown, 1957). The middle and right columns of Figure A.6 are an example of such a procedure in which each elementary process has its own statistical distribution. For sequential sources (top central column of Figure A.6) there is first a source distribution mechanism that occurs, then the results undergo a second distribution mechanism, and then these results follow a third mechanism, etc. The final counts, Y, found at the X-bin positions, produce a lognormal shape. Similarly, simultaneous source distributions (top right column of Figure A.6) will also produce Y numbers at the X positions that follow a lognormal shape. Lognormality is verified by a plot of the $\ln(X)$ versus Y (lower right corner of Figure A.6) showing a bell-shaped distribution. A family of lognormal distributions varies as in the lower left corner of Figure A.6.

The curves representing the number of hourly values of AE and Dst indices for various field levels in active and quiet times (Figure 3.58) are examples of lognormal shapes. Clearly, the field values used for these indices represent both the addition of simultaneous sources and sources that have occurred in sequence. The occurrence of Kp values (Figure 3.54) has a lognormal appearance; but in this case Julius Bartels, who originated the index in the 1930s, defined the pseudo-logarithmic values for the stations contributing to his index to keep the values on scale. This particular shape, with a maximum near $Kp = 2+$, was formed. There is some evidence that the lognormal form of the main and recovery phase of Dst may be due to the many contributing sources of negative field values occurring both simultaneously and in sequence to form a lognormal distribution (Figure 3.60). Let us consider the time into the storm to be the bin position of the lognormal distribution (in Figure A.6) and the negative Dst value to be the size of the measurement (accumulated fields) in each bin. Utilizing the symmetrical shape of the normal curve when the Dst is presented in log-time representation (Figure 3.60), the recovery phase of a storm can be predicted with reasonable accuracy from the growth in the storm main phase. Program **DSTDEMO** in Section C.4 illustrates the fitting.

A.10 Correlation of Paired Values

Often it is important to discover whether the values of two phenomena are linearly related. For this purpose a correlation coefficient, r, is such a measure which ranges from -1.0 to 0.0 to $+1.0$. In Figure A.8 three distributions of values of some phenomena X and corresponding values of another phenomena Y are displayed. In part a of the figure, the X and Y values have a high positive linear correlation (as X increases, Y increases), with a value between $r = +0.8$ and $r = +0.9$. In the center,

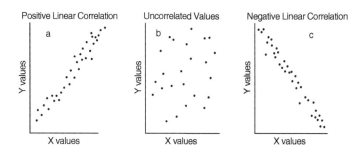

Figure A.8. Three distributions of X and Y values for which positive (a), zero (b), and negative (c) correlations are represented.

part b, of the figure, the X and Y values seem unrelated and show a correlation coefficient of $r = 0.0$. In part c of the figure, the X and Y values have a high negative linear correlaion (as X increases, Y decreases), with a value between $r = -0.8$ and $r = -0.9$. The slightly complicated formulation for r can be found in beginning probability and statistics textbooks. For your present use, see Appendix C.9 for the computer program **ANALYZ** which allows you to determine the linear correlation coefficient for two data sets that are to be compared. Also computed is the best fitting of a straight line to the data set (sometimes called the "least squares prediction") which minimizes the sum of squares of the deviations of the observed values from the fitted line.

Appendix B
Geomagnetic organizations, services, and bibliography

B.1 International Unions and Programs

The following quotation was taken, with permission, from *The National Geomagnetic Initiative* copyright 1993 by the National Academy of Sciences, courtesy of the National Academy Press, Washington, D.C. Revisions of this quoted material have been provided by J.H. Allen in order to modernize the statement to year 2002.

"The study of the Earth is intrinsically global. This was recognized by geologists, geodeticists, and geophysicists in the nineteenth century. During the past hundred years, the need for global collaboration in geosciences has become axiomatic; many mechanisms have been developed to encourage international cooperation in Earth sciences. Much international cooperation in science takes place under the non-governmental International Council for Science (ICSU).

By the latter part of the nineteenth century, international expeditions and exchange of data were common in the geosciences. This led to the development of international mechanisms for ongoing cooperation in geophysical and geological sciences. Seismic and magnetic observatories were being established worldwide. These de facto global networks of magnetic and seismic observatories led to international agreements on measurement standards and data exchange. These international activities led to the formation of an international organization that was the predecessor to the modern International Union of Geodesy and Geophysics (IUGG). The objectives of IUGG are the promotion and coordination of physical, chemical, and mathematical studies of the Earth and geospace environment. IUGG now consists of seven essentially autonomous associations: one of these, the International Association of Geomagnetism and Aeronomy (IAGA), is principally concerned with geomagnetism.

International collaboration in the study of the Earth received a major boost from the International Geophysical Year (IGY), which took place from 1957 to 1959. There were at least two major legacies of the IGY: first, the IGY demonstrated the possibility for international cooperation among all countries in a scientific endeavor of common interest and value. Second, the World Data Center System was established. Geomagnetism was a major program of the IGY.

From IGY to the present, geomagnetism has been a major or significant component of many international programs and organizations. Two bodies with major interests in geomagnetism were established in the 1960s: the ICSU Scientific Committee on Space Research (COSPAR) and the ICSU Scientific Committee on Solar–Terrestrial Physics (SCOSTEP). Both of these bodies sponsor programs involving geomagnetism. The International Magnetospheric Study (IMS) took place from 1976 to 1979. The Solar–Terrestrial Energy Program (STEP) of SCOSTEP was a significant program that ended in 1997. STEP was followed by four smaller programs: S-RAMP (STEP-Results, Applications, and Modeling Phase); ISCS (International Solar Cycle Study); EPIC (Equatorial Processes Including Coupling); and PSMOS (Planetary Scale Mesosphere Observing System). S-RAMP, the sequel to STEP, includes the working groups on space weather and climate impacts of solar variability. The future umbrella program of SCOSTEP will be CAWSES (Climate and Weather of the Sun–Earth System) in 2004–2008. Each of the current SCOSTEP programs is described in website http://www.ngdc.noaa.gov/stp/SCOSTEP/scostep.html.

The International Lithosphere Program (ILP), which was instituted by IUGG and IUGS (International Union of Geological Sciences) in 1980 as the successor to the International Geodynamics Project, seeks to elucidate the nature, dynamics, origin, and evolution of the lithosphere, with special attention to the continents and their margins.

The International Lithosphere Program (ILP) is being carried out under the guidance of the Inter-Union Commission on the Lithosphere (ICL), which was established under the auspices of ICSU. The ILP includes a broad array of topics that naturally call for collaborative efforts between geologists and geophysicists. Emphasis has been given to *key projects*. Geomagnetic problems play a role in several of the ILP key projects.

IUGG also established the SEDI (Studies of the Earth's Deep Interior) program. The SEDI program has major concern with the Earth's main magnetic field and dynamo processes."

IAGA assemblies are scheduled every two years on odd-numbered years. Alternate assemblies of IAGA are held as a part of the IUGG assembly that meets every four years on odd-numbered years (2003, 2007, 2011, . . .). Meeting sites are shifted between member countries.

B.2 World Data Centers for Geomagnetism

Geomagnetic data from world observatories are deposited at World Data Centers in a number of major countries. Researchers may obtain tabulations of hourly values, photocopies of analog records, and computer-compatible recordings of digitized records for the nominal cost of copying and mailing (or free of charge on an exchange basis). In addition, related information, such as activity indices, solar–terrestrial disturbance records, and global field models may be obtained. A Guide to the World Data Centers is provided at website http://www.wdc.r1.ac.uk/wdcmain/guide/gdsystem.html. Some WDCs of interest for our studies are listed below.

World Data Center for Solar–Terrestrial Physics, Boulder
    tel: 1 303 497 6761
    fax: 1 303 497 6513
 e-mail: wdcstp@ngdc.noaa.gov
website: http://www.ngdc.noaa.gov/stp/WDC/wdcstp.html
address: NOAA/NGDC E/GC2
           325 Broadway
           Boulder, Colorado 80305-3328
           USA

World Data Center for Solid Earth Geophysics, Boulder
    tel: 1 303 497 6521
    fax: 1 303 497 6513
 e-mail: wdcseg@ngdc.noaa.gov
website: http://www.ngdc.noaa.gov/seg/wdc/wdcseg.shtml
address: NOAA/NGDC E/GC1
           325 Broadway
           Boulder, Colorado 80305-3328
           USA

World Data Center for Satellite Information
    tel: 1 301 286 6695
    fax: 1 301 286 1771
 e-mail: request@nssdca.gsfc.nasa.gov
website: http://nssdca.gsfc.nasa.gov/about/about_wdc-a.html

address: NSSDC Code 633
NASA Goddard Space Flight Center
Greenbelt, Maryland 20771
USA

World Data Center for Rockets and Satellites, Obninsk
tel: 7 095 255 2012
fax: 7 095 255 2225
e-mail: sterin@meteo.ru
website: http://www.meteo.ru/
address: WDC for Rockets and Satellites
6 Korolev Str.
Obninsk
Kaluga Region 249020
Russia

World Data Center for Solar–Terrestrial Physics, Moscow
tel: 7 095 930 5619
fax: 7 095 930 5509
e-mail: kharin@wdcb.rssi.ru
website: http://www.wdcb.rssi.ru/WDCB/wdcb_stp.shtml
address: WDC for Solar–Terrestrial Physics
Molodezhnaya 3
Moscow 117296
Russia

World Data Center for Solid Earth Physics, Moscow
tel: 7 095 930 1762
fax: 7 095 930 5509
e-mail: sep@wdcb.rssi.ru
website: http://www.wdcb.rssi.ru/WDCB/wdcb_sep.shtml
address: WDC for Solid Earth Physics
Molodezhnaya 3
Moscow 117296
Russia

World Data Center for Geomagnetism, Copenhagen
tel: +45 39 157488
fax: +45 39 157460
e-mail: wdcc1@dmi.min.dk
website: http://www.dmi.dk/projects/wdcc1/
address: WDC for Geomagnetism
Lyngbyvej 100
DK-2100 Copenhagen
Denmark

World Data Centre for Geomagnetism, Edinburgh
        tel: +44 131 650 0234
        fax: +44 131 668 4368
     e-mail: d.kerridge@bgs.ac.uk
    website: http://ub.nmh.ac.uk
    address: WDC for Geomagnetism
                British Geological Survey
                Murchison House, West Mains Road
                Edinburgh EH9 3LA
                UK

World Data Center for Solar Activity, Meudon
        tel: +331 4507 7767 or 7801
        fax: +331 4507 7959
     e-mail: lantosp@mesioa.obspm.fr
    website: in preparation
    address: WDC for Solar Activity
                Observatoire de Meudon
                F-92195 Meudon Cedex
                France

World Data Centre for Solar Terrestrial Physics, Chilton
        tel: +44 1235 446579
        fax: +44 1235 445848
     e-mail: m.wild@rl.ac.uk
    website: http://www.wdc.rl.ac.uk
    address: Daresbury and Rutherford Appleton Laboratory R25
                Chilton
                Didcot
                Oxon OX11 0QX
                UK

World Data Center for Sunspot Index, Brussels
        tel: +32 3 373 02 76
        fax: +32 3 373 02 24
     e-mail: pierrec@oma.be
    website: http://sidc.oma.be
    address: WDC for Sunspot Index
                Observatorir Royal de Belgique
                7 Avenue Circulaire
                Brussels B1180
                Belgium

World Data Center for Aurora
      tel: +81 3 3962 6482
      fax: +81 3 3962 6482
    e-mail: ejiri@nipr.ac.jp
   website: http://www.nipr.ac.jp/english/aurora.html
   address: WDC for Aurora
              National Institute of Polar Research
              1910 Kaga, Itabashiku
              Tokyo 173
              Japan

World Data Center for Geomagnetism, Mumbai
      tel: +91 22 215 1607 or 1609
      fax: +91 22 218 9568
    e-mail: root@iig.iigm.res.in
   website: http://iig.iigm.res.in
   address: WDC for Geomagnetism
              Indian Institute of Geomagnetism
              Dr. Nanabhai Moos Marg, Colaba
              Mumbai 400005
              India

World Data Center for Geomagnetism, Kyoto
      tel: +81 75 753 3951
      fax: +81 75 722 7884
    e-mail: araki@kugi.kyoto-u.ac.jp
   website: http://swdcdb.kugi.kyoto-u.ac.jp
   address: WDC for Geomagnetism
              Data Analysis Center for Geomagnetism & Space
                 Magnetism
              Faculty of Science, Kyoto University
              Kyoto 606
              Japan

World Data Center for Solar Radio Emission
      tel: +81 267 98 4488
      fax: +81 267 98 2506
    e-mail: senome@nro.nao.ac.jp
   website: http://solar.nro.nao.ac.jp
   address: WDC for Solar Radio Emissions
              Nobeyama Radio Observatory
              Nagano 38413
              Japan

World Data Center for Solar–Terrestrial Activity
tel: +81 427 51 3911
fax: +81 427 59 4255
e-mail: hoshino@gtl.isas.ac.jp
website: http://www.isas.ac.jp
address: WDC for Solar–Terrestrial Activity
          Institute of Space and Astronautical Science
          311 Yoshinodai
          Sagamihara
          Kanagawa 229
          Japan

World Data Center for Solar–Terrestrial Science
tel: 61 02 9213 8000
fax: 61 02 9213 8060
e-mail: asfc@ips.gov.au
website: http://www.ips.gov.au/
address: WDC for Solar–Terrestrial Science
          IPS Radio and Space Services
          PO Box 1386
          Haymarket, NSW 1240
          Australia

World Data Center for Geophysics, Beijing
tel: +86 10 201 1118 ext. 325
fax: +86 10 203 1995
e-mail: none
website: none
address: WDC for Geophysics
          Institute of Geophysics
          Chinese Academy of Sciences
          Beijing 100101
          China

World Data Center for Space Sciences, Beijing
tel: 86 10 254 2551
fax: 86 10 254 2551
e-mail: duheng@sun20.cssar.ac.cn
website: none
address: WDC for Space Sciences
          Center for Space Science and Applied Research
          Chinese Academy of Sciences
          PO Box 8701
          Beijing 100080
          China

B.3 Special USGS Geomagnetics Website

The following description of the US Geological Survey Geomagnetic website at http://geomag.usgs.gov has been provided for this section by John Quinn of USGS.

"Selecting the above website address yields the GEOMAG Home Page, on which there are a dozen sections/topics listed that the user may access by single clicking on the HOT area surrounding the topic of interest.

Various charts of the Main Field magnetic components and their Secular Variations are contained in the section labeled Magnetic Charts. The user may select/examine IGRF charts for epoch 1995, US/UK World Magnetic Model (WMM) charts for epoch 2000, and Regional charts for the continental USA for epoch 1995 by single clicking on the desired topic. Clicking on WMM-2000 charts, for instance, yields a page that gives a verbal description of these charts and allows one to download a brochure either as a Postscript File or a PDF file. The brochure is for the 1995 epoch but, except for the coefficients, remains the same for one five-year epoch to the next. The brochure describes the mathematics behind the Spherical Harmonic Models that characterize the magnetic field. It also describes the computer algorithm that implements the mathematical formalism. Further down this page, one may view the magnetic field components in Mercator or Polar Stereographic projections. Clicking on the Mercator projection, for instance, bring up a new page which has small "thumbnail" views of each of the magnetic field components (D, I, F, H, and Z) for both the Main Field and its Secular Variation as color contour charts. To the right of these thumbnails, one can choose to download page-size versions of the charts in either Postscript or PDF formats. By clicking on a particular thumbnail, a full-size version of the chart will appear on the screen.

Clicking on the Magnetic Models section yields a page that discusses the difference between the International Geomagnetic Reference Field (IGRF) and WMM models and charts, their uses, and their data sources. Farther down the page are HOT zones which bring up charts of the magnetic coordinates as well as charts of the North and South geomagnetic pole movements from 1945 to 2000. When these pages are opened, one sees thumbnail charts and corresponding options to download the respective page-sized charts either as Postscript or as PDF files. Clicking on a particular thumbnail brings up the full page-sized chart on the computer screen. Returning to the Magnetic Models section and scrolling further down this page, one sees a diagram indicating how the magnetic components are defined with respect to a tangent plane attached to any point on the Earth's surface. Finally, at the bottom of this page is a HOT zone that takes the user to the magnetic models. The models may be used on-line, or they may be downloaded for free. The down-loaded program operates in Windows 2000/NT/95. The online version operates in Netscape or Microsoft Explorer. However, it is more easily read in Microsoft Explorer. The same set of models is available whether downloaded or used on-line. The magnetic field can be computed at any time between 1600 and 2005 by choosing the appropriate model. Magnetic values can be computed anywhere on/below/above the Earth's surface between altitudes of -100 km to $+1000$ km. The accuracy of the results get worse the farther back in time one goes. The downloaded version of the computational program is a collaboration between Interpx Corporation of Golden, Colorado, and the USGS. The program is called GEOMAGIX.

Another very interesting section of the USGS website is the Current Observatory Data Plots section. It presents a graphic that shows the location of all USGS operated magnetic observatories. Each observatory is designated by a red dot. Each red dot is a HOT zone. Clicking on a particular red dot for one observatory will bring up several tabs. A particular tab allows the user to view a full day's worth of data. Tabs begin with TODAY and go back one week for a total of eight tabs. The plots show the H, D, Z and F magnetic field components. The plots are shown in near real time with a twelve-minute delay. Note that the data being viewed is raw observatory data which still needs to be properly calibrated. The northern observatories tend to have more active plots because these observatories reside near the Auroral Zone. Also, note that there is sometimes a short delay (a minute or so) between the time a tab is clicked-on and the time that a plot actually appears on the screen.

The other sections are self-explanatory once they are clicked on. One section explains the INTERMAGNET program, one describes the USGS observatory network, and another describes the USGS National Geomagnetic Information Center (NGIC), which handles questions from students, scientific researchers, engineers, and military contractors. The Geomagnetic Information Node (GIN) section explains how one may acquire data via e-mail, not only from the USGS observatories but also from observatories operated by other countries which are also associated with the INTERMAGNET program."

B.4 Special Geomagnetic Data Sets

The WDC for Solar–Terrestrial Physics, Boulder, operated by the Solar–Terrestrial Physics Division of NGDC/NOAA, provides a Space Physics Interactive Data Resource (SPIDR) program at http://spidr.ngdc.

noaa.gov/spidr/. Eric Kihn, who maintains that website, provides this following description.

"SPIDR is designed to allow users to search, browse, retrieve, and display digital solar–terrestrial physics (STP) data. The four objectives of SPIDR design were: (1) to build a flexible and reliable database management system that delivers space and weather data to users through the world wide web and standard APIs, (2) to make full use of open-source software, (3) to allow users to preview all data prior to ordering and delivery, and (4) to provide data input mechanisms for numerical models of the atmosphere and space environment.

The SPIDR system currently handles the following data types for the STP division: DMSP satellite visible, infrared and microwave browse imagery, ionospheric parameters, geomagnetic 1.0-minute and hourly value data, radio telescope, satellite anomaly and city lights data sets. The goal is to manage and distribute all STP digital holdings through the SPIDR system providing comprehensive and authoritative on-line data services, analysis, and numerical modeling.

NGDC also participates in the SPIDR Mirror Project (SMP). The objective of SMP is to create an exact duplicate of the SPIDR system at other WDC sites allowing local users to enjoy greatly enhanced access and providing automatic mirroring of the international data in the NGDC archives. There are currently SPIDR sites in five locations: Japan, South Africa, Russia, Australia, and NGDC."

Users of SPIDR enter dates, parameters, and other search criteria through selection of HOT keys. There are about 200 geomagnetic stations' data, extending from 1900 to the present, available for one-minute or hourly display. One can also browse through the DMSP satellite pictures, display ionospheric global contour maps, and plot records at the user's desired scales. New data and imagery are continually added.

The following special list of international geomagnetic data internet locations has been provided by Helen Coffey of NGDC/NOAA:

Geomagnetic indices/SSC lists:
  http://www.ngdc.noaa.gov/seg/potfld/geomag.shtml
File transfer of indices:
  ftp://ftp.ngdc.noaa.gov/STP/GEOMAGNETIC_DATA/
  ftp://ftp.ngdc.noaa.gov/STP/SOLAR_DATA/RELATED_INDICES/
Storm sudden commencements and principal storms:
  ftp://ftp.ngdc.noaa.gov/STP/SOLAR_DATA/
    SUDDEN_COMMENCEMENTS/

IMAGE-27 Magnetometer stations for Estonia, Finland, Germany,

Norway, Poland, Russia, and Sweden to study auroral electrojets and
moving 2-D current systems:
    http://www.geo.fmi.fi/image

Canadian National Geomagnetism Program
Forecasts, indices, digital data for 12 stations:
    http://www.geolab.nrcan.gc.ca/geomag

Sodankyla Geophysical Observatory, Finland
Magnetometer 1914–present (no 1945), current 1-min geomag data,
ionosonde, and riometer:
    http://www.sgo.fi

Geophysical Observatory Tromso, Norway
10-sec data plots, K indices, daily mean values:
    http://Geo.phys.uit.no/geomag.html

Swedish Institute of Space Physics (IRF), Kiruna, Sweden:
    http://www.irf.se

Geomagnetism Group, British Geological Survey, Edinburgh, UK:
Magnetometers at Lerwick, Eskdalemuir, Hartland, Ascension Island,
Falkland Island, and Sable Island:
    http://www.geomag.bgs.ac.uk/observatories.html

B.5 Special Organization Services

Each location adds its special expertise of value to the particular national
interests of its home country.

Global Paleomagnetic Projects and Laboratories (GPPL)

N. Abrahamsen	E. Petrovsky
Dept. of Earth Sciences	Geophysical Institute
Aarhus University	Bocni II/1401
Finlandsgade 8	141 31 Praha 4
8200 Aarhus N	Czech Republic
Denmark	tel: 42-2-766-051
tel: 45-86-161-666	fax: 42-2-761-549
fax: 45-86-101-003	

International Service of Geomagnetic Indices (ISGI)
Institut de Physique du Globe
4 Place Jussieu, Tour 24-25, 2e etage
F-75252 Paris Cedex 05
France

Magnetic Activity Indices
Institut für Geophysik der Universität Göttingen
Postfach 2341
Herzberger Landstrasse 180
W-3400 Göttingen
Germany

Map Division (for US and global magnetic charts)
US Geological Survey
Mailstop 306, Box 25286
Federal Center
Denver, Colorado 80225
USA

US Magnetic Observatory Data
Geomagnetism Section
US Geological Survey
Mailstop 966, Box 25046
Denver, Colorado 80225-0046
USA
fax: 1-303-273-8506
web: http://www.geomag.usgs.gov

Quarterly Bulletin on Solar Activity (QBSA)
Solar Physics Division
National Astronomical Observatory
University of Tokyo
Mitaka, Tokyo 181
Japan

PAIGH Committee on Geomagnetism and Aeronomy
Magnetic observations in Argentina, Bolivia, Brasil, Colombia,
Costa Rica, Cuba, Mexico, and Peru.
contact Dr. L. M. Barreto at Observatorio Nacional
Rua General Jose Cristino 77
20921-400 Rio de Janeiro, RJ
Brasil
tel: 55-21-585-3215 ext. 214
fax: 55-21-580-6071

Institute of Earth Magnetism, Ionosphere, and Radio Waves, IZMIRAN
Russian Academy of Sciences
142092 Troitsk, Moscow Region
Russia
tel: (095) 334-0921
fax: (095) 334-0124

Tables and Maps of DGRF and IGRF
IAGA Bulletin Publications
Bureau of Publications SIIG-ISGI
CRPE, 4 Avenue de Neptune
94107 Saint-Maur-des Fosses Cedex
France
e-mail: aberthel@st-maur.cnet.fr

INTERMAGNET Office
Institute de Physique du Globe
4, place Jussier
F 75252 Paris cedex 05
France
tel: 33 2 3833 9501
fax: 33 2 3833 9504
e-mail: imso@ipgp.jussieu.fr

The INTERMAGNET data are collected at six Geomagnetic Information Nodes (GINs) described at http://www.intermagneti.gor/Gins_e.html. These GINs can be accessed and data transferred from them at the following e-mail addresses:

Edinburgh, UK: e_gin@mail.nmh.ac.uk
Golden, USA: gol_gin@ghtmail.cr.usgs.gov
Hiraiso, Japan: hiraiso_gin@crl.go.jp
Kyoto, Japan: kyoto-gin@swdcdb.kugi.kyoto-u.ac.jp
Ottawa, Canada: ottgin@geolab.nrcan.gc.ca
Paris, France: par_gin@ipgp.jussieu.fr

Annually produced CD-ROMs containing one-minute data from all INTERMAGNET observatories (starting with 1991) are available free of charge from the INTERMAGNET website or the office (address above).

The National Geophysical Data Center (NGDC) of NOAA is responsible for operating World Data Center, Boulder. There are two especially useful observatory publications available from NGDC: SE-50, "Bibliography of Historical Geomagnetic Main Field Survey and Secular Variations Reports," which is an inventory of historical reports pertaining to the Earth's main field; and SE-53, "A Report on Geomagnetic Observatories, 1995," which contains a listing of world geomagnetic observatories and their data coverage.

A CD-ROM (NGDC-05/01), "Solar Variability Affecting Earth," is available from NGDC for a nominal charge. This ROM contains over 100 solar–terrestrial data bases that have been accumulated until April 1990. The ROM includes solar activity records, IMF values, geomagnetic fields, indices, geomagnetic models, and ionospheric data.

B.6 Solar–Terrestrial Activity Forecasting Centers

Data collection and analysis centers about the Earth monitor the Sun and space environment for conditions of importance to determining the impact of solar–terrestrial disturbances on the Earth. Satellite operations, global communications, high-latitude electric power networks, etc. can be adversely affected by such solar-weather conditions (see Chapter 5). Regional Warning Centers (RWC) have been established by the International Space Environment Service (ISES) to provide notice of solar–terrestrial activity levels and disturbance forecasts to the public and to governmental bodies through a wide range of communication mechanisms (see website http://www.ises-spaceweather.org). Below is a list of the presently operating centers.

Australian Space Forecast Center
    IPS Radio and Space Services
    PO Box 1386
    Haymarket NSW 1240
    Australia
    tel: 61 2 921 38000
    fax: 61 2 921 38060
    e-mail: asfc@ips.gov.au

Canadian RWC
    Geomagnetism Laboratory
    Geological Survey of Canada
    7 Observatory Crescent
    Ottawa K1A 0Y3
    Canada
    tel: 1-613-837-4241
    fax: 1-613-824-9803
    e-mail: forecast@geolab.nrcan.gc.ca

Chinese RWC
    Beijing Astronomical Observatory
    Chinese Academy of Sciences
    Beijing 100080
    China
    tel: 255-1968
    fax: 28-1261
    email: jlwang@bepc2.ihepc.ac.cn

Czech RWC
    Czech Academy of Sciences
    Geophysical Institute

Bocni II
141 31 Praha 4
Czech Republic
tel: 42-276-2548
fax: 42-276-2528
e-mail: ion@cspgig11.bitnet

Russia RWC
Hydrometeorological Service
6 Pavlika Morozova St.
Moscow
Russia
e-mail: geophys@sovamsu.sovusa.com

Indian RWC
National Physical Laboratory
Hillside Road
New Delhi 110012
India
tel: 91-11-572-1436 (or -572-6570)
fax: 91-11-575-2678
e-mail: npl@sirnetd.ernet.in

Polish RWC
Space Research Center
Ordona
2101-237 Warsaw
Poland
tel: 403-766
fax: 121-273
e-mail: cbkpan@plearn.bitnet

Western Hemisphere RWC
Space Environment Center
NOAA, R/E/SE
325 Broadway
Boulder, Colorado 80305-3328
USA
tel: 1-303-497-3171
fax: 1-303-497-3137
e-mail: sesc@sel.noaa.gov

Japanese RWC
Space Science Division
Communications Research Laboratory

4-2-1 Nukuikita-machi, Koganie-shi
Tokyo 184
Japan
tel: 81-29-265-9719
fax: 81-29-265-9721
e-mail: soltech@crl.go.jp

Swedish RWC
Swedish Institute of Space Physics
Scheelev 17
SE-223 70 Lund
Sweden
tel: +46-46-2862120
fax: +46-46-129879
e-mail: henrik@irfl.lu.se

Belgian RWC
Royal Observatory of Belgium
Av. Circulaire, 3
B-1180 Brussels
Belgium
tel: +32 (0) 237 30276
fax: +32 (0) 237 30224
e-mail: pierre.cugnon@oma.be

The British Geomagnetism Information and Forecast Service (GIFS)
supplies 27-day forecasts of Ap index, 10.7-cm solar flux, and activity
indices from UK observatories.

Geomagnetism Group
British Geological Survey
Murchison House, West Mains Road
Edinburgh EH9 3LA
UK
tel: 44-31-667-1000
fax: 44-31-668-4368

The Space Environment Center of NOAA, which operates the
Western Hemisphere RWC, provides the latest solar and geophysical information in a 40-second message updated every 3 hours.
Solar–terrestrial information (indices, conditions for the last 24 hours,
and forecasts for the next 24 hours) are available on a telephone recording
((303) 497-3235 at country code 1) and radio broadcasts both on station
WWV (at 2.5, 5.0, 10.0, and 20.0 MHz) at 18 minutes past the hour and

on station WWVH (at 2.5, 5.0, 10.0, and 15.0 MHz) at 45 minutes past the hour. Space environment plots as well as most recent solar images can be obtained from the website http://www. sec.noaa.gov (select "Space Weather Now" then select "Customer Services").

B.7 Bibliography for Geomagnetism

The following is my list of recommended reading for those who desire a historical perspective and greater details on geomagnetism subjects than can be found in this introductory textbook.

Akasofu, S.-I., and S. Chapman, *Solar–Terrestrial Physics*, 889 pp., Clarendon Press, Oxford, 1972.

Brown, G.C., and A.E. Mussett, *The Inaccessible Earth*, 235 pp., George Allen & Unwin Press, London, 1981.

Campbell, W.H., *Earth Magnetism: A Guided Tour through Magnetic Fields*, Harcourt Academic Press, San Diego, 151 pp., 2000.

Chapman, S., and J. Bartels, *Geomagnetism*, 2 vols., Clarendon Press, reprinted, 1,049 pp., Oxford University Press, Oxford, 1940.

Davies, K., *Ionospheric Radio*, 580 pp., Peter Peregrinus Ltd., London, 1990.

Davis, N., *The Aurora Watchers Handbook*, 230 pp., University of Alaska Press, Fairbanks, 1992.

Fowler, C.R.M., *The Solid Earth: An Introduction to Global Geophysics*, Cambridge University Press, New York, 472 pp., 1990.

Gilbert, W., *De Magnete*, 1600. Translated by P. Fleury Mottelay in 1893; reprinted, 368 pp., Dover Publications Inc., New York, 1958.

Hoyt, D.V. and K.H. Schatten, *The Role of the Sun in Climate Change*, Oxford University Press, Oxford, 280 pp., 1997.

Jacobs, J.A., ed., *Geomagnetism*, four vols., 2,545 pp., Academic Press, New York, 1987–1991.

Jeanloz, R., The nature of the Earth's core, *Ann. Rev. Earth Planet. Sci.*, **18**, 357–386, 1990.

Kan, J.R., T.A. Potemra, S. Kokubun, and T. Iijima, *Magnetospheric Substorms, Geophysical Monograph 64*, 476 pp., AGU, Washington, D.C., 1991.

Livingston, J.D., *Driving Force, the Natural Magic of Magnets*, Harvard University Press, Cambridge, USA, 311 pp., 1996.

Matsushita, S., and W.H. Campbell, eds., *Physics of Geomagnetic Phenomena*, 2 vols., 1,398 pp., Academic Press, New York, 1967.

Mayaud, P.N., *Derivation, Meaning, and Use of Geomagnetic Indices*, 154 pp., AGU, Washington, D.C., 1980.

McElhinny, M.W. and P.L. McFadden, *Paleomagnetism: Continents and Oceans*, Academic Press, San Diego, 386 pp., 2000.

Melchoir, P., *The physics of the Earth's Core, an Introduction*, Pergamon Press, Oxford, 256 pp., 1986.

Parkinson, W.D., *Introduction to Geomagnetism*, 433 pp., Scottish Academic Press Ltd., Edinburgh, 1983.

Piper, J.D.A., *Palaeomagnetism and the Continental Crust*, 434 pp., Open University Press, Milton Keynes Pub., 1987.

Stacey, F.D., *Physics of the Earth*, 513 pp., Brisbane Brookfield Press, Brisbane, 1992.

Song, P., H.J. Singer, and G.L. Siscoe, eds., *Space Weather, American Geophysical Union*, Washington, 449 pp, 2001.

Tauxe, L., *Paleomagnetic Principles and Practice*, Kluwer Academic Publishers, Boston, 299 pp., 1998.

B.8 Principal Scientific Journals for Geomagnetism

Geomagnetism and Aeronomy
Geomagnetizm i Aeronomiya
English translation from Russian
INTERPERIODICA
PO Box 1831
Birmingham, AL 35201-1831
USA

Geophysical Journal International
Royal Astronomical Society
Burlington House
Piccadilly, London WQ1J 0BQ
UK

Geophysical Research Letters
American Geophysical Union
2000 Florida Avenue, NW
Washington, DC 20009
USA

Geophysics
Society of Exploration Geophysics
PO Box 702740
Tulsa, Oklahoma 74170
USA

International Journal of Geomagnetism and Aeronomy
American Geophysical Union
2000 Florida Avenue, NW
Washington, DC 20009
USA

Journal of Atmospheric and Solar–Terrestrial Physics
Elsevier Science
PO Box 945
New York, NY 10159-0945
USA

Journal of Earth, Planets and Space
Terra Scientific Publication Company
2003 Sansei Jiyugaoka Haimu
5-27-19 Okusawa, Setagaya-ku
Tokyo 158-0083
Japan

Journal of Geophysical Research
American Geophysical Union
2000 Florida Avenue, NW
Washington, DC 20009
USA

Physics of the Earth and Planetary Interiors
Elsevier Science Publishers
Journal Department, PO Box 211
1000 AE Amsterdam
Netherlands

Planetary and Space Science
Elsevier Science
PO Box 945
New York, NY 10159-0945
USA

Pure and Applied Geophysics
Birkhauser Verlag publishers
PO Box 133, CH-4010
Basel
Switzerland

Reviews of Geophysics
American Geophysical Union
2000 Florida Avenue, NW
Washington, DC 20009
USA

Appendix C
Utility programs for geomagnetic fields

Below is a description of computer program files, mentioned throughout this textbook, that may be obtained by the readers in two ways: (1) free of charge at the website prepared by Susan McLean of NGDC/NOAA (http://www.ngdc.noaa.gov/seg/potfld/geomag.shtml) or (2) on a high-density (1.4 Mb) floppy disk sold by NGDC/NOAA.

At the website main page, select the highlighted "useful computer programs." The executable files are designed for a DOS-compatible, personal computer. The programs were designed to assist the user in understanding the subject of geomagnetic fields; no claim is made regarding the suitability of the software for any other purpose. No restriction has been placed on the sharing of these programs; also, no warranty (expressed or implied), no endorsement, no guarantee of accuracy, and no responsibility for the program's functioning, can be made by the program authors, the author or publisher of this book, or by the National Geophysical Data Center. The files ending in .EXE are the executable programs; all the other files provide necessary input for some of the programs and must also be copied for proper operation of the set. If the files have been downloaded from the NGDC website, then copy all of them to a disk on your computer before running.

C.1 Geomagnetic Coordinates 1940–2005

The **GMCORD** program provides a determination of the geomagnetic coordinates for any selected global location. The program uses a polynomial fit (see the **POLYFIT** program below) of the dipole

Gauss coefficients from the geomagnetic reference field models (see file **ALL-IGRF.TAB** below). The fitting allows specification of any date from the first day of January 1940 to the projected values on the last day of December 2005. In addition to the computation of the dipole coordinates, the geomagnetic pole locations, the dipole moment, and station dipole field values are displayed on your screen. This program uses a computational method devised by Fraser-Smith (1987) and coded by W.H. Campbell. See Section 1.5.

To run the program, place the disk in the computer, select the drive letter that is being used for the disk, and enter **GMCORD**. Enter a name for your selected station location, enter its latitude and longitude in geographic coordinates (decimal degrees with + for North and East, – for South and West), and answer the date questions (months are to be given as numbers 1 to 12). The displayed output may be saved by using the "print screen" key on your computer.

C.2 Fields from the IGRF Model

The **GEOMAG** program provides field values computed from the geomagnetic reference field model (IGRF) applicable to a user-selected date from 1900 through 2004 (see **ALL-IGRFTAB** file below). This program, provided by S. McLean of NGDC, is an abbreviated version of the *Geomagnetic Field Models and Synthesis Software* package for field models starting from 1900 that can be obtained, with detailed documentation, at nominal cost from the World Data Center for Solid Earth Geophysics, NGDC/NOAA, 325 Broadway, Boulder, Colorado 80305, USA. That organization notes, "If the programs are incorporated into other software, a statement identifying them may be required under 17 U.S.C. 403 to appear with any copyright notice." See Section 1.4.

To run the program, insert the disk in the computer, select the drive letter that is being used for the disk, and enter **GEOMAG**. There is an explanatory text at the start of the program. Use the file **igrf.dat** with its location (e.g. **A:\igrf.dat**) when asked for the model. The program will prompt you for the date as well as the location latitude and longitude of interest. Entry of location is either in decimal degrees or in degrees, minutes, and seconds. Entry of the date is in decimal year or in year, month, and day. With the decimal year entry use .0 for 1 January and .5 for midyear (e.g., 2000.0 for 1 January 2000, and 2000.5 for 2 July 2000).

C.3 Quiet-Day Field Variation, Sq

The **SQ1MODEL** program by W.H. Campbell produces an estimate of the daily variation of geomagnetic field for any global location.

The model was derived using data gathered at a global distribution of observatories, on the quietest days in the extremely quiet year 1965. Yearly variations were obtained using a Fourier analysis of the daily Fourier coefficients for selected quiet days (see the **FOURSQI** program below). These coefficients were smoothed in latitude and longitude to form the incremental global data source file SQGLDATA.DAT on the disk. The data, arranged in geomagnetic coordinates, are simply extended to other years (1940 to 2005) by adjustment to the changed geomagnetic coordinates (using a version of the **GMCORD** program) at the prescribed location on the selected date. It must be realized that on solar-active years the entire Sq system can be modified somewhat from the quietest-year behavior. Because of the scarcity of observatories in the polar region at the time the data were gathered, there is less confidence in the values that the program produces at very high latitudes. See Section 2.6.

To run the program, you can elect to have the list of values for each increment of minutes (either local or universal time) directed to your printer or placed in a directory file with a name of your selection. Longitude values should be positive for east and negative for west; latitude values should be positive for north, negative for south. The field values, in both *HDZ* and *XYZ* field component directions, are given in universal time and local meridian time.

C.4 The Geomagnetic Disturbance Index, Dst

During a typical geomagnetic storm the magnetic field is depressed (*H* component is negative) everywhere in the middle and lower latitudes of the Earth. The **DSTDEMO** program by W.H. Campbell and E.R. Schiffmacher illustrates the change of these midlatitude fields during a storm. The program also shows how the shape of the Dst index, during geomagnetic storms, can be fitted to lognormal distribution curves. For convenience in this display, all the storm values (obtained from the *H* component of field) are presented as positive magnitudes. The user can either select from ten sample geomagnetic storms or enter hourly *magnitudes* of the *H* component of field through the main phase and recovery phase of a storm (i.e., enter the storm-time depression of *H* as positive values).

A lognormal distribution becomes a normal (symmetric) distribution when the amplitudes are plotted using the logarithm of the *x*-axis values. This symmetry allows the prediction of storm decay (geomagnetic storm recovery phase) from the storm value rise to maximum (geomagnetic storm main phase). The program computes the field decay from the hourly values given from the start of the main phase of the storm until one hour past the peak magnitude of the storm-time Dst index.

Screen graphics show the comparison between the predicted recovery and the observed values. See Section 3.15.

Try one of the sample Dst storms first to see how **DSTDEMO** works. The program needs to read the accompanying sample data files so you are asked to give the location where you copied those files. The program asks questions about the printer so that a hard copy of the results can be an option. When the program is run with user-supplied hourly Dst index or H-component field, negative input values will be made positive for analysis purposes. The projection technique requires that values must be given up to an hour past the peak of the storm. If more past-peak values are entered, they will be shown as an overplot on the prediction of recovery. To run the program, insert the disk in the computer, select the drive letter, and enter **DSTDEMO**.

C.5 Location of the Sun and Moon

The **SUN-MOON** program, prepared by E.R. Schiffmacher, gives the position of the Sun and Moon for any observer latitude/longitude and selected time (in universal time, UT). The solar and lunar local transit, rise and set times, and location with respect to the observer at any given hour and minute (UT) are specified. The solar zenith angle, χ, given in this program, is 90° minus the elevation angle of the Sun. The square root of the cosine of this solar zenith angle (called the Chapman function; see Section 2.3) is used when discussing the ionospheric E-region ionization and electrical conductivity for the quiet-time dynamo currents of thermospheric wind and when discussing tidal origin of fields. To run the program, insert the disk in the computer, select the drive letter for the disk, enter **SUN-MOON**, and follow the directions.

C.6 Day Number

The **DAYNUM** program determines the number of the day in the year (1 January is number 1) for any day, month, and year selected by the user. Adjustments for leap years are made. The day number is not to be confused with the Julian Day, which is the count of the number of days starting with the first of January 4713 BC.

To start the program, simply insert the disk into the computer drive bay, set the computer for that drive, and enter **DAYNUM**. The day, month, and year must be entered as whole numbers. The year must be entered as a four-digit number greater than 1000. Many scientific data (and most satellite records) are organized by year and day number rather than year, month, and day.

C.7 Polynomial Fitting

The **POLYFIT** program creates a polynomial equation (see Appendix A.1) portrayal of data points (independent variable X, dependent variable Y). Such useful representations of related phenomena have the form:

$$Y = c + a_1 X^1 + a_2 X^2 + a_3 X^3 + \cdots a_n X^n, \qquad (C.1)$$

where c, a_1, a_2, a_3, ... a_n are constants, and the "degree" of the polynomial, n, is at least two less than the number of data points. To run the program, insert the disk into the computer drive bay, set the computer for that drive, and enter **POLYFIT**. Answer the questions regarding the number of data points and desired degree of the fitting. Then enter the pairs of points, X and Y, separated by a comma. The program gives the values of the constant, the polynomial coefficients, the correlation coefficient, and the standard error of estimate. Also, an opportunity is provided to use the polynomial to determine any Y value for a given X.

For many geomagnetic computations, an extrapolation to future predictions uses a polynomial fitting of the known past behavior. The geomagnetic coordinate computations are extended to 10 years past the present using a polynomial fitting of previous main-field Gauss coefficients.

C.8 Quiet-Day Spectral Analysis

The **FOURSQ1** program computes the Fourier harmonic spectral components (cf. Figure 1.12) of quiet-field records (cf. Figures 2.23 and 2.24) along with a determination of the daily mean level and trend (slope). One can select the number of desired harmonics, up to twenty-four. A special iterative technique of W.H. Campbell is employed to extract the linear trend, embedded in the daily harmonics, to avoid improper trend contributions from sine wave harmonics. An input of field value data at evenly spaced intervals (with a point at the beginning and at the end) is required and a test is used to determine whether the number of points is sufficient for the highest selected harmonic. The output is to the computer screen and to a user-selected file. To start the program, place the disk in the drive bay, set the computer for that drive, and enter **FOURSQ1**. Answer the questions at the prompt. The daily quiet-time Sq field variations are analyzed using this program.

C.9 Median of Sorted Values

The program **SORTVAL** is designed to sort a list of unordered data values and place them in ascending size. A median value (or range), for which the number of entries with lower value is equal to the number with higher value, is determined from the ordered list. The presence of outliers in a data set are often first indicated when the median of a data set differs greatly from the mean (average) value of the set.

SORTVAL is run from the computer DOS mode; simply enter the program name and touch ENTER on your keyboard. You will be asked to enter the total number of values in your list and then enter the data. For an odd number of data samples, the exact median is computed; for an even number of data samples, the boundary values for the median are found . The results of the program are displayed on your screen. However, you are given the opportunity also to have the results sent to your printer.

C.10 Mean, Standard Deviation, and Correlation

The **ANALYZ** computer program provides two choices for the user: (1) to determine the sample mean and standard deviation for a data set of single values, and (2) to find the linear regression Correlation Coefficient for a set of paired (x and y coordinate) data values. The program displays the values of a best fitting linear representation of the data pairs along with the Standard Error of Estimate. You are given the opportunity to use the linear regression line to determine the corresponding value of y for any given x.

The **ANALYZ** program is run in the computer DOS mode; simply enter the program name and touch ENTER on your keyboard. You are asked to select between the two types of study (single data set and paired values set). Follow the directions for entering the number of values (or paired values) and your data. You are given the opportunity to correct an input error. Both the input and results are displayed on the screen. For the best fitting line representing the paired values, you can select any x value of your choosing and have the value of corresponding y shown.

C.11 Demonstration of Spherical Harmonics

The **SPH** program is provided by Phil McFadden of Geoscience Australia (formerly Australian Geological Survey Organisation). It is a graphical representation of the spherical harmonics (cf. Figure 1.15) used in global modeling of the Earth's main field (see Section 1.4). To run the program, place your disk in the floppy drive and enter the letters **SPH**. The screen prompt will ask for the degree (n) of the polynomial that is desired; enter any value less than seventeen. Next, the prompt will ask you to enter the order (m) of the desired polynomial; select any value less than or equal to the n you have already selected. The program allows you to view the spherical surface at different rotations and tilts. The next prompt asks for the step size (in degrees) that you wish to have for rotation of the figure about the pole (e.g., twenty). Then you are asked for the step size that you want for tilting the pole of the figure toward you (e.g., twenty).

The rotation of the small slash in the upper left corner of your screen indicates that the computation is being processed. The displayed spherical polynomial surface is colored light green for values above the base-level sphere and dark green for those below. Use the **S** key to add (or remove) a grid of the base sphere in red line color. The **up, down, right**, and **left** arrow keys of your computer can be used to rotate or tilt the figure in degree steps of the increment size you selected at the start. The **space bar** can be used to select a new polynomial figure. The **ESC** key ends the program. The spherical harmonic, Schmidt-normalized Legendre polynomials of degree n and order m are used to fit the global measurements of the main geomagnetic field producing the IGRF and DGRF Gauss coefficients g and h listed in **ALL-IGRF.TAB** below. A considerably more detailed form of this program along with a computation of associated field components can be obtained by writing to Phil McFadden at pmcfadden@pcug.org.au for his new version of **SPH.**

C.12 Table of All Field Models

The **ALL-IGRF.TAB** file is a table of all the spherical harmonic Gauss coefficients, g and h, of the International Geomagnetic Reference Fields (IGRF) and Definitive Geomagnetic Reference Fields (DGRF) to cover the years from 1900 to 2005 (cf. Table 1.1). This table was adopted by the Working Group 8 of IAGA Division V.

For each five-year epoch (column) the g and h Gauss coefficients (in nanotesla units) are given (in ordered rows) up to values of degree (n) 10 and order (m) 10. The **SV** column to the right represents the secular variation in units of nanotesla (gamma) per year and is used to extend the 2000 epoch values to the year 2005. These coefficients are used to determine the geomagnetic potential function V (from which the XYZ or HDZ field component values are derived) with the formulation:

$$V = a \sum_{n=1}^{n} (a/r)^{n+1} \left[g_n^m \cos(m\phi) + h_n^m \sin(m\phi) \right] P_n^m(\cos(\theta)), \qquad (C.2)$$

where a is the mean Earth radius, r is the radial distance from the center of the Earth, ϕ is the longitude measured eastward from Greenwich, θ is the geocentric colatitude, and P_n^m is the Schmidt-normalized associated Legendre polynomial of degree n and order m. The maximum number of n is N ($N = 10$ in the table) and m is less than or equal to n. See Equation (1.39).

References

Adam, A., Geothermal effects in the formation of electrically conducting zones and temperature distribution in the Earth, *Phys. Earth Planet Int.*, **17**, 21–28, 1978.

Adam, A., Relation of the graphite and fluid bearing conducting dikes to the tectonics and sesimicity (Review of the Transdanubian crustal conductivity anomaly), *Earth Planets Space*, **53**, 903–918, 2001.

Aitchison, J., and J.A.C. Brown, *The Lognormal Distribution with Special Reference to its Use in Economics*, 176 pp., Cambridge University Press, New York, 1957.

Akasofu, S.-I., The development of the auroral substorm, *Planet. Space Sci.*, **12**, 273–282, 1964.

Akasofu, S.-I., and S. Chapman, *Solar-Terrestrial Physics*, 889 pp., Clarendon Press, Oxford, 1972.

Albertson, V.D., and J.A. vanBaalen, Electric and magnetic fields at the Earth's surface due to auroral currents, *IEEE Trans. Power Apparatus and Systems, PAS-*89, 578–584, 1970.

Allen, J.H., L. Frank, H. Sauer, and P. Reiff, Effects of the March 1989 solar activity, *Eos Trans. AGU*, **70**(46), 1486–1488, 1989.

Allen, J.H., and D.C. Wilkinson, Solar-terrestrial activity affecting systems in space and on Earth, *NOAA/NGDC internal report*, 33 pp., 1992.

Baker, D.N., R.D. Belian, P.R. Higbie, R.W. Klebesadel, and J.B. Blake, Deep dielectric charging effects due to high energy electrons in Earth's outer magnetosphere, *J. Electrostat.*, **20**, 3, 1987.

Baker, D.N., S. Kanekal, J.B. Blake, B. Klecker, and G. Rostoker, Satellite Anomalies linked to electron increase in the magnetosphere, *Eos Trans. AGU*, **75**(35), 401–405, 1994.

Baker, R.R., J.G. Mather, and J.H. Kennaugh, Magnetic bones in human sinuses, *Science*, **301**, 78–80, 1983.

Barreto, L.M., Methods of azimuth determination at a magnetic repeat station: a comparison, *J. Geomag. Geoelectr.*, **48**, 1523–1530, 1996.

Bartels, J., Solar eruptions and their ionospheric effects – a classical observation and its new interpretation, *Terr. Mag. Atmos. Electr.*, **42**, 235–239, 1937.

Basu, S., S. Basu, C.E. Valladares, H.-C. Yeh, S.-Y. Su, E. MacKenzie, P.J. Sultan, J. Aarons, F.J. Rich, P. Doherty, K.M. Groves, and T.W. Bullett, Ionospheric effects of major magnetic storms during the International Space Weather Period of September and October 1999: GPS observations, VHF/UHF scintillations, and in situ density structures at middle and equatorial latitudes, *J. Geophys. Res.*, **106**, 30389–30413, 2001.

Becker, R.O., C.H. Bachman, and H. Friedman, Relation between natural magnetic field intensity and increase of psychiatric disturbances in the human population, paper presented at the International Conference on High Magnetic Fields, Mass. Inst. Tech., Cambridge, Mass., 3 Nov. 1961.

Blais, G., and P. Metsa, Operating the Hydro-Quebec grid under magnetic storm conditions since the storm of 13 March 1989, Proceedings of the Solar-Terrestrial Predictions-IV Workshop, Ottawa, Canada, 18–22 May 1992, *NOAA/Dept. of Commerce Pub.* vol. 1., pp. 108–130, 1993.

Blakemore, R., and R.B. Frankel, Magnetic navigation in bacteria, *Sci. Amer.*, **245**, 58–65, 1981.

Bureau, Y.R.J., and M.A. Persinger, Geomagnetic activity and enhanced mortality in rats with acute (epileptic) limbic lability, *Int. J. Biometeorology*, **36**, 226–232, 1992.

Campbell, W.H., Correlation of sunspot numbers with quantity of S. Chapman publications, *Eos Trans. AGU*, **49**, 609–610, 1968.

Campbell, W.H., The field levels near midnight at low and equatorial geomagnetic stations, *J. Atmos. Terr. Phys.*, **35**, 1127–1146, 1973.

Campbell, W.H., An interpretation of induced electric currents in long pipelines caused by natural geomagnetic sources of the upper atmosphere, *Surveys Geophys.*, **8**, 239–259, 1986.

Campbell, W.H., Geomagnetic storms, the Dst ring-current myth, and lognormal distributions, *J. Atmos. Terr. Phys.*, **58**, 1171–1187, 1996.

Campbell, W.H., A misuse of public funds: UN support for geomagnetic forecasting of earthquakes and meteorological disasters, *EOS, Trans. Amer. Geophys. Un.*, **79**, 463–465, 1998.

Campbell, W.H., B.R. Arora, and E.R. Schiffmacher, Polar cap field response to IMF *By* sector changes on quiet days at a longitude line of observatories, *J. Geomag. Geoelectr.*, **46**, 735–746, 1994.

Campbell, W.H., C.E. Barton, R.H. Chamalaun, and W. Wesh, Quiet-day ionospheric currents and their application to upper mantle conductivity in Australia, *Earth Planets Space*, **50**, 347–360, 1998.

Campbell, W.H., and T.C. Thornberry, Propagation of Pc1 hydromagnetic waves across North America, *J. Geophys. Res.*, **77**, 1941–1950, 1972.

Campbell, W.H., and J. Young, Auroral-zone observations of infrasonic pressure waves related to ionospheric disturbances and geomagnetic activity, *J. Geophys. Res.*, **68**, 5909–5916, 1963.

Carapiperis, L.N., The Etesian winds, III, Secular changes and periodicity of the Etesian winds, *Upomnemata Tou Ethnikon Asterskopeion Athenon*, Ser. II (Meteorology), no. 11, 1962.

Chapman, S., *The Earth's Magnetism*, 117 pp., John Wiley & Sons, New York, 1936.

Chapman, S., and J. Bartels, *Geomagnetism*, reprint, 2 vols., 1049 pp., Oxford University Press, Oxford, 1940.

Chapman, S., and J.C.P. Miller, Analysis of lunar variations, *Mon. Not. Roy. Astr. Soc., Geophys. Suppl.*, **649**, 1940.

Christiansen, F., V.O. Papitashvili, and T. Neubert, Seasonal variations of the high-latitude field-aligned currents inferred from Oersted and Magsat observations, *J. Geophys. Res.*, **107**, A2, JA900104, 2002.

Clayton, H.H., *World Weather*, MacMillan, New York, 1923.

Cliver, E.W. and H.V. Cane, Forum: Impulsive and gradual SEP events., *EOS, Trans. Amer. Geophys. Un.*, **83**, 62–69, 2002.

Cliver, E.W., Y. Kamide, A.G. Ling, and N. Yokoyama, Semidiurnal variation of the geomagnetic Dst index: evidence for a dominant nonstorm component, *J. Geophys. Res.*, **106**, 21297–21304, 2001.

Courtillot, V., J.L. LeMouel, J. Ducruix, and A. Cazenave, Geomagnetic secular variation as a precursor of climatic change, *Nature*, **297**(5865), 386–387, 1982.

Dawson, E., and L.R. Newitt, The magnetic poles of the Earth, *J. Geomag. Geoelectr.*, **34**, 225–240, 1982.

Dubrov, A.P., *The Geomagnetic Field and Life, Geomagnetobiology*, English ed., translated from Russian by F.L. Sinclair, 318 pp., Plenum Press, New York, 1978.

Eather, R.H., *Majestic Lights, The Aurora in Science, History, and the Arts*, 323 pp., American Geophysical Union, Washington, D.C., 1980.

Eddy, J.A., Historical evidence for the existence of the solar cycle, in *The Solar Output and Its Variation*, O.R. White, ed., pp. 51–71, Colorado Associated University Press, Boulder, 1977.

Eddy, J.A., *A New Sun; The Solar Results From Skylab*, 198 pp., Marshall Space Flight Center, NASA, Washington, D.C., 1979.

Feynman, R.P., R.B. Leighton, and M. Sands, *The Feynman Lectures on Physics*, 3 vols., Addison-Wesley, Reading, Mass., 1965.

Fraser-Smith, A.C., Centered and eccentric geomagnetic dipoles and their poles, 1600–1985, *Rev. Geophys.*, **25**, 1–16, 1987.

Friedman, II., R.O. Becker, and C.H. Bachman, Geomagnetic parameters and psychiatric hospital admissions, *Nature*, **200**, 626–628, 1963.

Friis-Christensen, E., Y. Kamide, A.D. Richmond, and S. Matsushita, Interplanetary magnetic field control of high-latitude electric fields and currents determined from Greenland magnetometer data, *J. Geophys. Res.*, **90**, 1325–1338, 1985.

Friis-Christensen, E., and K. Lasssen, Length of the solar cycle: an indicator of solar activity closely associated with climate, *Science*, **254**, 698–700, 1991.

Fuller-Rowell, T.J., M.V. Codrescu, R.J. Moffett, and S. Quegan, Response of the thermosphere and ionosphere to geomagnetic storms, *J. Geophys. Res.*, **99**, 3893–3914, 1994.

Gauss, C.F., Allgemeine Theorie des Erdmagnetismus, Resultate aus den Beobachtungen des Magnetischen Vereins im Jahre 1838, eds. C.F. Gauss and W. Weber. English translation by E. Sabine and R. Taylor in *Scientific Memoirs Selected from the Transactions of Foreign Academies and Learned Societies and from Foreign Journals*, pp. 184–251, J. and R.E. Taylor Pub., London, 1848, 1838.

Geller, R.J., Thinking about the unpredictable, *EOS, Trans. Amer. Geophys. Un.*, **78**, 63–67, 1997.

Gilbert, W., *De Magnete*, Original 1600 book tanslated by P. Fleury Mottelay in 1893, republished by Dover Publications, New York, 1958.

Gnevyshev, M.N., K.F. Novikova, A.I. Ol', and N.V. Tokareva, Sudden death from cardiovascular diseases and solar activity, in *Effects of Solar Activity on the Earth's Atmosphere and Biosphere*, N.M. Gnevyshev and A.I. Ol', eds., pp. 201–210, Acad. Sci. U.S.S.R, English trans., Israel Prog. for Sci. Trans., Jerusalem, 1977.

Gnevyshev, M.N., and A.I. Ol', eds., *Effects of Solar Activity on the Earth's Atmosphere and Biosphere*, Academy of Science, U.S.S.R, English trans., 290 pp., Israel Program for Sci. Trans., Jerusalem, 1977.

Gold, J.L., J.L. Kirschvink, and K.S. Deffeyes, Bees have magnetic remanence, *Science*, **201**, 1026–1028, 1978.

Gopalswamy, N., A. Lara, S. Yashiro, M.L. Kaiser, and R.A. Howard, Predicting the 1-AU arrival times of coronal mass ejections, *J. Geophys. Res.*, **106**, 29207–29217, 2001.

Gorney, D.J., H.C. Koons, and R.L. Waltersheid, Some prospects for artificial intelligence techniques in solar-terrestrial predictions, Solar-Terrestrial Predictions-IV Workshop, Ottawa, Canada, 18–22 May 1992, vol. 2, *NOAA Dept. of Commerce Pub.*, 550–564, 1993.

Goubau, W.M., T.D. Gamble, and J. Clarke, Magnetotelluric data analysis: removal of bias, *Geophys.*, **43**, 1913–1925, 1989.

Gough, D.I., and M.R. Ingham, Interpretation methods for magnetometer arrays, *Rev. Geophys.*, **21**, 805–827, 1983.

Groom, R.W., and R.C. Bailey, Decomposition of magnetotelluric impedance tensors in the presence of local three-dimensional galvanic distortion, *J. Geophys. Res.*, **94**, 1913–1925, 1989.

Haekkinen, L.V.T., T.I. Pulkkinen, H. Nevanlinna, R.J. Pirjola, and E.I. Tanskanen, Effects of induced currents on Dst and on magnetic variations at midlatitude stations, *J. Geophys. Res.*, **107**, JA900130, 2002.

Hayakawa, M., and Y. Fujinawa, eds., *Electromagnetic Phenomena Related to Earthquake Prediction*, 678 pp., Terra Scientific, Tokyo, 1994.

Heirtzler, J.R., J.H. Allen, and D.C. Wilkinson, Ever-present South Atlantic anomaly damages spacecraft, *EOS Trans. Amer. Geophys. Un.*, **83**, 165 & 169, 2002.

Heirtzler, J.R., X. LePichon, and J.G. Brown, Magnetic anomalies over the Reykjanes Ridge, *Deep Sea Res.*, **13**, 427–443, 1966.

Herman, J.R., and R.A. Goldberg, Sun, weather, and climate, *NASA Pub. SP 426*, 360 pp., Washington D.C., 1978.

Higuchi, T. and S.I. Ohtani, Automatic identification of large-scale field-aligned current structures, *J. Geophys. Res.*, **105**, 25305–25315, 2000.

Hines, C.O., A possible mechanism for the production of sun-weather correlations, *J. Atmos. Sci.*, **31**, 589–591, 1974.

Hoffman, K.A., Ancient magnetic reversals: Clues to the geodynamo, *Sci. Amer.*, **258**(5), 76–83, 1988.

Hoffman, R.A., and P.A. Bracken, Higher-order ring currents and particle energy storage in the magnetosphere. *J. Geophys. Res.*, **72**, 6039–6049, 1967.

Hoffman, R.A., K.W. Ogilvie, and M.H. Acuna, Fleet of satellites and ground-based instruments probes the Sun-Earth system, *Eos Trans. AGU*, **77**, 149–150, 1996.

Hofmann-Wellenhof, B., H. Lichtenegger, and J. Collins, *Global Positioning System: Theory and Practice*, 3rd ed., 355 pp., Springer-Verlag, Wien, 1994.

Hones, E.W., Magnetic reconnection in the Earth's magnetotail, *Aus. J. Phys.*, **38**, 981–997, 1985.

Horbury, T.S., D. Burgess, M. Fraenz, and C.J. Owen, Prediction of Earth arrival times of interplanetary southward magnetic field turnings, *J. Geophys. Res.*, **106**, 30001–30009, 2001.

Hoyt, D.V. and K.H. Schatten, *The Role of the Sun in Climate Change*, Oxford University Press, Oxford, UK, 280 pp, 1997.

Hruska, J., R. Coles, H.L. Lam, and G.J. van Beek, *The Major Magnetic Storm of 13–14 March*, 1989: *Its Character in Canada and Some Effects*, Geophys. Div., Geolog. Survey Canada report, Ottawa, 1990.

Hundhausen, A.R., An interplanetary view of coronal holes, in *Coronal Holes and High Speed Wind Streams*, J.B. Zirker, ed., Chapter VII, pp. 225–329, Colorado Associated Press, Boulder, 1977.

Iijima, T., and T.A. Potemra, Field-aligned currents in the dayside cusp observed by TRIAD, *J. Geophys. Res.*, **81**, 5971–5979, 1976.

Jacobs, J.A., ed., *Geomagnetism*, 4 vols., Academic Press, New York, 1987–1991.

Jankowski, J., and C. Sucksdorff, *Guide for Magnetic Measurements and Observatory Practice*, Int. Assoc. Geomagnetism and Aeronomy, 120 pp, 1997.

Jeanloz, R., The nature of the Earth's core, *Ann. Rev. Earth Planet. Sci.*, **18**, 357–386, 1990.

Judge, D.L., D.R. McMullin, P. Gangopadhyay, H.S. Ogawa, F.M. Ipavich, A.B. Galvin, E. Moebius, P. Bochsler, P. Wurtz, M. Hilchenbach, H. Gruenwaldt, D. Hovestadt, P. Klecker, and F. Gliem, Space weather observations using SOHO CELIAS complement of instruments, *J. Geophys. Res.*, **106**, 29963–29968, 2001.

Kan, J.R., T.A. Potemra, S. Kokubun, and T. Iijima, *Magnetospheric Substorms*, Geophysical Monograph 64, 476 pp., American Geophysical Union, Washington, D.C., 1991.

Kappenman, J.L., Geomagnetic disturbances and power system effects, Proceedings of the Solar-Terrestrial Predictions-IV Workshop, Ottawa, 18–22 May 1992, *NOAA/Dept. of Commerce Pub.*, vol. l, pp. 131–141, 1993.

Kappenman, J.L., and V.D. Albertson, Bracing for geomagnetic storms, *IEEE Spectrum*, **27**(3), 27–33, 1990.

King, J.W., Sun-weather relationships, *Astronaut. Aeronaut.*, **13**, 10–19, 1975.

Kirschvink, J.L., Rock magnetism linked to human brain magnetite, *Eos Trans. AGU*, **75**, 178, 1994.

Kirschvink, J.L., D.S. Jones, and B.J. MacFadden, eds., *Magnetite Biomineralization and Magnetoreception in Organisms: A New Biomagnetism*, 682 pp., Plenum Press, New York, 1985.

Kleusberg, A., The global positioning system and ionospheric conditions, Solar predictions-IV Workshop, Ottawa, Canada, 18–22 May 1992, *NOAA/Dept. of Commerce Pub.*, vol. 1, pp. 142–146, 1993.

Kurtz, R.D., J.M. DeLaurier, and J.C. Gupta, Electrical conductivity distribution beneath Vancouver Island: A region of active plate subduction, *J. Geophys. Res.*, **95**, 10,920–10,946, 1990.

Lakhina, G.S., B.T. Tsurutani, H. Kojima, and H. Matsumoto, Broadband plasma waves in the boundary layers, *J. Geophys. Res.*, **105**, 27791–27831, 2000.

Langel, R.A., The main field in *Geomagnetism*, Vol. 1, J.A. Jacobs, ed., Chap. 4, pp. 249–512, Academic Press, New York, 1987.

Langel, R.A., and R.T. Baldwin, Satellite data for geomagnetic modeling, in *Types and Characteristics of Data for Geomagnetic Field Modeling, NASA Conf. Pub.* 3153, R.A. Langel and R.T. Baldwin, ed., pp. 75–135, NASA, Greenbelt, Md., 1992.

Lanzerotti, L.J., and L.V. Medford, Geomagnetic disturbance and long-haul telecommunications cables, *Proc. Electr. Power Res. Inst. Conf. on Geomag. Induced Currents*, EPRI, 3412 Hillview Ave., Palo Alto, Calif. 94304, Nov. 1989.

Lassen, K. and E. Friis-Christensen, Variability of the solar cycle length during the past five centuries and the apparent association with terrestrial climate, *J. Atmos. Terr. Phys.*, **57**, 835, 1995.

Lawrence, E.N., Terrestrial climate and the solar cycle, *Weather*, **20**, 334–343, 1965.

Lay, T., T.J. Ahrens, P. Olson, J. Smyth, and D. Loper, Studies of the Earth's deep interior: Goals and trends, *Phys. Today*, **43**(10), 44–52, 1990.

Lett, J.T., W. Atwell, and M.J. Golightly, Radiation hazards to humans in deep space: a summary with special reference to large solar particle events, 140–158, in *Solar-Terrestrial Predictions: Proc. of a Workshop at Leura, Australia*, October 16–20, 1989, vol. 1, 1990.

Levy, R.H., H.E. Petschek, and G.L. Siscoe, Aerodynamic aspects of magnetospheric flow, *Amer. Inst. Aeronaut. Astron.*, **2**, 2065–2076, 1964.

Livingston, J.D., *Driving Force, the Natural Magic of Magnets*, Harvard University Press, 250 pp, 1996.

Lui, A.T.Y., R.W. McEntire, and S.M. Krimigis, Evolution of the ring current during two geomagnetic storms, *J. Geophys. Res.*, **92**, 7459–7470, 1987.

Main, I., Long odds on prediction, *Nature*, **385**, 19–20, 1997.

Matsushita, S., Solar quiet and lunar daily variation fields, in *Physics of Geomagnetic Phenomena*, S. Matsushita and W.H. Campbell, eds., Chapter III-1, pp. 301–424, Academic Press, New York, 1967.

Matsushita, S., and W.H. Campbell, eds., *Physics of Geomagnetic Phenomena*, 2 vols., 1398 pp., Academic Press, New York, 1967.

Matsushita, S., and W.H. Campbell, Lunar semidiurnal variations of the geomagnetic field determined from 2.5-min data scalings, *J. Atmos. Terr. Phys.*, **34**, 1187–1200, 1972.

Matsushita, S., and W.Y. Xu, Seasonal variations of L equivalent current systems, *J. Geophys. Res.*, **89**, 285–294, 1984.

Maugh, T.H., II, Magnetic navigation an attractive possibility, *Science*, **215**, 1492–1493, 1982.

Maxwell, J.C., *Treatise on Electricity and Magnetism*, Cambridge University Press, Cambridge, 1873.

McCormac, B.M., ed., *Weather and Climate Responses to Solar Variations*, 626 pp., Colorado Associated University Press, Boulder, 1983.

McElhinny, M.W., *Palaeomagnetism and Plate Tectonics*, 358 pp., Cambridge University Press, Cambridge, 1973.

Melchoir, P., *The Physics of the Earth's Core: an Introduction*, Pergamon Press, Oxford, 256 pp, 1986.

Merrill, R.T., and P. McFadden, Paleomagnetism and the nature of the geodynamo, *Science*, **248**, 345–350, 1990.

Mursula, K., T. Braeysy, K. Niskala, and C.T. Russell, Pc1 pearls revisited: structured elecromagnetic ion cyclon waves on Polar satellite and on ground, *J. Geophys. Res.*, **106**, 29543–29553, 2001.

Neubert, T., M. Mandea, G. Hulot, R. von Frese, E. Primdahl, J.L. Jorgensen, E. Friis-Christensen, P. Stauning, N. Olsen, and T. Risbo, Oersted satellite captures high-precision geomagnetic data, *EOS, Trans. Amer. Geophys. Un.*, **82**, 81–88, 2001.

Newitt, L.R., C.E. Barton, and J. Bitterly, *Guide for Magnetic Repeat Station Surveys*, Int. Assoc. Geomagnetism and Aeronomy, 112 pp, 1996.

Nikolaev, Yu. S., Ya.Ya. Rudakov, S.M. Mansurov, and L.G. Mansurova, Interplanetary magnetic field sector structure and disturbances of the central nervous system activity, *Preprint N 17a*, Acad. Sci. U.S.S.R, 29 pp., IZMIRAN, Moscow, 1976.

Norton, R.B., The middle latitude F region during some severe ionospheric storms, *Proc. IEEE*, **57**, 1147, 1969.

Novikova, K.F., and B.A. Ryvkin, Solar activity and cardiovascular diseases, in *Effects of Solar Activity on the Earth's Atmosphere and Biosphere*, M.N. Gnevyshev and A.I. Ol', eds., pp. 184–200, Acad. Sci. U.S.S.R, English trans., Israel Prog. Sci. Trans., Jerusalem, 1977.

Ohtani, S., P.T. Newell, and K. Takahashi, Dawn-dusk profile of field-aligned currents on May 11, 1999: a familiar pattern driven by an unusual cause, *Geophys. Res. Let.*, **27**, 3777–3780, 2000.

Ohtani, S., M. Nose, G. Rostoker, H. Singer, A.T.Y. Lui, and M. Nakamura, Storm-substorm relationship: contribution of the tail current to Dst, *J. Geophys. Res.*, **106**, 21199–21209, 2001.

Ohtani, S., R. Yamaguchi, M. Nose, H. Kawano, M. Engebretson, and K. Yumoto, Quiet time magnetotail dynamics and their implications for the substorm trigger, *J. Geophys. Res.*, **107**, SMP 6.1–6.10, 2002.

Oldenburg, D.W., Inversion of electromagnetic data: An overview of new techniques, *Surveys Geophys.*, **11**, 231–270, 1990.

Onwumechilli, A., Geomagnetic variations in the equatorial zone, in *Physics of Geomagnetic Phenomena*, S. Matsushita and W.H. Campbell, eds., Chapter III-2, pp. 425–507, Academic Press, New York, 1967.

Opdyke, N.D. and J.E.T. Channell, *Magnetic Stratigraphy*, Academic Press, San Diego, 346 pp, 1996.

Persinger, M.A., Increased geomagnetic activity and the occurrence of bereavement hallucinations, *Neurosci. Lett.*, **88**, 271–274, 1988.

Pfaff, R.P., M.H. Acuna, P.A. Marionni and N.B. Trivedi, DC polarization field, current density, and plasma density measured in the daytime equatorial electrojet, *Geophs. Res. Let.*, **24**, 1667–1670, 1997.

Piper, J.D.A., *Palaeomagnetism and the Continental Crust*, 434 pp., Halsted Press, New York, 1987.

Quinn, T.P., Evidence for celestial and magnetic compass orientation in lake migrating sockeye salmon fry, *J. Comp. Physiol.*, **137**, 243–248, 1980.

Rajaram, M., and S. Mitra, Correlations between convulsive seizure and geomagnetic activity, *Neurosci. Lett.*, **24**, 187–191, 1981.

Randall, W., and S. Randall, The solar wind and hallucinations - a possible relation to magnetic disturbances., *Bioelectromagnetics*, **12**, 67–70, 1991.

Rich, F.J., and M. Hairston, Large-scale convection patterns observed by DMSP, *J. Geophys. Res.*, **99**, 3827–3844, 1994.

Roberts, W.O., and R.H. Olson, New evidence for the effects of variable solar corpuscular emission on weather, *Rev. Geophys. Space Phys.*, **11**, 731–740, 1973.

Roberts, W.O., and R.H. Olson, Great Plains weather, *Nature*, **254**, 380, 1975.

Roble, R., The Thermosphere, in *The Upper Atmosphere and Magnetosphere*, Chapter 3, pp. 57–71, National Research Council, National Academy of Sciences, Washington, D.C., 1977.

Roederer, J.G., Are magnetic storms hazardous to your health?, *Eos Trans. AGU*, **76**, 441–445, 1995.

Rosen, J., The seasonal variation of geomagnetic disturbance amplitudes, *Bull. Astron. Inst. Neth.*, **18**, 295–305, 1966.

Rostoker, G., Current flow in the magnetosphere during magnetospheric substorms, *J. Geophys. Res.*, **79**, 1994–1998, 1974.

Rostoker, G., Magnetospheric substorms – their phenomenology and predictability, Solar-Terrestrial Predictions-IV Workshop, Ottawa, Canada, 18–22 May 1992, *NOAA Dept of Commerce Pub.*, vol. 3, pp. 21–35, 1993.

Rostoker, G., Phenomenology and physics of magnetospheric substorms, *J. Geophys. Res.*, **101**, 12995–12973, 1996.

Rufenach, C.L., R.L. McPherron, and J. Schaper, The quiet geomagnetic field at geosynchronous orbit and its dependence on solar wind dynamic pressure, *J. Geophys. Res.*, **97**, 25–42, 1992.

Satori, G., and B. Zieger, Spectral characteristics of Schumann resonances observed in Central Europe, *J. Geophys. Res.*, **101**, 29663–29669, 1996.

Schmucker, U., An introduction to induction anomalies, *J. Geomag. Geoelectr.*, **22**, 9–33, 1970.

Schunk, R.W., ed., Handbook of Ionospheric Models, *SCOSTEP*, Boulder, Colorado, 350 pp, 1996.

Schuster, A., The diurnal variation of terrestrial magnetism, *Phil. Trans. Roy. Soc. London, Ser. A.*, **180**, 467–518, 1889.

Schuster, A., The diurnal variation of terrestrial magnetism, *Phil. Trans. Roy. Soc. London, Ser. A.*, **208**, 163–204, 1908.

Shapka, R., Geomagnetic effects on modern pipeline systems, Solar-Terrestrial Predictions-IV Workshop, Ottawa, Canada, 18–22 May 1992, *NOAA Dept. of Commerce Pub.*, vol. 1, pp. 163–170, 1993.

Sidorenkov, N.S., Solar corpuscular streams and weather on the Earth (English trans.), *Rept. No. JPRS* 62197, p. 21, Joint Publications Research Service, Natl. Tech. Info. Ser., Springfield, Va., 1974.

Smith, E.J., B.T. Tsurutani, and R.L. Rosenberg, Observations of the interplanetary sector structure up to heliographic latitudes of 16°: Pioneer 11, *J. Geophys. Res.*, **83**, 717–724, 1978.

Stewart, B., On the great magnetic disturbance which extended from August 28 to September 7, 1859, as recorded at the Kew observatory, *Phil. Trans. R. Soc.*, **151**, part 2, 423–430, 1861.

Stewart, B., Hypothetical views regarding the connection between the state of the Sun and terrestrial magnetism, *Encyclopedia Britannica*, 9th ed., vol. **16**, 159–184, 1883.

Svalgaard, L., Solar activity and weather, *SUIPR Rept. No.* 526, Institute for Plasma Research, Stanford Univ., Stanford, Calif., 1973.

Takahashi, K., B.A. Toth, and J.V. Olson, An automated procedure for near-real-time Kp estimates, *J. Geophys. Res.*, **106**, 21017–21032, 2001.

Tarling, D.H., *Principles and Applications of Paleomagnetism*, 164 pp., Chapman and Hall, London, 1971.

Thompson, R.J., A technique for predicting the amplitude of the solar cycle, *Solar Physics*, **148**, 383–388, 1993.

Thompson, S.M., and M.G. Kivelson, New evidence for the orgin of giant pulsations, *J. Geophys. Res.*, **106**, 21237–21253, 2001.

Trichtchenko, L. and D.H. Boteler, Specification of geomagnetically induced electric fields and currents in pipelines, *J. Geophys. Res.*, **106**, 21039–21048, 2001.

Tsurutani, B.T., X.Y. Zhou, V.M. Vasyliunas, G. Haerendel, J.K. Arballo, and G.S. Lakhina, Interplanetary shocks, magnetopause boundary layers, and dayside auroras: the importance of a very small magnetospheric region, *Surveys Geophys.*, **22**, 101–130, 2001.

Tsyganenko, N.A., Global quantitative models of the geomagnetic field in the cis lunar magnetosphere for different disturbance levels, *Planet. Space Sci.*, **35**, 1347–1358, 1987.

Tsyganenko, N.A., A magnetospheric magnetic field model with a warped tail current sheet, *Planet. Space Sci.*, **37**, 5–20, 1989.

Vaivadds, A., W. Baumhohann, E. Georgescu, G. Haerendes, R. Nakamura, R.R. Lessard, P. Eglitis, L.M. Kistler, and R.E. Ergun, Correlation studies of compressional Pc5 pulsations in space and Ps6 pulsations on the ground, *J. Geophys. Res.*, **106**, 29797–29806, 2001.

Van Allen, J.A., Charged particles in the magnetosphere, *Rev. Geophys.*, **7**, 233–255, 1969.

Vennerstroem, S., B. Sieger, and E. Friis-Christensen, An improved method of inferring interplanetary sector structure, 1905 present, *J. Geophys. Res.*, **106**, 16011–16020, 2001.

Veroe, J., Solar cycle effect on Pc3 geomagnetic pulsations, *J. Geophys. Res.*, **101**, 2461–2465, 1996.

Veroe, J., B. Zieger, and H. Luehr, Upstream waves and surface geomagnetic pulsations, in *Geophysical Monograph 81*, Amer. Geophys. Union Pub., 1994.

Villoresi, G., T.K. Breus, N. Iucci, L.I. Dorman, and S.I. Rapoport, The influence of geophysical and social effects on the incidence of clinically important pathologies (Moscow 1979–1981), *Physica Medica*, **X**, 79, 1994.

Vine, F.J. and D.H. Mathews, Magnetic anomalies over ocean ridges, *Nature*, **199**, 947–949, 1963.

Wait, J.R., *Geo-Electromagnetism*, 268 pp., Academic Press, New York, 1982.

Walpole, P.H., G.M. Mason, J.E. Mazur, D.J. Mabry, J.E. Stephens, R. Whitley, and D.C. Welch, High voltage power supply anomalies on the SAMPEX/LICA instrument associated with geomagnetic activity, paper presented at AGU Fall Meeting, San Francisco, American Geophysical Union, December 1995.

Webb, D.F., J.C. Johnson, and R.R. Radick, Solar Mass Ejection Imager (SMEI): a new tool for space weather, *EOS, Trans. Amer. Geophys. Un.*, **83**, 33–39, 2002.

Wegener, A., The Origins of Continents and Oceans, 4th ed., 1929. (English translation by J. Biram, Methuen, London, 1967).

Wiens, R.C., D.S. Burnett, M. Neugebauer, C. Sasaki, D. Sevilla, E. Stansbery, B. Clark, N. Smith, L. Oldham, B. Barraclough, E.E. Dors, J. Steinberg, D.R. Reinsenfeld, J.E. Nordholt, A. Jurewicz, and K. Cyr, Genesis mission to return solar wind samples to Earth, *EOS, Trans. Amer. Geophys. Un.*, **83**, 229–234, 2002.

Wilcox, J.M., P.H. Scherrer, and J.T. Hoeksema, Interplanetary magnetic field and tropospheric circulation, in *Weather and Climate Responses to Solar Variations*, B.M. McCormac, ed., pp. 365–379, Colorado Associated Press, Boulder, 1983.

Wilcox, J.M., P.H. Scherrer, L. Svalgaard, W.O. Roberts, R.H. Olson, and R.L. Jenne, Influence of solar magnetic sector structure on terrestrial atmospheric vorticity, *SUIPR Rept. No.* 530, Institute for Plasma Research, Stanford Univ., Stanford, Calif., 1973.

Williamson, S.J., L. Kaufman, and D. Brenner, Magnetic fields of the human brain, *U.S. Naval Research Rev.*, 1–18 October 1977.

Wiltschko, W. and R. Wiltschko, Magnetic orientation in birds, *Current Ornithology*, **5**, 67–121, 1988.

Wollin, G., W.B.F. Ryan, D.B. Ericson, and J.H. Foster, Paleoclimate, paleomagnetism and the eccentricity of the Earth's orbit, *Geophys. Res. Lett.*, **4**, 267–274, 1977.

Wrenn, G.L., Conclusive evidence for internal dielectric charging anomalies on geosynchronous communications spacecraft, paper presented at Users Conference, Space Environment Laboratory NOAA, Boulder, Colo., 2–5 May 1994.

Yacob, A., Day-time enhancement of geomagnetic disturbances at the magnetic equator in India, *Indian J. Meteorol. Geophys.*, **17**, 271–276, 1966.

Zanetti, L.J., T.A. Potemra, B.J. Anderson, R.E. Erlandson, D.B. Holland, M.H. Acuna, J.G. Kappenman, R. Lesher, and B. Feero, Ionospheric currents correlated with geomagnetic induced currents: Freja magnetic field measurement and the sunburst monitor system, *Geophys. Res. Lett.*, **21**, 1867–1870, 1994.

Zhou, X., and B.T. Tsurutani, Interplanetary shock triggering of nightside geomagnetic activity: substorms, pseudobreakups, and quiescent events, *J. Geophys. Res.*, **106**, 18957–18967, 2001.

Zhou, X., B. Tsurutani, and W.D. Gonzalez, The solar wind depletion (SWD) event of 26 April 1999: triggering of an auroral "pseudobreakup" event, *Geophys. Res. Let.*, **27**, 4025–4028, 2000.

Index